# ELEMENTARY QUANTUM MECHANICS

Holden-Day Series in Physics

# ELEMENTARY
# QUANTUM
# MECHANICS

## David S. Saxon

*University of California, Los Angeles*

---

## HOLDEN-DAY

*San Francisco, Cambridge, London, Amsterdam*

Library of Congress Catalog Card Number 68–16996.
Printed in the United States of America.

# *Preface*

This book is based on lectures given by the author in an intensive undergraduate course in quantum mechanics which occupies a central role in the physics curriculum at UCLA. It is a required course for all third-year physics and astrophysics students, but it is taken by some seniors and many graduate students, both in physics and in related fields.

Students enrolling in the course are expected to have had an introduction to elementary Hamiltonian mechanics, to the extent of knowing, for simple systems, what the Hamiltonian function is and what the Hamiltonian equations are. Students are also expected to have had training in mathematics through differential equations and Fourier series and to have at least seen many of the special functions of mathematical physics. In an effort to keep the mathematics as simple as possible, however, the first two-thirds of the book is largely confined to the consideration of one-dimensional systems.

The stress throughout is on the formulation of quantum mechanics and not on its applications. At UCLA the applications follow in immediately subsequent courses selected from atomic, nuclear, solid state and elementary particle physics. The last chapter is intended to pave the way for these applications; in it a number of relatively advanced topics are somewhat briefly presented.

The coverage is rather broad and not everything is treated in depth. Wherever the text is frankly introductory, however, references to a complete treatment are given. In all other respects the book is self-contained. Experience with a preliminary edition has shown that it is accessible to students and that they can learn from it largely by themselves. To a considerable degree the teacher is thus left free to illuminate the subject in his own way.

One hundred and fifty problems are presented, and these play an important pedagogical role. The problems are not exclusively illustrative of material presented in the text; they also amplify it. A significant number are intended to broaden the scope of the course by pointing the way to new topics and new points of view. Many problems are too difficult for the student to master in his first attempt. He is encouraged to return to them again and again as his understanding grows. Eventually he should be able to handle any and all of them. Answers or complete solutions to some fifty representative problems are given in Appendix III. About forty exercises are scattered throughout the text. These are mostly concerned with the working out of details, but not all of them are trivial.

At UCLA the material in the text is presented in a sequence covering two quarters. However, the text is also intended for use in a one-semester course; any, or all, of the starred sections in the table of contents can be omitted without harm to the logical development. If it is desired, on the other hand, to use the text for a one-year course, some supplementation would be desirable. The Heisenberg and interaction representations, and transformation theory in general, are topics which at once come to mind. At the applied level, the Zeeman and Stark effects, Bloch waves, the Hartree–Fock and Fermi–Thomas methods, simple molecules and isotopic spin are a suitable list from which to choose.

The author has benefited from numerous criticisms and suggestions from a host of colleagues and students. To each of them, he expresses his deep gratitude and especially to Dr. Ronald Blum for his meticulous reading of both the preliminary edition and the final manuscript. The author will be equally grateful for additional comments and for the correction of misprints and errors.

*David S. Saxon*

*November, 1967*

# Contents

* For a one-semester course, any or all of the starred sections can be omitted without harm to the logical development (see Preface).

## APPENDICES

"*And now reader,—bestir thyself—for though we will always lend thee proper assistance in difficult places, as we do not, like some others, expect thee to use the arts of divination to discover our meaning, yet we shall not indulge thy laziness where nothing but thy own attention is required; for thou art highly mistaken if thou dost imagine that we intended when we begun this great work to leave thy sagacity nothing to do, or that without sometimes exercising this talent thou wilt be able to travel through our pages with any pleasure or profit to thyself.*"

HENRY FIELDING

# I

# *The dual nature of matter and radiation*

## 1. THE BREAKDOWN OF CLASSICAL PHYSICS*

In the latter part of the 19th century, most physicists believed that the ultimate description of nature had already been achieved and that only the details remained to be worked out. This belief was based on the spectacular and uniform success of Newtonian mechanics, combined with Newtonian gravitation and Maxwellian electrodynamics, in describing and predicting the properties of macroscopic systems which ranged in size from the scale of the laboratory to that of the cosmos. However, as soon as experimental techniques were developed to the stage where atomic systems could be studied, difficulties appeared which could not be resolved within the laws, and even concepts, of classical physics. The necessary new laws and new concepts, developed over the first quarter of the 20th century, are those of quantum mechanics.

The difficulties encountered were of several kinds. First, there were difficulties with some of the predictions of the beautiful and general classical equipartition theorem. Straightforward applications of this theorem gave the wrong, and even a nonsensical, black-body radiation spectrum and gave wrong results for the specific heats of material systems. In both cases, the empirical result implies that only certain of the degrees of freedom participate fully in the energy exchanges leading to statistical equilibrium, while others participate little or not at all.

Second, there were difficulties in explaining the structure, and indeed the very existence, of atoms as systems of charged particles. For any such system, static equilibrium is impossible under purely electro-

---

* For a detailed discussion of the experimental and historical background of quantum mechanics, see references [1] through [5] in the selected list of references given in Appendix II.

magnetic forces, while dynamic equilibrium, for example, in the form of a miniature solar system, is equally impossible. Particles in dynamic equilibrium are accelerated and, classically, accelerated charges must radiate, thus causing rapid collapse of the orbits, whatever their precise nature might be. Accepting the fact that atoms somehow do manage to exist, there is still the problem of explaining atomic spectra, the characteristic radiation caused by the acceleration of the charged constituents of an atom when it is disturbed from its equilibrium configuration. Classically, one would expect such spectra to consist of the harmonics of a few fundamental frequencies. The observed spectra instead satisfy the Ritz combination law, which states that the frequencies are expressible as differences between a relatively few basic frequencies, or terms, and not as multiples.

A third, and more special, class of difficulties is illustrated by the photo-electric effect. Photo-emission of electrons from an illuminated surface takes place under circumstances which permit no classical explanation. The essential difficulty is this: the *number* of emitted electrons is proportional to the intensity of the incident light and thus to the electromagnetic energy falling on the surface, but the *energy* transferred to the individual photo-electrons does not depend at all upon the intensity of the illumination. Instead this energy depends upon the *frequency* of the light, increasing linearly with frequency above a certain threshold value, characteristic of the surface material. For frequencies below this threshold, photo-emission simply does not occur. Otherwise stated, at frequencies below threshold, no photo-electrons are emitted even if a relatively large amount of electromagnetic energy is being transmitted into the surface. On the other hand, at frequencies above threshold, no matter how weak the light source, some photo-electrons are always emitted and always with the full energy appropriate to the frequency.

The explanation of these various difficulties began in 1901, when Planck assumed the existence of energy quanta in order to obtain the desired modification of the equipartition theorem. The implication that electromagnetic radiation therefore had corpuscular aspects was emphasized, and indeed first recognized, in 1905 in Einstein's direct and simple predictions of the characteristics of photo-electric emission. It was also Einstein who first realized, two years later, that the low-temperature behavior of the specific heats of solids could be explained by quantizing the vibrational modes of internal motion of a material object according to Planck's rules. The first understanding of atomic structure and spectra came in 1913, when Bohr introduced the revolutionary idea of *stationary states* and gave quantum conditions for their determination. These conditions were subsequently generalized by Sommerfeld and Wilson, and the resultant theory accounted almost perfectly for the spectrum and structure of atomic hydrogen. But the Bohr theory encountered increasingly serious difficulties as attempts were made to apply it to more complex problems and to more complex systems. The helium atom, for example, proved to be completely intractable. The first indication of the ultimate solution to these problems came in 1924, when de Broglie suggested that, just as light waves exhibit particle-like behavior, so do particles exhibit wave-like behavior. Following up this suggestion, Schrödinger developed, in 1926, the famous wave equation which bears his name. Slightly earlier, and from a very different point of view, Heisenberg had arrived at a mathematically equivalent statement in terms of matrices. At about the same time, Uhlenbeck and Goudsmit introduced the idea of electron spin, Pauli enunciated the exclusion principle, and the formulation of nonrelativistic quantum mechanics was substantially completed.

## 2. QUANTUM MECHANICAL CONCEPTS

The laws of quantum mechanics cannot be *derived*, any more than can Newton's laws or Maxwell's equations. Ideally, however, one might hope that these laws could be deduced, more or less directly, as the simplest logical consequence of some well-selected set of experiments. Unfortunately, the quantum mechanical description of nature is too abstract to make this possible; the basic constructs of quantum theory are one level removed from everyday experience. These constructs are the following:

*State Functions.* The description of a system proceeds through the specification of a special function, called the state function of the system, which cannot itself be directly observed. The information contained in the state function is inherently *statistical* or *probabilistic*.

*Observables.* Specification of a state function implies a set of observations, or measurements, of the physical properties, or attributes, of the system in question. Properties susceptible of measurement, such as energy, momentum, angular momentum, and other dynamical variables, are called observables. Observations or observables are represented by abstract mathematical objects called *operators*.

The process of observation requires that some interaction take place between the measuring apparatus and the system being observed. Classically, such interactions may be imagined to be as small as one pleases. Normally they are taken to be infinitesimal, in which case the system is left undisturbed by an observation. On the quantum level, however, the interaction is discrete in character, and it cannot be decreased beyond a definite limit. The act of observation thus introduces certain irreducible and uncontrollable disturbances into the system. The observation of some property $A$, say, will produce unpredictable changes in some other related observable $B$. The existence of an absolute limit to an interaction or a disturbance permits an absolute meaning to be given to the idea of size. A system may be thought of as large or small, and treated as classical or quantum mechanical, to the extent that a given irreducible interaction can be safely regarded as negligible or not.

The notion that precise observation of one property makes a second property (called complementary to the first) unobservable is a completely quantum mechanical idea with no counterpart in classical physics. The attributes of being wave-like or particle-like furnish one example of a pair of complementary properties. The wave-particle duality of quantum mechanical systems is a statement of the fact that such a system can exhibit either property, depending upon the observations to which it has been subjected. A second and more quantitative example of a pair of complementary observables is furnished by the dynamical variables, position and momentum. Observing the position of a particle, say by looking at it, which means by shining light on it, will necessarily produce a finite disturbance in its momentum. This follows because of the corpuscular nature of light; a measurement of position requires at least one photon to strike the particle, and it is this collision which produces the disturbance. One immediate consequence of this relationship between measurement and disturbance is that precise particle trajectories cannot be defined at the quantum level. The existence of a precise trajectory implies precise knowledge of both position and momentum *at the same time*. But simultaneous knowledge of both is not possible if measurement of one produces a significant and uncontrollable disturbance in the other, as is the case for quantum mechanical systems. We emphasize that these mutual disturbances or uncertainties are not a matter of experimental technique; they follow instead as an inevitable consequence of measurement or observation. The necessary existence of such effects in a pair of complementary variables was first enunciated by Heisenberg in his statement of the famous *uncertainty principle*.

We shall return to these questions later, but now we want to begin our development of the laws of quantum mechanics. Our approach, which is not the historical one, will proceed in the following way. First, in the remainder of this chapter, we shall try to make plausible some of the ideas of quantum mechanics, and particularly the ideas of complementarity and uncertainty. We shall do this by considering some experiments and observations which emphasize that matter is dual in nature and that, as one immediate consequence, the precise particle trajectories of Newtonian mechanics do not exist. This at once poses the problem of how the state of motion of a quantum mechanical system is to be characterized and how such systems are to be described. In Chapter II we answer this question by introducing the *state function* of a system, and we then discuss its probabilistic interpretation. In Chapter III we consider the general properties of *observables* and *dynamical variables* in quantum mechanics and give rules for obtaining their abstract operator representations. Next, in Chapters IV and V, we complete the first stage of our formulation by introducing *Schrödinger's equation,* which governs the time development of quantum systems. Methods of solving Schrödinger's equation for the simplest possible system, the motion of a single particle in one dimension, are discussed in Chapters VI and VII. Only in the final four chapters are we ready to treat the general problem of systems of interacting particles in three dimensions, thus making contact with the real world. Throughout our development we shall continually use the principle that the predictions of the quantum laws must correspond to the predictions of classical physics in the appropriate limit. As we shall see, this *principle of correspondence* plays a key role in determining the form of the quantum mechanical equations.

The emphasis throughout will be on the quantum mechanical properties of material systems. Because of its complexity, no corresponding systematic development of the quantum properties of electromagnetic fields will be presented, although relevant quantum properties will occasionally be asserted and perhaps even made plausible.[1]

## 3. THE WAVE ASPECTS OF PARTICLES

The experiment which most nearly isolates the basic elements of the quantum mechanical description of nature is the scattering of a beam of electrons by a metallic crystal, first performed by Davisson and Germer in 1927. Their experiment was designed to test the prediction of de Broglie that, by analogy with the already well-established corpuscular properties of light, there is associated with a particle of momentum $p$ a wave of wavelength $\lambda$, now called the *de Broglie wavelength,* given in terms of the momentum by

$$\lambda = h/p.$$

The universal constant $h$ is *Planck's constant* or the *quantum of action.* Motivating de Broglie was the desire to provide a basis for understanding, in terms of fitting an integral number of half-wavelengths into a Bohr orbit, Bohr's apparently arbitrary quantization condition. In any case,

---

[1] Specifically, in Section 5 of the present chapter, the corpuscular nature of light is invoked to account for the nature of black-body radiation and of Compton scattering. We shall not refer to radiation again until Section 6, Chapter VII, when its emission and absorption is presented heuristically and semiclassically. Finally, in Section 4, Chapter XI, we briefly discuss the motion of a charged particle in a classical, externally prescribed electromagnetic field.

Davisson and Germer observed that the electrons of momentum $p$ scattered by the crystal were indeed distributed in a diffraction pattern, exactly as would be x-rays of the same wavelength scattered by the same crystal; and thus they directly, conclusively and quantitatively verified de Broglie's hypothesis.

The quantum of action is seen to have the dimensions of momentum-length or, equivalently, of energy-time, and its numerical value is

$$h = 6.625 \times 10^{-27} \text{ erg-sec.}$$

In most quantum mechanical applications it turns out to be more convenient to use the quantity $h/2\pi$, which is abbreviated as $\hbar$ and is called "$h$ bar." It has the numerical value

$$\hbar \equiv h/2\pi = 1.054 \times 10^{-27} \text{ erg-sec.}$$

In terms of $\hbar$, the de Broglie relation can be rewritten in the form

$$\lambdabar \equiv \lambda/2\pi = \hbar/p,$$

where we have introduced the reduced wavelength $\lambdabar$ (called "lambda bar"), which is physically a more significant length characterizing the wave than is the wavelength itself. It is also convenient to define the *wave number* $k$ (strictly speaking, the *reduced* wave number) as the reciprocal of $\lambdabar$. Thus we can also write the de Broglie relation in the form

$$p = \hbar k.$$

To collect these relations in a single expression let us write, finally,

$$p = h/\lambda = 2\pi\hbar/\lambda = \hbar/\lambdabar = \hbar k. \tag{1}$$

The de Broglie hypothesis, and the Davisson–Germer experiment, are in sharp conflict with classical physics in that *both* particle and wave properties are assigned to the same entity. The nature and implications of the conflict can be made much clearer by imagining the experiment to be performed with a beam of electrons so limited in intensity that only a single electron is scattered by the crystal and recorded at a time. In that event, no diffraction pattern at all would be observed at first; a given electron would be scattered in some direction or other in an apparently random way. However, as time went on and the slowly accumulating number of scattered electrons mounted into the thousands and millions, it would become increasingly clear that more electrons are scattered in some directions than in others, and thus the diffraction pattern would gradually emerge.

The following conclusions can be drawn from the results of the Davisson–Germer experiment:

(a) Electrons exhibit both particle and wave properties. The quantitative connection between these is expressed by the de Broglie relation, equation (1).

(b) The exact behavior of a given electron cannot be predicted, only its probable behavior.

(c) Precisely defined trajectories do not exist at the quantum level.

(d) The probability that an electron is observed to be in a given region is proportional to the intensity of its associated wave field.

(e) The superposition principle applies to de Broglie waves, just as it does to electromagnetic waves.

Conclusions (a) and (b) require no further comment. Conclusion (c) follows from (b), because classically a particle moves along a unique trajectory under the influence of specified forces for given initial conditions. Conclusion (d) is inferred from the parallelism between the x-ray and electron diffraction patterns from a given crystal. Finally, conclusion (e) follows from the fact that the diffraction pattern is produced by interference of secondary waves generated at each atomic site in the crystal, that is, by a *linear combination* or *superposition* of these scattered waves.

These conclusions are the starting point for our whole development of quantum mechanics. They have been reached without reference to the specific character of the interaction between electrons (or x-rays either, for that matter) with the atoms in the crystal and without reference to the details of the diffraction pattern formed as a result of that interaction. This is no oversight, however, for our argument is based entirely on the behavior of a crystal as a three-dimensional diffraction grating, calibrated by observation of its effects upon x-rays of known properties. Nonetheless, it is a little unsatisfying, pedagogically speaking, to have reached such significant conclusions without exploring all the details. Unfortunately, these details require an understanding of the interaction of an electron with the atoms in a crystalline solid, and this interaction cannot be understood before we understand quantum mechanics itself. For that reason we shall now consider two highly idealized "crucial" experiments which will force us to essentially the same conclusions in a more or less transparent way. These experiments are one-dimensional versions of scattering and diffraction, and they involve nothing but the simplest kinds of systems. However, as will shortly become apparent, our experiments are actually performable only in principle and not in practice.

In the first experiment, as shown in Figure 1(a), a particle of positive charge $e$ and mass $m$ is sent with momentum $p$ down the axis of a long drift tube, the walls of which are at ground potential. Aligned with the first drift tube, and infinitesimally separated from it is a second drift tube at a higher potential $V_0$.

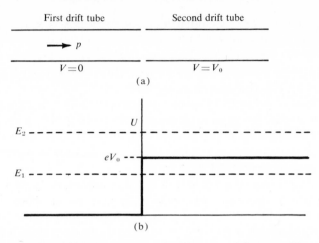

FIGURE 1. (a) The drift tube system. (b) The potential energy $U$ as a function of distance along the axis of the drift tube system. For simplicity, we have taken $U$ to change discontinuously. A classical particle is reflected if its energy is $E_1$, transmitted if its energy is $E_2$.

Suppose first that the energy of the particle is $E_1 = p_1{}^2/2m$ and that $E_1$ is less than $eV_0$, as shown in Figure 1(b). Classically, the resulting motion is such that the particle is reflected at the interface and returns along the axis of the first drift tube with its momentum unchanged in magnitude. Next suppose the energy is increased to a value $E_2$, which exceeds $eV_0$, as is also shown in Figure 1(b). The classical prediction is that the electron will be decelerated at the interface and will proceed into the second drift tube with momentum $\bar{p}$ such that

$$\bar{p}^2/2m = E_2 - eV_0.$$

The results of such an experiment agree with the classical prediction in the first instance, but not in the second. For $E_2$ somewhat greater than $eV_0$, the particle is not always transmitted as predicted but is sometimes reflected. However, as $E_2$ increases, the likelihood of reflection decreases until, eventually, the particle is almost never reflected and the classical prediction becomes correct. If we define the transmission co-

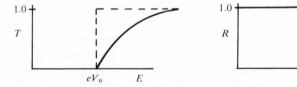

FIGURE 2. Transmission and reflection coefficients as a function of energy for the drift tube of Figure 1. The dotted lines are the classical predictions.

efficient $T$ as the relative number of times the particle is transmitted, and the reflection coefficient $R$ as the relative number of times it is reflected, with $T + R = 1$, the results are shown in Figure 2. The classical prediction is the dotted line, and the experimental result is the solid curve, which is clearly impossible to explain on classical grounds. Note that, over the energy region where either reflection or transmission can occur, there is no way of predicting the precise behavior of, or assigning a precise trajectory to, a given incident particle. The best one can do is to say that a particle will be reflected with probability $R$ or, equivalently, transmitted with probability $T = 1 - R$.

We now go on to a second idealized and still more revealing experiment in which a third drift tube at ground potential is aligned with the second. The potential $U$ then is as shown in Figure 3. The length of the

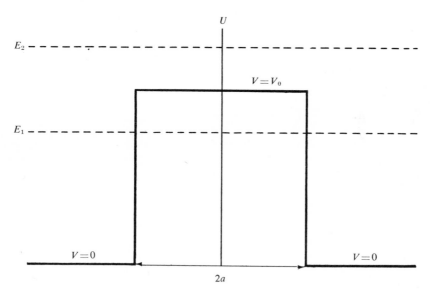

FIGURE 3.   The repulsive square well potential.

middle drift tube is $2a$, and the origin has been taken halfway along the middle tube. A potential such as that in Figure 3 is called a repulsive square well potential; if $V_0$ were negative, it would be attractive.

The classical prediction is, of course, that the particle will be reflected if its energy is less than $eV_0$, say $E_1$ in Figure 3, and will be transmitted past the barrier if its energy exceeds $eV_0$, say $E_2$ in the figure. Again this classical prediction is wrong, but now it is wrong in both instances, if the barrier is sufficiently thin. *Whatever the sign* of $E - eV_0$, provided this difference is not too large, some fraction of the particles is transmitted and some fraction reflected. Defining reflection and transmission coefficients as before, the experimental transmission coefficient as a function of energy is plotted in Figure 4. For comparison, the classical prediction is also shown.

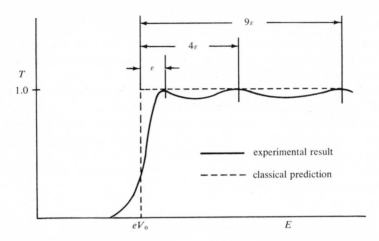

FIGURE 4.   Transmission coefficient for the repulsive square well potential.

The results are quite remarkable and unexpected. Particularly astonishing is the fact that the particle is sometimes transmitted through the barrier when its energy is too small for the particle to cross over it, that is, when its kinetic energy would be *negative* if the particle were inside the barrier. Classically, no meaning can be assigned to a negative kinetic energy, and motion in such a region is impossible. We thus have the paradox that the particle somehow appears on the other side of a region through which it cannot pass. This is commonly called the tunnel effect, because the particle appears to have tunneled through the potential barrier. For the moment, we merely remark that this is further evidence that the idea of a classical trajectory loses its meaning where quantum effects are important.

We now focus our attention on the oscillations in the transmission coefficient. If the first maximum occurs at an energy $\epsilon$ above the barrier height, the second is observed to occur at $4\epsilon$, the third at $9\epsilon$ and so on. If the experiment is repeated for different barrier widths, the value of $\epsilon$ is found to vary inversely with the square of the barrier width. We thus deduce that the energy $E_n$ of the $n$th maximum is such that $\sqrt{E_n - eV_0}$ is proportional to $n/a$. Introducing the momentum $\bar{p}$ of the particle while passing over the barrier, we see that the momentum $\bar{p}_n$ of the $n$th maximum satisfies the simple relation

$$\bar{p}_n = h \frac{n}{a},$$

where the constant of proportionality turns out to be just Planck's constant. Otherwise stated, when the width of the barrier, $2a$, is a half-integral multiple of $h/p$, the transmission achieves its maximum value of unity (and the reflection coefficient becomes zero), so that the barrier becomes perfectly transparent only for these special values.

This behavior is exactly analogous to that for the transmission of light through a thin dielectric slab or film, where the reflection coefficient vanishes whenever the thickness of the film is a half-integral number of wavelengths. This makes clear that what is being observed is a wave phenomenon and, more explicitly, that associated with a particle of momentum $p$ is a wave of wavelength $\lambda$, in precise agreement with de Broglie's prediction and the results of the Davisson–Germer experiment.

Our explanation of the observations is then something like this. We associate with the incident particle in the first drift tube a wave, which we shall henceforth call a *de Broglie wave,*

$$\psi = e^{2\pi i x/\lambda} = e^{ipx/\hbar}. \tag{2}$$

When this wave impinges on the first face of the potential barrier, part of it is transmitted into the barrier, part of it is reflected. The transmitted wave inside the barrier has the form

$$\psi \simeq e^{i\bar{p}x/\hbar}.$$

This wave is, in turn, partially transmitted out of the barrier and partially reflected at the second interface. The reflected wave travels back toward the first interface where part is again reflected and part transmitted, and so on. The wave eventually transmitted to the right is thus a superposition of a multiply reflected set of waves. The condition for these to interfere constructively to give a maximum in the transmission is that the barrier be a half-integral number of wavelengths thick. Implicit in this explanation is the idea that the intensity of the final transmitted and reflected waves is to be associated with the probabilities for transmission and reflection of the particle.

On the basis of this interpretation, note that negative kinetic energy, or imaginary momentum, is no longer nonsensical. For imaginary momentum the de Broglie wavelength is also imaginary, and hence the corresponding waves are *attenuated* rather than *propagating* waves. But such waves exist and make sense. Indeed, the tunnel effect can be qualitatively explained on this basis. That portion of the incident wave which is transmitted into the barrier becomes an attenuated wave. It reaches the second interface diminished in amplitude, but upon transmission through the second interface becomes a propagating wave again.

If the barrier is thick, the attenuation becomes very great and the transmission drops exponentially to zero, in agreement with observation.[2]

## 4. NUMERICAL MAGNITUDES AND THE QUANTUM DOMAIN

It is quite instructive to examine the magnitudes of the de Broglie wavelength for some representative cases:

    (a)   Electron of energy $E$ (electron volts)

$$\lambda = \frac{\hbar}{p} = \frac{\hbar}{\sqrt{2mE}} \simeq 10^{-8}\,E^{-1/2}\ \text{cm}$$

    (b)   Proton of energy $E$ (electron volts)

$$\lambda \simeq 5 \times 10^{-10}\,E^{-1/2}\ \text{cm}$$

    (c)   One gm mass moving at one cm/sec

$$\lambda \simeq 10^{-27}\ \text{cm}.$$

These numbers tell us at once why quantum effects manifest themselves only at the atomic level. On the macroscopic level all dimensions are so enormous, compared to the de Broglie wavelength, that wave aspects are undetectable. In the atomic and subatomic domain, the dimensions become comparable to the de Broglie wavelength and the wave aspects dominate.

These numbers also make clear the difficulty of actually performing our idealized drift tube experiments in the laboratory. For simplicity, we assumed the potentials to change discontinuously. In actuality the potentials will change over some distance, say $b$. This complicates the analysis but does not change the *qualitative* features of the results. However, the *magnitude* of the quantum effects are crucially dependent on the size of $b$. Only if $b$ is rather smaller than, or at most comparable with, the wavelength, will the effects be appreciable. Looking at the most favorable case, that of the electron, we see that the gap between drift tubes would have to be at most a few angstroms, that is, a few atom diameters.

There are, however, analogs of our experiment on the atomic scale. Thus thermonic emission of electrons from a metal corresponds to our first experiment. Field emission, where tunneling plays a dominant role, corresponds to the second. So does nuclear alpha decay. The passage of an externally incident electron through an atom also corresponds roughly to our second experiment. Resonances in the transmission are indeed observed, as in our experiment, and are known as the Ramsauer

---

[2] A detailed treatment is presented in Section 7 of Chapter VI.

effect. Unfortunately, all of these involve complex physical systems whose relevant properties cannot be fully understood before we understand quantum mechanics itself.

## 5. THE PARTICLE ASPECTS OF WAVES

In the preceding, we have demonstrated that classical particles have a dual nature in that they also exhibit wave properties. We now briefly describe some experiments which, conversely, demonstrate that electromagnetic waves have particle properties. The first indication of this arose in connection with the spectral properties of the radiation from a perfectly absorbing, or a black, body. An approximation to such a body is obtained as follows. Imagine a container to be constructed with walls opaque to electromagnetic radiation and suppose it to have an infinitesimal hole in its surface. Radiation which enters the hole will not, with appreciable probability, find its way out again, and the hole is thus a black body. The radiation field in the interior of the container, and in thermal equilibrium with it at a given temperature $T$, is then black-body radiation. It can be studied experimentally by examining the radiation which leaks out through the infinitesimal hole. Its spectral distribution and volume density turn out to depend only upon the temperature and not upon the detailed properties of the walls or of anything else. We note that it is just this freedom from dependence upon detail which makes black-body radiation such an important testing ground for our understanding of the energy interchange between matter and radiation in thermal equilibrium. Classical physics gives an unambiguous and almost totally wrong answer to the question of what the spectrum of this radiation ought to be. The argument is as follows.

The electromagnetic field in the interior of a cavity can be completely described as a superposition of the characteristic modes of harmonic vibration of the field in the given cavity. The amplitude of each mode is independent and may, in principle, be arbitrarily assigned. Thus each mode represents a degree of freedom of the radiation field, and these degrees of freedom are vibrational in character. According to the equipartition theorem of classical statistical mechanics, each vibrational degree of freedom has the same mean energy $kT$ in thermal equilibrium. Now it is not hard to show that the number of modes in the frequency interval between $\nu$ and $(\nu + d\nu)$ is given by $(8\pi/c^3)\, V\nu^2 d\nu$, where $V$ is the volume of the cavity. Thus we obtain the paradoxical result that the energy density spectrum of the black-body radiation is given by $(8\pi/c^3)\, kT\nu^2 d\nu$, which means that the density of radiation with frequency between $\nu$ and $\nu + d\nu$ increases indefinitely with the square of the frequency and that the total electromagnetic energy in the cavity is infinite.

**Exercise 1.** Consider a cubical box of volume $V$ with perfectly conducting walls.

(a)   Show that the number of modes with frequency between $\nu$ and $\nu + d\nu$ is given by $(8\pi/c^3) V\nu^2 \, d\nu$ (reference [3]).

(b)   Would such a box, even with the proverbial speck of dust in it, actually behave like a black body at *all* frequencies? In particular, what are its properties at very low frequencies?

We have described the classical result, which is known as the Rayleigh–Jeans Law, as almost totally wrong; however, the low frequency part of the spectrum is, in fact, accurately predicted by this relation. At higher frequencies, the observed spectrum is less intense than that predicted classically, and eventually it falls exponentially to zero. To put it another way, the degrees of freedom associated with the higher frequencies do not participate fully in the sharing of energy, and the highest not at all.

The mystery of the non-participation of some degrees of freedom was first penetrated by Planck when he proposed that the energy of a vibrational mode of frequency $\nu$ could take on only discrete values, and could not vary continuously as it would classically.[3] In particular, he assumed that the energy could increase from zero only in equal steps or jumps of magnitude proportional to the frequency. The proportionality constant is just Planck's constant, of course, so that the energy of a quantum of frequency $\nu$, or angular frequency $\omega$, is

$$E = h\nu = \hbar\omega \tag{3}$$

and the energy of an oscillator would then have as its only permissible values $0, \hbar\omega, 2\hbar\omega, \ldots$ .

It is easy to see that Planck's idea is at least qualitatively correct. For sufficiently low frequency modes, the energy steps are very small compared to thermal energies, and the classical equipartition theorem is unaffected. For sufficiently high frequency modes, on the other hand, the energy steps are very large compared to thermal energies, and these modes do not participate in the energy sharing process. Specifically, it turns out that the mean energy of a vibrational degree of freedom of frequency $\nu$ at temperature $T$ is

$$\bar{E} = \frac{h\nu}{e^{h\nu/kT} - 1} = \frac{\hbar\omega}{e^{\hbar\omega/kT} - 1}, \tag{4}$$

---

[3] We present the argument from a modern point of view. Planck actually ascribed quantum characteristics only to the material oscillators, which he introduced to represent the properties of the walls of the enclosure, and not to the modes of the electromagnetic field. It was Einstein who first realized that the *radiation field* is also necessarily quantized.

which is seen to take on the classical value $kT$ when $\hbar\omega/kT \ll 1$ and to be exponentially small for $\hbar\omega/kT \gg 1$. The corresponding energy density of black-body radiation of frequency between $\nu$ and $\nu + d\nu$ is then

$$E(\nu)\ d\nu = \frac{8\pi}{c^3}\ \frac{h\nu^3}{e^{h\nu/kT} - 1}\ d\nu, \tag{5}$$

which is the Planck radiation law. It is in excellent agreement with experiment and historically it furnished the first, and quite accurate, determination of $\hbar$.

---

**Exercise 2.** (See reference [3].)

(a)   Derive equation (4) and the Planck radiation law, equation (5).

(b)   Denoting the wavelength at the maximum of the black-body spectrum by $\lambda_m$, show that $\lambda_m T =$ constant (Wien's displacement law).

(c)   Show that the total energy radiated by a black body at temperature $T$ is proportional to $T^4$ (Stefan's Law).

---

Although Planck gave a completely successful solution to the difficulties of black-body radiation, his work attracted little attention.[4] Indeed, it was not even taken very seriously before 1905, when Einstein applied the quantum idea to the explanation of the phenomenon of photoelectric emission by explicitly introducing the corpuscular properties of electromagnetic radiation. These corpuscular properties are even more explicitly demonstrated in the Compton effect. When x-rays of a given frequency are scattered from (essentially) free electrons at rest, the frequency of the scattered x-rays is not unaltered but decreases in a definite way with increasing scattering angle. This effect is precisely described by treating the x-rays as relativistic *particles* of energy $\hbar\omega$ and momentum $\hbar\omega/c$, and applying the usual energy and momentum conservation laws to the collision.

---

**Exercise 3.**  Show that for Compton scattering

$$\lambdabar' - \lambdabar = 2\lambdabar_c \sin^2 \frac{\phi}{2},$$

where $\lambdabar_c = \hbar/mc$ is the so-called *Compton wavelength, m* is the mass of the electron, $\lambdabar$ is the wavelength of the incident x-rays and $\lambdabar'$ is the wavelength of x-rays scattered through the angle $\phi$. The Compton wavelength plays the role of a fundamental length associated with a particle

---

[4] E. U. Condon, in *Physics Today*, Vol. 15, No. 10, p. 37, Oct. 1962.

of mass $m$. What is its approximate numerical value for an electron? For a proton? For a $\pi$-meson? For a billiard ball? (See reference [3].)

## 6. COMPLEMENTARITY

We have now established a certain symmetry in nature between particles and waves which is totally lacking in classical physics, where a given entity must be exclusively one or the other. But this has come at the price of great conceptual difficulty. We must somehow accommodate the classically irreconcilable wave and particle concepts. This accommodation involves what is known as the *principle of complementarity,* first enunciated by Bohr. The wave-particle duality is just one of many examples of complementarity.

The idea is the following: Objects in nature are neither particles nor waves; a given experiment or measurement which emphasizes one of these properties necessarily does so at the expense of the other. An experiment properly designed to isolate the particle properties, such as Compton scattering or the observation of cloud chamber tracks, provides no information on the wave aspects. Conversely, an experiment properly designed to isolate the wave properties, for example, diffraction, provides no information about the particle properties. The conflict is thus resolved in the sense that irreconcilable aspects are not simultaneously observable *in principle.* Other examples of complementary aspects are the position and linear momentum of a particle, the energy of a given state and the length of time for which that state exists, the angular orientation of a system and its angular momentum, and so on. We shall elaborate on these various aspects in due course. We are now, however, in a position to give a reasonably general statement of the principle of complementarity. The quantum mechanical description of the properties of a physical system is expressed in terms of pairs of mutually complementary variables or properties. Increasing precision in the determination of one such variable necessarily implies decreasing precision in the determination of the other.

## 7. THE CORRESPONDENCE PRINCIPLE

Thus far we have been concentrating our attention on experiments which defy explanation in terms of classical mechanics and which, at the same time, isolate certain aspects of the laws of quantum mechanics. We must not lose sight, however, of the fact that there exists an enormous domain, the domain of macroscopic physics, for which classical physics works and works extremely well. There is thus an obvious requirement which quantum mechanics must satisfy—namely, that in the appropriate or classical limit, it must lead to the same predictions as does classical mechanics. Mathematically, this limit is that in which $\hbar$ may be regarded as small. For the electromagnetic field, for example, this means that the number of quanta in the field must be very large. For particles it means that the de Broglie wavelength must be very small compared to all relevant lengths. Of course, the statements of quantum mechanics are probabilistic in nature, we have argued, while those of classical mechanics are completely deterministic. Thus, in the classical limit, the quantum mechanical probabilities must become practical certainties; fluctuations must become negligible.

This principle, that in the classical limit the predictions of the laws of quantum mechanics must be in one-to-one correspondence with the predictions of classical mechanics, is called the *correspondence principle*. Its requirements are sufficiently stringent that, starting with the idea of de Broglie waves and their probabilistic interpretation, the laws of quantum mechanics can be more or less completely determined from the correspondence principle, as we shall eventually demonstrate.

---

**Problem 1.** Calculate, to two significant figures, the de Broglie wavelengths of the following:

(a)  An electron moving at $10^7$ cm/sec.

(b)  A thermal neutron at room temperature, that is, a neutron in thermal equilibrium at 300°K and moving with mean thermal energy.

(c)  A 50 *MeV* proton.

(d)  A 100 gm golf ball moving at 30 meters/sec.

**Problem 2.** Consider an electron and proton each with the same kinetic energy, $T$. Calculate the de Broglie wavelength of each, to one significant figure, in the following cases:

(a)  $T = 30\ eV$.

(b)  $T = 30\ keV$.

(c)  $T = 30\ MeV$.

(d)  $T = 30\ GeV = 30{,}000\ MeV$.

NOTE: To sufficient accuracy, the rest energy of an electron is 0.5 *MeV*, and of a proton it is one *GeV*. Note also that the relation between kinetic energy, momentum and rest mass can be expressed as

$$E = T + mc^2 = \sqrt{(mc^2)^2 + (pc)^2}.$$

---

# II

# *State functions and their interpretation*

## 1. THE IDEA OF A STATE FUNCTION; SUPERPOSITION OF STATES

We have been led to the idea that the description of the behavior of a particle requires the introduction of de Broglie waves. These waves exhibit characteristic interference, and the intensity of these waves in a given region is associated with the probability of finding the particle in that region.

We now seek to generalize these ideas, and at the same time to make them more definite. To simplify the mathematical features, we shall consider the motion of a single particle in one dimension under the influence of some arbitrary, but prescribed, external force. As a first step we ask how the state of motion of such a particle is to be described at some given instant. In classical mechanics, a description is normally given by specifying the position and momentum of the particle at the instant in question. Newton's laws then furnish a prescription for determining the development of the state of motion in time. But we have emphasized that such a description will not do in quantum mechanics, since particle trajectories are not well defined. We must start somewhere, however, and we shall make the minimum assumption that the state of a particle at time $t$ is completely describable, at least as completely as possible, by some function $\psi$ which we shall call the *state function* of the particle or system.

We must then address ourselves to the following questions:

(1) How is $\psi$ to be specified? That is, what variables does it depend on?

(2)  How is $\psi$ to be interpreted? That is, how are the observable properties of a system to be inferred from $\psi$?

(3)  How does $\psi$ develop in time? That is, what is the equation of motion for the system?

As a tentative answer to the first question we shall make the simplest possible assumption, namely, that the state function of a structureless[1] one-dimensional particle at a given time $t$ can be expressed in terms of space coordinates alone, $\psi = \psi_t(x)$, where the subscript $t$ denotes the instant at which the description applies. Putting this in more conventional notation, we write

$$\psi = \psi(x, t), \tag{1}$$

where $t$ plays the role of a parameter. Our assumption that $\psi$ can be so expressed for a structureless particle turns out to be correct. This means that *any* physical state can be specified in terms of an appropriate $\psi$ of the form of equation (1). What about the converse? Does every arbitrarily chosen $\psi$ correspond to some physical state? The answer is no. Only a certain class of state functions, which we shall call *physically admissible,* correspond in fact to realizable physical states. For example, it turns out that $\psi$ must be single-valued and bounded, in a sense to be defined later, if it is to be physically admissible.

Proceeding now to the second question, which is the main business of the present chapter, we first give a precise meaning to the probabilistic aspects of the quantum mechanical state function. We shall make the plausible and physically necessary assumption that the probability of finding a particle in a given region of space is large where $\psi$ is relatively large and small where $\psi$ is relatively small. Since probabilities can never be negative, and since $\psi$ itself takes on both positive and negative values (and indeed turns out to be a complex function), the simplest association we can make is to take the relative probability proportional to the absolute value squared of $\psi$, which is analogous to the intensity of an ordinary wave field. More precisely, if $P(x, t)\ dx$ is the *relative* probability of finding the particle at time $t$ in a volume element $dx$ centered about $x$, we write

$$P(x, t)\ dx = |\psi(x, t)|^2\ dx = \psi^*(x, t)\psi(x, t)\ dx \geq 0,$$

where $\psi^*$ denotes the complex conjugate of $\psi$. We can convert to *absolute* probabilities $\rho(x, t)\ dx$ by writing

$$\rho(x, t)\ dx = \frac{P(x, t)\ dx}{\int P(x, t)\ dx}$$

or

---

[1] By a structureless particle we mean a conventional mass point. The description must be modified for a particle with internal degrees of freedom, such as spin, as we shall see.

$$\rho(x, t) = \frac{\psi^*(x, t)\psi(x, t)}{\int \psi^*(x, t)\psi(x, t) \, dx}, \qquad (2)$$

where the integral extends over all space. That $\rho \, dx$ is indeed an absolute probability follows from the fact that, evidently,

$$\int \rho \, dx = 1.$$

This means that the probability of finding the particle *somewhere* in space, anywhere, correctly has the value unity. The quantity $\rho$ is called the *probability density*. If the probability density is not to lose its meaning, the integral in the denominator must be bounded. Hence all physically admissible state functions must be square integrable.[2]

Note that, according to equation (2), $\rho$ is unchanged if $\psi$ is multiplied by any arbitrary space-independent factor, that is to say, by an arbitrary factor $c(t)$, which may be complex. In that sense, $\psi$ is undetermined up to such a factor. It is generally convenient to choose this multiplicative factor in such a way that

$$\int \psi^*\psi \, dx = 1, \qquad (3)$$

which can always be done for physically admissible state functions. This condition is called the normalization condition and state functions which satisfy it are called *normalized*. For normalized state functions $\psi^*\psi$ is itself the probability density,

$$\rho(x, t) = \psi^*(x, t)\psi(x, t), \qquad (4)$$

and $\psi$ can then be interpreted as a probability *amplitude*.

The actual normalization procedure is the following: Suppose $\psi$ to be some given physically admissible state function. Evaluate $\int \psi^*\psi \, dx$ and denote the result by $M$, a real number. Then

$$\psi \equiv \sqrt{M} \, e^{i\delta} \, \psi'$$

defines the normalized state function $\psi'$ for arbitrary $\delta$. We emphasize that normalization is a matter of convenience and that no physical significance is to be attributed to the *absolute* numerical magnitude of a state function. Only relative magnitudes are important. Otherwise stated, a state function which is *everywhere* increased by an order of magnitude is physically unchanged. This is in sharp contrast to the situation in classical physics. An increase by a similar factor in the amplitude of the

---

[2] While this statement is correct, physicists frequently find it convenient to work with idealized state functions which satisfy the weaker condition, or others equivalent to it,

$$\int \psi^*(x, t)\psi(x, t) \, e^{-\alpha|x|} \, dx = M(\alpha, t),$$

where $M$ is finite for arbitrarily small but non-zero $\alpha$. We shall shortly see some examples.

pressure in an acoustical wave, for example, results in a significantly altered physical situation, readily apparent to even the most casual observer.

It is important to understand the precise nature of the probabilistic quantities we have introduced. We are discussing a system which consists of a one-dimensional particle moving under the influence of some prescribed external force. Imagine now an ensemble of such systems, identical to one another and satisfying identical initial conditions. Suppose that at some instant $t$ the coordinates of the particle in each system in the ensemble are measured. The measured values will not all be the same, as would be the case classically, but instead will be distributed over some range of coordinate values. The quantity $\rho(x, t)\ dx$ then gives the fraction of the systems in the ensemble for which the measured coordinates lie between $x$ and $x + dx$.

One important property of state functions must still be emphasized. The existence of interference, the observation of which led us to associate wave properties with particles in the first place, implies that if $\psi_1$ describes one possible state of the system and if $\psi_2$ describes a second possible state, then

$$\psi_3 = a_1\psi_1 + a_2\psi_2,$$

with $a_1$ and $a_2$ arbitrary, also describes a possible state of the system. By extension, we see that an arbitrary superposition of any set of possible state functions is also a possible state function. This is called the *principle of superposition.* That this principle applies is one of our basic assumptions; its applicability sharply differentiates the probablistic aspects of quantum mechanics from those of classical statistical mechanics.

To make the relationship between interference and the superposition principle clear, consider the probability density corresponding to the particular superposition $\psi_3$ defined above. We have

$$\psi_3{}^*\psi_3 = |a_1|^2\psi_1{}^*\psi_1 + |a_2|^2\psi_2{}^*\psi_2 + a_1a_2{}^*\psi_1\psi_2{}^* + a_1{}^*a_2\psi_1{}^*\psi_2.$$

The first two terms give just the sum of the individual probabilities for each state, weighted by the extent to which each is present in the superposition, exactly as would be the case classically. The last two terms are the interference terms. These terms are not expressible solely in terms of the individual probabilities associated with each state, but are simultaneously and mutually a property of both states. Their sign is determined by the relative phase of $a_1\psi_1$ and $a_2\psi_2$, and it can be either positive or negative corresponding to constructive or destructive interference in the probabilities. The radical nature of this behavior must not be overlooked. It means that a set of states, each of which independently describes the occurrence of some event with finite probability, can be combined in such a way that the given event cannot occur at all!

An interesting example is the famous double slit experiment, in which the interference pattern of a beam of particles incident upon a double slit system in an opaque screen is studied. The experiment is shown schematically in Figure 1(a). The first screen contains identical slits at

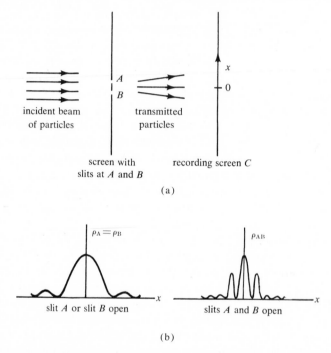

(a)

(b)

FIGURE 1. The double slit experiment. (a) Schematic experimental arrangement. (b) Distribution of particles recorded on the screen $C$.

$A$ and at $B$, either of which can be opened or closed. Any electrons passing through the slit system are recorded on the distant screen $C$. In Figure 1(b), the distribution of particles recorded when either $A$ or $B$ is open is shown on the left, that when both $A$ and $B$ are open is shown on the right. In the former case, the result is the typical Fraunhofer pattern, in the latter this pattern is modulated by interference and is clearly *not* the superposition of the *probability* for transmission through either slit alone. To relate this to the superposition principle, let $\psi_A$ denote the state function of an electron for $A$ open and $B$ closed, $\psi_B$ that for $B$ open and $A$ closed, and $\psi_{AB}$ that for both $A$ and $B$ open. Let $\rho_A$, $\rho_B$ and $\rho_{AB}$ denote the corresponding probability densities. Then, to good approximation, we have

$$\psi_{AB} = \psi_A + \psi_B,$$

whence

$$\rho_{AB} \equiv |\psi_{AB}|^2 = |\psi_A|^2 + |\psi_B|^2 + \psi_A{}^*\psi_B + \psi_B{}^*\psi_A.$$

Since $\rho_A = \rho_B$, we thus have, in sharp contrast to the classical result $\rho_{AB} = \rho_A + \rho_B = 2\rho_A$,

$$\rho_{AB} = 2\rho_A[1 + \cos \delta(x)],$$

where $\delta(x)$ is the phase of $\psi_B$ relative to $\psi_A$,

$$\psi_B = \psi_A e^{i\delta}.$$

The phase factor $\delta$ increases linearly with distance from 0 along the recording screen, and interference minima occur whenever $\delta$ is an odd multiple of $\pi$. We here see quite explicitly how superposition leads to interference. Note, in particular, that when both slits are open the probability of an electron arriving at the screen at an interference minimum is zero, even though its probability of reaching the same point on the screen is quite finite when only one slit is open!

One further aspect of this experiment deserves comment. The electron's particle nature manifests itself in the fact that an electron is, after all, a localizable entity. When detected or recorded in any way, it is always observed as just that; one never sees only a part of an electron. Thus an electron passing through the first screen must pass through one slit or the other. If it passes through $A$, how can it know about $B$ and thus somehow adjust its behavior to give the experimental result? The answer is that, in just this respect, the electron is not localized; it also has attributes which are distributed in space like a wave. In short, it exhibits *both* particle and wave properties. The complementary aspects of this duality are emphasized by introducing an additional detector that permits one to observe through which of the two slits a given electron actually passes. This can be done, and sure enough, each electron is always observed to pass through one slit or the other. However, the act of observation necessarily involves an interaction of some kind between the measuring apparatus and the electron, and this interaction produces an uncontrollable disturbance which destroys the phase relationship necessary for interference. In other words, as one observes which slit the electron passes through, one forces it to act entirely like a particle, and thus the wave-generated interference pattern disappears and the classical result appropriate to classical particles is observed.

## 2. EXPECTATION VALUES

Given our probabilistic interpretation of the state function $\psi(x, t)$, we now show how to extract information from it concerning the behavior of a particle. Specifically, recalling that $\rho(x, t)$ refers to the distribution of measured values of the particle coordinate for an ensemble of systems, we see that the (ensemble) average, or *expectation value*, of the position, written $\langle x \rangle$, is simply

$$\langle x \rangle = \int x \rho(x, t) \, dx, \tag{5}$$

where the integral extends over all space. We emphasize that this follows just because $\rho(x, t) \, dx$ is that fraction of the measured values of position which lies between $x$ and $x + dx$. Suppose now that we are concerned about some function of the position of the particle, $f(x)$. Then $\rho(x, t) \, dx$ is the fraction of the times the measured value of $f(x)$ would lie between $f(x)$ and $f(x + dx)$. Hence we have, for the (ensemble) average or expectation value of $f(x)$, in the same notation,

$$\langle f(x) \rangle = \int f(x) \rho(x, t) \, dx. \tag{6}$$

As an example, if a particle is moving in a potential $V(x)$, and its probability density function is $\rho(x, t)$, then its mean potential energy can be computed according to equation (6), with $f(x) = V(x)$.

Let us express these expectation values in terms of the state function $\psi(x, t)$. We have at once

$$\langle f(x) \rangle = \frac{\int \psi^*(x, t) f(x) \psi(x, t) \, dx}{\int \psi^* \psi \, dx} \tag{7}$$

or, if the state function is normalized,

$$\langle f(x) \rangle = \int \psi^*(x, t) f(x) \psi(x, t) \, dx. \tag{8}$$

Of course, the order of the factors in the integrand of these expressions is a matter of indifference. We could equally well have written $f\psi^*\psi$ or $\psi^*\psi f$, both of which are less complicated-looking than the form obtained by inserting $f$ between $\psi^*$ and $\psi$. We have chosen this last, however, for reasons of future convenience.

We have seen how to calculate the quantum analog of the position of a particle (or any function of its position). What about the remaining one-dimensional dynamical variable, the momentum? One way of proceeding might be thought to be the following. In general, since $\psi = \psi(x, t)$, the expectation value of $x$ is a function of time, $\langle x \rangle = f(t)$. Hence the quantity $m d\langle x \rangle / dt$ can certainly be calculated if the time dependence of $\psi$ is known. This quantity ought then to correspond to the momentum, at least in the classical limit. There are two difficulties with this approach. The first is a fundamental one having to do with the nature of the momentum as a dynamical variable. Classically, the existence of a trajectory gives a precise meaning to the mathematical operations involved in evaluating the quantity $m\,dx/dt$. In quantum mechanics, no precise trajectories exist and the quantity $dx/dt$ must presently be regarded as undefined. It thus makes no sense at this stage to talk about $p$, if it is merely defined as $m\,dx/dt$, that is, as a purely kinematical quantity. On the other hand, $p$ must certainly have a dynamical meaning, quite independently of trajectories. Regarded as a dynamical variable, on the same footing as the position variable, we must make sense out of the momentum and out of such related quantities as its expectation value $\langle p \rangle$, and indeed this is our next task.[3]

The second difficulty referred to above is more of a practical kind. To compute a quantity like $d\langle x \rangle/dt$, we must know the answer to the third question asked at the beginning of this chapter: How do state functions develop in time? We are not yet prepared to answer that question. Indeed, once we understand the momentum as a quantum mechanical dynamical variable, we shall make use of the correspondence principle requirements

$$\langle p \rangle = \frac{m\,d\langle x \rangle}{dt}$$

and

$$\frac{d\langle p \rangle}{dt} = -\left\langle \frac{dV(x)}{dx} \right\rangle$$

in order to establish the time dependence of state functions.

## 3. COMPARISON BETWEEN THE QUANTUM AND CLASSICAL DESCRIPTIONS OF A STATE; WAVE PACKETS

Our discussion has been rather far removed from classical physics, in which we are accustomed to prescribing the precise position and velocity of a particle at some instant and not a probability distribution, much less an intrinsically unobservable probability amplitude. Since quantum mechanics is intended to be more general than classical mechanics, which it must contain as a limiting case, we now discuss the sense in which we can, in fact, recover the classical description, starting from the concept of a quantum mechanical state function. Our task is not a difficult one. A classical trajectory is nothing more than some curve in space which evolves in time in some definite way. The quantum mechanical state function has all of space and time as its domain. Although it thus

---

[3] Classically, the description which places position and momentum on an equal footing as dynamical variables is the Hamiltonian description. We thus anticipate that the Hamiltonian function will bear closely on the formulation of quantum mechanical laws.

appears to be an inherently non-localized entity, it can certainly be used to describe a trajectory if it is simply chosen to be a very special and localized space-time function—namely one which vanishes everywhere except in the infinitesimal neighborhood of the trajectory in question.

Such localized, or sharply peaked, state functions are called *wave packets*. They play a key role in the isolation of many physical effects, and particularly, of course, in understanding the relationship between classical and quantum mechanics. An example of a wave packet at some given instant is the Gaussian function,

$$\psi = A \, \exp\left[-(x - x_0)^2/2L^2\right]. \tag{9}$$

Noting that the relative probability distribution is then

$$\psi^*\psi = |A|^2 \exp\left[-(x - x_0)^2/L^2\right], \tag{10}$$

we see that we have here a state localized about the point $x = x_0$ within a neighborhood of dimension $L$. The smaller $L$ is, the more localized the state function; the classical limit of absolute precision corresponds to the limit in which $L$ approaches zero.

Specification of the state function at a given instant is entirely analogous to the classical specification of the initial position of a particle. If one seems more vague and mysterious than the other, it is only because, at the classical level, we are accustomed to the establishment of initial conditions through our own direct and personal involvement, at least in imagination, as when we throw a piece of chalk or set into motion a mechanism that fires a satellite. At both levels, the details by which initial conditions are established are irrelevant to the subsequent developments; all we need to know is what the initial conditions in fact are. That we are not yet able to discuss *how* a well-defined quantum mechanical initial state is actually prepared thus need not be a source of difficulty. To repeat, we need to know only what the initial state is, not where it came from.

Given some initial state, its time development is, of course, determined by the equations of motion, both classically and quantum mechanically.[4] Suppose the classical equations of motion upon integration yield the trajectory

$$x = f(t).$$

It is then tempting to guess that a suitable form for the corresponding quantum mechanical probability function in the classical limit is

---

[4] The initial position and momentum must *both* be specified, of course, in the classical case. In the quantum mechanical case, both *cannot* be prescribed with arbitrary precision. Information about the momentum is *implicity* contained in the state function. How to extract that information is the subject of the following chapter.

$$\psi^*\psi = |A|^2 \exp\{-[x - f(t)]^2/L^2\}$$

for sufficiently small $L$. This expression represents a wave packet of width $L$ moving along the classical trajectory in accordance with the classical equations of motion. This intuitive supposition can be explicitly tested for the special case of the motion of a free particle. For such a particle, of mass $m$, say, starting from the origin with initial momentum $p_0$, we have classically,

$$x = p_0 t/m,$$

and we thus are supposing that the quantum mechanical probability distribution might be given by the moving wave packet

$$\psi^*\psi = |A|^2 \exp[-(x - p_0 t/m)^2/L^2]. \tag{11}$$

The actual result, obtained in Chapter IV (equation IV-22) by integration of the quantum mechanical equations of motion, is precisely this, except that the constant width $L$ is replaced by the time dependent width

$$L(t) = \sqrt{L^2 + (\hbar^2 t^2/m^2 L^2)}\,.$$

Thus the correct result reveals that, in actuality, the wave packet grows in size from its initial width $L$. However, for *macroscopic* particles the second term under the square root sign is readily seen to remain negligible over *cosmological* time intervals,[5] hence equation (11) deviates undetectably from the correct result and our intuitive expectations are substantially correct. We shall return to this subject again in Chapter IV.

---

**Problem 1.** Consider a particle described by a Gaussian wave packet,

$$\psi = A \exp[-(x - x_0)^2/2a^2].$$

(a)   Calculate $A$ if $\psi$ is normalized.

(b)   Calculate $\langle x \rangle$.

(c)   Calculate the mean square deviation in the particle's position, $\langle (x - \langle x \rangle)^2 \rangle$.

(d)   Suppose the particle is moving in a potential $V(x)$. Calculate $\langle V \rangle$ for $V = mgx$; for $V = \frac{1}{2}kx^2$. See Appendix I for the evaluation of Gaussian Integrals.

**Problem 2.**

(a)   The same as Problem 1, except for the state function

$$\psi_1 = A \exp[i(x - x_0)/a] \exp[-(x - x_0)^2/2a^2].$$

(b)   Consider the superposition state

[5] This follows because $\hbar$ is so small in macroscopic terms.

$$\psi_{\pm} = c_{\pm} \, [\psi_1 \pm \psi]$$

where $\psi$ is the wave packet of Problem 1, $\psi_1$ that of part (a) above. Evaluate $c_{\pm}$. Plot and compare the probability density for the four cases $\psi^*\psi$, $\psi_1^*\psi_1$, $\psi_+^*\psi_+$, $\psi_-^*\psi_-$.

# III

# *Linear momentum*

## 1. STATE FUNCTIONS CORRESPONDING TO A DEFINITE MOMENTUM

We have now come to understand some of the properties of state functions and have seen that our next task is to understand linear momentum as a quantum mechanical dynamical variable. The essential clue is provided by the de Broglie description of a free particle of definite momentum $p$. Associated in some way with such a particle, we have argued, is a wave of reduced wavelength $\lambda = \hbar/p$. We now make this vague relationship explicit by assuming that *the de Broglie wave itself is the state function of the particle.* Specifically, we write

$$\psi(x, t) = \exp[i(x/\lambda) - i\omega t]$$

or, expressing $\lambda$ in terms of $p$,

$$\psi_p(x, t) = \exp[i(px/\hbar) - i\omega(p)t], \tag{1}$$

where we have attached a subscript $p$ to $\psi$ to denote that this state function describes a particle which is moving with definite, fixed linear momentum $p$. The frequency $\omega$ of de Broglie waves has not yet received any special attention, and in writing equation (1) we have therefore taken $\omega$ to be some characteristic, but as yet unknown, function of $p$.

The identification of that particular state function which describes a particle with definite momentum is an absolutely crucial step in our method of development. It is offered here as a reasonably direct, but hardly unambiguous, deduction from the Davisson–Germer experiment. So there will be no misunderstanding, we state as emphatically as possible that the quantum mechanical rabbit is already in the hat, once equation (1) is accepted and understood. Except for spin and the exclusion principle, *all else follows from the correspondence principle alone.*

Because of the importance of this result, we shall comment on it in some detail. Note first that we have written $\psi_p$ as a *complex* exponential function. This choice requires elaboration because a traveling wave can certainly be represented by a real trigonometric function as easily as by an exponential. Indeed, all classical wave fields are actually represented by such real functions, even if complex notation is used for convenience. That this striking property is essential for quantum mechanical state functions can be made plausible by the following argument: For a free particle, all points in space are physically equivalent. In particular, the choice of origin is irrelevant; the state of the system cannot depend in any essential way on this choice. Suppose, now, that the origin is shifted to the left through some arbitrary distance $b$, by which we mean that $x$ is replaced by $x + b$. Then, as defined by equation (1), $\psi_p$ is merely multiplied by the physically undetectable constant phase factor $e^{ipb/\hbar}$. As required, the description of the state is seen to contain no *physically* significant reference to the origin. This would not be the case were a real trigonometric function used to represent $\psi_p$. In fact, if we had started with the most general possible traveling wave,

$$\psi = A \cos (px/\hbar - \omega t) + B \sin (px/\hbar - \omega t),$$

the demand that $\psi$ reduce to a multiple of itself under an arbitrary translation then would at once have led us to the exponential form of equation (1).[1]

---

**Exercise 1.** Prove this last assertion.

---

Still another feature of $\psi_p$ requires comment. It is a state function corresponding to a total absence of localization in space. The relative probability density is

$$\psi_p{}^*\psi_p = 1,$$

which means that the particle is just as likely to be found in any one volume element as in any other. As an immediate consequence, the state function $\psi_p$ is not physically admissible, except in the weak sense referred to in the footnote following equation (II-3). Nonetheless, because $\psi_p$ does correspond to a precise value of the momentum $p$ it is a useful idealization, as we shall at once show.

---

[1] A more conventional, and perhaps more convincing, argument can be given in terms of the requirement that the probability of finding the particle *somewhere* in space must be unity *for all times*. We shall return to the subject in Section 7 of Chapter IV.

## 2. CONSTRUCTION OF WAVE PACKETS
## BY SUPERPOSITION

We now give an important and instructive example of the utility of these idealized non-physical states by combining them to form a wave packet, the most intuitively physical kind of state function. We do this by constructing a general superposition of momentum states $\psi_p$. Since there is a continuum of possible values of $p$, the superposition takes the form of an *integral* rather than a sum and we write

$$\psi(x, t) = \frac{1}{\sqrt{2\pi\hbar}} \int_{-\infty}^{\infty} \phi(p) \, \exp[i(px/\hbar) - i\omega(p)t] \, dp \qquad (2)$$

where the factor $1/\sqrt{2\pi\hbar}$ has been introduced for reasons of future convenience. In this superposition, the amplitude of the state function $\psi_p$ corresponding to momentum $p$ is denoted by $\phi(p)$. For the present we shall not be concerned with the time dependence of state functions or with the relationship between $\omega$ and $p$. We shall consider instead only the description at some fixed instant, which we take to be $t = 0$ for simplicity. We thus write, in place of equation (2),

$$\psi(x) = \frac{1}{\sqrt{2\pi\hbar}} \int_{-\infty}^{\infty} \phi(p) \, e^{ipx/\hbar} \, dp, \qquad (3)$$

where now

$$\psi(x) \equiv \psi(x, t = 0).$$

It is perhaps helpful at this stage to give an example, even if a purely mathematical one, of how a physically admissible normalizable state $\psi(x)$ can in fact be obtained by superposition of the idealized inadmissible momentum states $\exp[ipx/\hbar]$. To particularize to a very simple case,

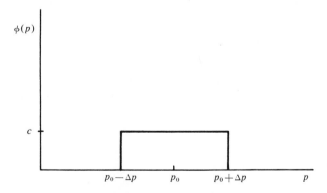

FIGURE 1.   The momentum distribution of equation (4).

let us suppose that $\phi(p)$ is constant in some interval of width $\Delta p$ on each side of some fixed momentum $p_0$, and is identically zero outside of that interval. In other words, we choose $\phi(p)$ to be a square distribution,

$$\phi(p) = \begin{cases} c, & |p - p_0| \leq \Delta p \\ 0, & |p - p_0| > \Delta p, \end{cases} \tag{4}$$

where $c$ is an arbitrary constant. This means that the state we are considering is one in which the momentum does not have some precisely determined numerical value but is instead distributed uniformly over a band of width $2\Delta p$ centered about $p_0$, as illustrated in Figure 1.

With this choice for $\phi(p)$, equation (3) becomes

$$\psi(x) = \frac{1}{\sqrt{2\pi\hbar}} \int_{p_0 - \Delta p}^{p_0 + \Delta p} c\, e^{ipx/\hbar}\, dp$$

$$= \sqrt{\hbar/2\pi}\, \frac{c}{ix} \left[ e^{i(p_0 + \Delta p)x/\hbar} - e^{i(p_0 - \Delta p)x/\hbar} \right]$$

or, factoring out the term $e^{i p_0 x/\hbar}$ and simplifying,

$$\psi(x) = c\sqrt{2\hbar/\pi}\, \frac{\sin \Delta px/\hbar}{x}\, e^{ip_0 x/\hbar}. \tag{5}$$

This example has thus yielded a state function which is a de Broglie wave corresponding to the momentum $p_0$, but modulated by the factor $(1/x) \sin (\Delta px/\hbar)$. This factor makes $\psi(x)$ normalizable and thus physically admissible. To carry the example one step further, let us fix the constant $c$ in equation (4) and equation (5) by actually normalizing $\psi$. We have

$$1 = \int_{-\infty}^{\infty} \psi^*(x)\psi(x)\, dx = \frac{2\hbar|c|^2}{\pi} \int_{-\infty}^{\infty} \frac{\sin^2 \Delta px/\hbar}{x}\, dx$$

whence, introducing the new integration variable $u = \Delta px/h$, we obtain

$$|c| = 1/\sqrt{2\Delta p}, \tag{6}$$

where we have made use of the known result

$$\int_{-\infty}^{\infty} \frac{\sin^2 u}{u^2}\, du = \pi. \tag{7}$$

To summarize, the particular momentum distribution of equation (4) yields by superposition the normalized wave packet of equation (5), provided $c$ satisfies equation (6).

Observe that in the limit as $\Delta p$ approaches 0, we recover a pure de Broglie wave of momentum $p_0$, that is to say

$$\lim_{\Delta p \to 0} \psi(x) = \sqrt{\Delta p/\pi\hbar}\, e^{ip_0 x/\hbar}.$$

The appearance of the factor $\sqrt{\Delta p}$, which means that the amplitude of $\psi$ becomes infinitesimal, is a consequence of the fact that the state function is not normalizable in this limit. We thus cannot proceed to the limit in the usual way. If, however, the relevant physical dimensions of our physical system are say $L$, then the state function need be considered only over a domain of that dimension. Hence, if $\Delta p L/\hbar \ll 1$, the normalizable wave packet state deviates undetectably from the pure de Broglie state. This means that the above limit is in fact *physically* achievable, which is to say that $\Delta p$ can be regarded as *effectively* zero, provided it is much smaller than $\hbar/L$. We have here an illustration of the way in which the non-physical and non-normalizable pure momentum states can nonetheless serve as useful idealizations of true physical states.

Returning now to the general case, we seek to interpret the amplitude $\phi(p)$ which appears in the superposition integral defined in equation (3). Suppose we had been less ambitious and had considered the superposition of only two momentum state functions,

$$\psi(x) = a_1 e^{ip_1 x/\hbar} + a_2 e^{ip_2 x/\hbar}.$$

Evidently, this combination corresponds to a state in which the momentum is either $p_1$ or $p_2$, with relative probability amplitudes $a_1$ and $a_2$, respectively. The more general state of equation (3) is a state in which all possible momenta are present with probability determined by $\phi(p)$. It is natural to assume that $\phi(p)$ is proportional to the momentum probability amplitude or the probability amplitude in momentum space. If $\rho(p)$ denotes the corresponding probability density, we thus take the probability that the particle has momentum between $p$ and $p + dp$ to be

$$\rho(p)\ dp = \frac{\phi^*\phi\ dp}{\displaystyle\int_{-\infty}^{\infty} \phi^*(p)\phi(p)\ dp},$$

where, as indicated, the integral extends over all momentum space. If $\phi(p)$ is normalized so that

$$\int \phi^*(p)\phi(p)\ dp = 1,$$

then we have simply

$$\rho(p) = \phi^*\phi, \tag{9}$$

and $\phi(p)$ gives the probability amplitude directly.

Accepting this interpretation, we now have a first answer to our question about the momentum as a dynamical variable. If $\phi(p)$ is given or known, we have, in exact analogy to our procedure in $x$-space, or configuration space,

$$\langle p \rangle = \int p\rho(p) \, dp$$
$$= \int \phi^*(p)p\phi(p) \, dp,$$

where $\langle p \rangle$ is to be interpreted as the average momentum for an ensemble of identical systems identically prepared. More generally, for any function of the momentum $f(p)$, we have

$$\langle f(p) \rangle = \int \phi^*(p)f(p)\phi(p) \, dp.$$

As a particular example, the expectation value of the kinetic energy is

$$\left\langle \frac{p^2}{2m} \right\rangle = \int \phi^*(p) \, \frac{p^2}{2m} \, \phi(p) \, dp.$$

We now see how to handle the linear momentum $p$ in a manner analogous to that used for the coordinate $x$. Some questions still remain, however. If $\phi(p)$ is given, we know that the corresponding $\psi(x)$ is uniquely determined by equation (3). What about the converse—how is $\phi(p)$ to be determined if $\psi(x)$ is given? Note, moreover, that the unique relationship between $\phi(p)$ and $\psi(x)$ implied by equation (3) means that if one of these functions is normalized, there is no freedom left to normalize the other. Hence as a test of the consistency of our formulation and its interpretation, we must demand that if $\phi(p)$ is normalized so must $\psi(x)$ be, and conversely. In the particular example of equations (4) and (5), it is readily verified that this consistency requirement is indeed satisfied. Recall that $\psi(x)$ is normalized if $|c| = 1/\sqrt{2\Delta p}$ and hence, from equation (4),

$$\int \phi^*\phi \, dp = \frac{1}{2\Delta p} \int_{p_0-\Delta p}^{p_0+\Delta p} dp = 1,$$

so that $\phi$ is also normalized. Of course, we must demonstrate that this is true in general and not merely for such special examples.

To answer these and similar questions, we now make a brief mathematical digression into the properties of Fourier integrals, as integrals of the form of equation (3) are called.

## 3. FOURIER TRANSFORMS; THE DIRAC DELTA FUNCTION[2]

Recall that a function $f(\theta)$, piecewise continuous in the interval $-\pi \leqslant \theta \leqslant \pi$, can be represented by a Fourier series. Writing such a series in exponential form we have

---

[2] See items [6] through [13] in the selected list of references given in Appendix II.

$$f(\theta) = \sum_{-\infty}^{\infty} A_n \, e^{in\theta},$$

where

$$A_n = \frac{1}{2\pi} \int_{-\pi}^{\pi} f(\theta) \, e^{-in\theta} \, d\theta$$

Now replace $\theta$ by $\pi x/L$. We then obtain

$$f(x) = \sum_{-\infty}^{\infty} A_n \, e^{in\pi x/L}$$

$$A_n = \frac{1}{2L} \int_{-L}^{L} f(x) \, e^{-in\pi x/L} \, dx.$$

We next want to consider the limit of these expressions when $L$ tends to infinity. To prepare the way, we write

$$k_n \equiv n\pi/L$$

$$\Delta k \equiv k_{n+1} - k_n = \pi/L$$

whence, also,

$$k_n = n\Delta k.$$

We next write

$$A_n \equiv (1/L) \, \sqrt{\pi/2} \, g(k_n) = (1/L) \, \sqrt{\pi/2} \, g(n\Delta k).$$

Then, putting this all together,

$$f(x) = \sum_{-\infty}^{\infty} \frac{\sqrt{\pi/2}}{L} \, g(n\Delta k) \, e^{in\Delta kx} = \sqrt{1/(2\pi)} \sum_{-\infty}^{\infty} g(n\Delta k) \, e^{in\Delta kx} \, \Delta k$$

$$g(n\Delta k) = \sqrt{1/(2\pi)} \int_{-L}^{L} f(x) \, e^{-in\Delta kx} \, dx$$

and letting $L \to \infty$, so that $\Delta k \to 0$ while $n\Delta k \to k$, we have, finally, utilizing the elementary definition of the integral as the limit of a sum,

$$f(x) = \frac{1}{\sqrt{2\pi}} \int_{-\infty}^{\infty} g(k) \, e^{ikx} \, dk$$

$$g(k) = \frac{1}{\sqrt{2\pi}} \int_{-\infty}^{\infty} f(x) \, e^{-ikx} \, dx. \tag{11}$$

The symmetrically related pair of functions $f(x)$ and $g(k)$ are called *Fourier transforms,* one of the other, and the expressions of equation (11) are called *Fourier integral representations.*

Equation (11) tells us how to calculate $f(x)$, given $g(k)$, and conversely. One interesting aspect of these relations is the following. Consider some arbitrary $f(x)$ and suppose $g(k)$ to be determined from the second of the equations in (11). Substituting the resulting expression for $g(k)$ in terms of $f(x)$ back into the first equation, we then have the identity

$$f(x) = \frac{1}{2\pi} \int_{-\infty}^{\infty} dk \, e^{ikx} \int_{-\infty}^{\infty} f(x') \, e^{-ikx'} \, dx',$$

where we have introduced $x'$ as the dummy integration variable in the integral representation of $g(k)$. Now take the mathematically risky step of interchanging the order of integration. The result can be written in the form

$$f(x) = \int_{-\infty}^{\infty} dx' \, f(x')\delta(x - x'), \tag{12}$$

where we have introduced the abbreviation

$$\delta(x - x') \equiv \frac{1}{2\pi} \int_{-\infty}^{\infty} e^{ik(x-x')} \, dk. \tag{13}$$

The function $\delta(x - x')$, first introduced by Dirac, is called the *Dirac delta function*.[3] Since $f(x)$ is arbitrary within wide bounds, the delta function evidently has very strange properties. These properties are readily identified from the form of equation (12), which, after all, simply states that the product of an arbitrary function $f$ and the $\delta$-function, when integrated over all space, yields as an answer the value assumed by $f$ at that particular isolated point where the argument of the $\delta$-function vanishes. In other words, $\delta(z)$, say, singles out in the integration only the value of $f(z)$ at the point $z = 0$. The behavior of $f(z)$ everywhere else is irrelevant. This implies that $\delta(z)$ *vanishes everywhere except at the point $z = 0$.* At $z = 0$, it becomes indefinitely large, but in such a way that it remains integrable. This last is easily made explicit by observing that, if $f(x)$ is chosen to be a constant, then equation (12) at once yields

---

[3] Replacing $x - x'$ by $z$, and $k$ by $y$, our definition can be restated in the form

$$\delta(z) \equiv \frac{1}{2\pi} \int_{-\infty}^{\infty} e^{iyz} \, dy.$$

This means, for example, that in analogy to equation (13) we also have

$$\delta(k - k') = \frac{1}{2\pi} \int_{-\infty}^{\infty} e^{i(k-k')x} \, dx.$$

$$\int_{-\infty}^{\infty} \delta(x - x') \, dx' = 1, \tag{14}$$

so that the integral of a $\delta$-function is normalized to unity.[4]

To summarize, the significant properties of the Dirac $\delta$-function are defined by equations (14) and (12), respectively, and equation (13) is its Fourier integral representation. A list of additional useful properties are the following:

$$\delta(-x) = \delta(x) \tag{15}$$

$$a\delta(\pm ax) = \delta(x), \qquad a > 0 \tag{16}$$

$$\delta(x^2 - a^2) = \frac{1}{2a} \left[ \delta(x - a) + \delta(x + a) \right] \tag{17}$$

$$\int_{-\infty}^{\infty} f(x) \, \frac{d\delta(x - a)}{dx} \, dx = -\frac{df}{dx}\bigg|_{x=a} . \tag{18}$$

The proof of these relations is left to the problems.

As a final mathematical point we now mention the convolution theorem. Suppose $f_1(x)$ and $f_2(x)$ are given arbitrary functions with Fourier transforms $g_1(k)$ and $g_2(k)$, respectively. This theorem states that

$$\int_{-\infty}^{\infty} e^{-ikx} f_1(x) f_2(x) \, dx = \int_{-\infty}^{\infty} g_1(k') g_2(k - k') \, dk'. \tag{19}$$

The proof is not difficult, but it provides an instructive exercise in the manipulation of Fourier integrals. Substitution of the Fourier integral representations for $f_1(x)$ and $f_2(x)$ gives

$$\int dx \, e^{-ikx} f_1(x) f_2(x) = \frac{1}{2\pi} \int dx \, e^{-ikx} \int dk' \, g_1(k') \, e^{ik'x}$$

$$\times \int dk'' \, g_2(k'') \, e^{ik''x}.$$

Noting that the dependence on $x$ of the right side is explicit, we interchange the order of integration and evaluate the integral over $x$ first. Specifically, we write

---

[4] An alternative version of our argument can be constructed as follows. Consider equation (12) for fixed $x$, say $x = b$. Suppose $f(x)$ to be altered by some arbitrary amount $\eta(x)$ in the infinitesimal neighborhood of any point $a \neq b$. The left side of equation (12) is unchanged; it remains $f(b)$, and hence the additional contribution on the right must yield zero. This at once implies $\delta(a - b) = 0$, $b \neq a$, in agreement with our conclusion above. Incidentally, it may help the student to visualize the properties of the $\delta$-function if he considers it to be the limit of a sharply peaked but perfectly well-behaved function, for example, a Gaussian, as discussed in the problems.

$$\int dx \ e^{-ikx} f_1(x) f_2(x) = \int\int_{-\infty}^{\infty} dk' \ dk'' \ g_1(k') g_2(k'')$$

$$\times \frac{1}{2\pi} \int dx \ e^{i(k''+k'-k)x}.$$

The last factor is now recognized as $\delta(k'' - k + k')$ according to equation (13). Finally, evaluation of the integral over $k''$ gives the stated result. As a special case we note that

$$\int_{-\infty}^{\infty} f^*(x) f(x) \ dx = \int_{-\infty}^{\infty} g^*(k) g(k) \ dk. \tag{20}$$

---

**Exercise 2.** Prove equation (20).

---

## 4. MOMENTUM AND CONFIGURATION SPACE

We are now in a position to provide a precise relationship between wave functions in configuration space, $\psi(x)$, and in momentum space, $\phi(p)$. We had as equation (3)

$$\psi(x) = \frac{1}{\sqrt{2\pi\hbar}} \int_{-\infty}^{\infty} \phi(p) \ e^{ipx/\hbar} \ dp$$

whence, according to equation (11),

$$\phi(p) = \frac{1}{\sqrt{2\pi\hbar}} \int_{-\infty}^{\infty} \psi(x) \ e^{-ipx/\hbar} \ dx. \tag{21}$$

Note that only in a system of units in which $\hbar = 1$, often used by physicists, is $\phi$ the Fourier transform of $\psi$. More generally,

$$\overline{\phi}(p/\hbar) \equiv \sqrt{\hbar} \ \phi(p)$$

is its transform, the transform variable being $p/\hbar = 2\pi/\lambda \equiv k$, where $k$ is recalled to be the (reduced) wave number. With this identification, equation (20) becomes, with $f = \psi$, $g = \sqrt{\hbar} \ \phi$,

$$\int \psi^* \psi \ dx = \int \phi^* \phi \ dp$$

and the condition that $\psi$ and $\phi$ be simultaneously normalizable is indeed satisfied.

We have now established two completely equivalent representations of, or ways of writing, a given state function, one in configuration space, one in momentum space. Neither contains more information than the other, neither has special claims to distinction. Together, they permit us to treat position and linear momentum on an equal footing as dynamical variables.

## 5. THE MOMENTUM AND POSITION OPERATORS

Given some state function, $\psi(x)$, we now know how to calculate expectation values of any function of position, or of any function of momentum. In the latter case, we must calculate the state function in momentum space, or briefly, we must transform to momentum space. This procedure is clearly rather roundabout and inefficient. Can we not instead calculate expectation values of momentum from $\psi(x)$ directly? More than a matter of convenience is actually involved, since our present technique will not permit us to calculate expectation values of mixed functions of position and momentum, such as angular momentum, for example.

To answer our question, consider first

$$\langle p \rangle = \int \phi^*(p) p \phi(p) \, dp.$$

Using equation (17) to express $\phi(p)$ in terms of $\psi(x)$ and similarly for $\phi^*(p)$ in terms of $\psi^*(x)$, we obtain

$$\langle p \rangle = \frac{1}{2\pi\hbar} \int\int\int dp \, dx \, dx' \, \psi^*(x') \, e^{ipx'/\hbar} \, p \, \psi(x) \, e^{-ipx/\hbar}. \qquad (22)$$

Now the dependence on $p$ is explicit and we therefore seek to perform the $p$ integration first. This integration becomes very simple if the factor $p$ in the integrand is eliminated. To do so use the fact that

$$p \, e^{-ipx/\hbar} = -\frac{\hbar}{i} \frac{d}{dx} e^{-ipx/\hbar}, \qquad (23)$$

as is readily verified by performing the indicated differentiation on the right-hand side. Introducing this result into equation (22), we obtain

$$\langle p \rangle = \frac{1}{2\pi\hbar} \int\int\int dp \, dx \, dx' \, \psi^*(x') \, e^{ipx'/\hbar} \, \psi(x) \left( -\frac{\hbar}{i} \frac{d}{dx} e^{-ipx/\hbar} \right).$$

Next integrate with respect to $x$ by parts. The integrated part is proportional to the value of $\psi(x)$ at infinity and hence is zero for physically admissible state functions because all such functions necessarily vanish at infinity. Thus the result of the integration by parts is

$$\langle p \rangle = \frac{1}{2\pi\hbar} \int\int\int dp \, dx \, dx' \, \psi^*(x') [e^{ip(x'-x)/\hbar}] \frac{\hbar}{i} \frac{d\psi(x)}{dx}.$$

The $p$ integration is now easily performed and we obtain

$$\langle p \rangle = \int\int\int dx \, dx' \, \psi^*(x') \, \frac{\hbar}{i} \frac{d\psi(x)}{dx} \, \delta(x' - x)$$

whence, finally,

$$\langle p \rangle = \int dx \, \psi^*(x) \, \frac{\hbar}{i} \frac{d}{dx} \psi(x). \qquad (24)$$

Next consider the expectation value of $p^n$ for arbitrary $n$. We have, using the same technique,

$$\langle p^n \rangle = \int \phi^*(p) p^n \phi(p) \, dp$$

$$= \frac{1}{2\pi\hbar} \int\int\int dp \, dx' \, dx \, \psi^*(x') \, e^{ipx'/\hbar} \, p^n \, \psi(x) \, e^{-ipx/\hbar}$$

or

$$\langle p^n \rangle = \frac{1}{2\pi\hbar} \int\int\int dp \, dx' \, dx \, \psi^*(x') \, e^{ipx'/\hbar} \, \psi(x) \left( -\frac{\hbar}{i} \frac{d}{dx} \right)^n e^{-ipx/\hbar} \qquad (25)$$

where, in the last step, we have used the following generalization of equation (23):

$$p^n \, e^{-ipx/\hbar} = \left( -\frac{\hbar}{i} \frac{d}{dx} \right)^n e^{-ipx/\hbar} \equiv \left( -\frac{\hbar}{i} \right)^n \frac{d^n(e^{-ipx/\hbar})}{dx^n}. \qquad (26)$$

Integration of equation (25) by parts with respect to $x$, $n$ times repeated, then yields, since the integrated part always vanishes,

$$\langle p^n \rangle = \frac{1}{2\pi\hbar} \int\int\int dp \, dx' \, dx \, \psi^*(x') \, e^{ip(x'-x)/\hbar} \left( \frac{\hbar}{i} \frac{d}{dx} \right)^n \psi(x).$$

Again the $p$ integration yields $2\pi\hbar \, \delta(x' - x)$ and hence, after the integration over $x'$ is performed, we obtain as the final result

$$\langle p^n \rangle = \int dx \, \psi^*(x) \left( \frac{\hbar}{i} \frac{d}{dx} \right)^n \psi(x) \equiv \int dx \, \psi^*(x) \left( \frac{\hbar}{i} \right)^n \frac{d^n \psi(x)}{dx^n}. \qquad (27)$$

To generalize still further, consider any function $f(p)$ which can be expanded as a power series in $p$. By an obvious extension of the argument we then have

$$\langle f(p) \rangle = \int \psi^*(x) f\left( \frac{\hbar}{i} \frac{d}{dx} \right) \psi(x) \, dx. \qquad (28)$$

We intend to draw a deep and important conclusion from this result, so let us first make sure we understand precisely what it says. By hypothesis, $f(p)$ is expressible as a power series in $p$. Imagine $f$ to be so expressed,

$$f(p) = \sum_n a_n \, p^n.$$

Then equation (28) can be written in the form

$$\langle f(p) \rangle = \sum_n a_n \langle p^n \rangle,$$

each term of which can be evaluated using equation (27). Evidently this is not a very useful or practical way to deal with complicated functions of the momentum; it is much simpler to do so by transforming to momentum space. Indeed, we remark parenthetically, for functions which cannot be expanded in a power series, no alternative to this latter procedure is possible.

Our concern at present, however, is more formal than practical. Specifically, we observe that we can rewrite equation (28) in the following way,

$$\langle f(p) \rangle = \int_{-\infty}^{\infty} \psi^*(x) f(p) \psi(x) \ dx, \tag{29}$$

*provided we understand that the momentum p in the integrand is to be replaced by the differential operator $(\hbar/i) \, d/dx$ which acts on the function* $\psi(x)$ *standing to its right.* This replacement can be restated as follows: In configuration space, the linear momentum is represented by the operation of space differentiation multiplied by $\hbar/i$,

$$p = \frac{\hbar}{i} \frac{d}{dx}, \tag{30}$$

by which we mean that

$$p\psi(x) = \frac{\hbar}{i} \frac{d\psi}{dx} \tag{31}$$

or, more generally, that

$$p^n\psi(x) = \left(\frac{\hbar}{i} \frac{d}{dx}\right)^n \psi(x) = \left(\frac{\hbar}{i}\right)^n \frac{d^n\psi(x)}{dx^n}. \tag{32}$$

Strictly viewed, our argument appears to admit this interpretation of the momentum as a *differential operator* in configuration space only insofar as the calculation of expectation values is concerned, that is to say, only in an expression such as that of equation (29). We now extend this interpretation by accepting equation (30) as a *general* statement, the operational content of which is expressed by equations (31) and (32) as they stand, quite independently of the calculation of expectation values. This generalization is a conceptual one, and an unusually useful one as we shall abundantly show, but we emphasize that no additional *physical* assumptions are implied by it. This follows because every observable or measurable consequence of quantum mechanics is ultimately expressed in terms of expectation values, and hence in terms of integrals over quadratic functionals of the state function, as equation (29) illustrates.

We have thus far confined our attention to the properties of the momentum in configuration space. What can we say about the properties of the position in momentum space? By exactly parallel arguments, which are readily traced through, it can be shown that the position is represented by the operator of differentiation in momentum space multiplied by $(-\hbar/i)$,

$$x = -\frac{\hbar}{i}\frac{d}{dp}. \tag{33}$$

---

**Exercise 3.** Starting from

$$\langle x \rangle = \int \psi^*(x) x \psi(x) \ dx$$

and using equation (3) to express $\psi(x)$ in terms of $\phi(p)$, deduce equation (33).

---

This means that, in complete analogy with the configuration space results,

$$x\phi(p) = -\frac{\hbar}{i}\frac{d\phi(p)}{dp} \tag{34}$$

or, more generally, that

$$x^n\phi(p) = \left(-\frac{\hbar}{i}\frac{d}{dp}\right)^n \phi(p) \equiv \left(-\frac{\hbar}{i}\right)^n \frac{d^n\phi(p)}{dp^n}, \tag{35}$$

and, therefore, that in momentum space

$$\langle f(x) \rangle = \int \phi^*(p)f(x)\phi(p) \ dp = \int \phi^*(p)f\left(-\frac{\hbar}{i}\frac{d}{dp}\right)\phi(p) \ dp. \tag{36}$$

To restate our results in general form, we see that in quantum mechanics *dynamical variables are not merely numbers, but are represented by operators which act upon state functions.* The particular form of the operators depends upon the representation. In configuration space, the position variable is just the number $x$, the momentum variable is the differential operator (30). In momentum space, conversely, the position variable is the differential operator (33), the momentum variable is a number. These results are summarized in Table 1.

| Representation | Dynamical variable | |
|---|---|---|
| | Position | Momentum |
| Position space | $x$ | $\frac{\hbar}{i}\frac{d}{dx}$ |
| Momentum space | $-\frac{\hbar}{i}\frac{d}{dp}$ | $p$ |

TABLE I.   Operator representations of position and momentum.

Our characterization of dynamical variables at the quantum level has led us to introduce the concept of an operator. We shall encounter such entities with increasing frequency as our development continues. Before going on, therefore, let us briefly summarize their relevant properties.

Any unambiguous rule by which one function is changed to some other function is called an *operation*, and the abstract representation of this process is called an *operator*. Symbolically, we thus write

$$Af(x) = g(x), \qquad (37)$$

where the operator $A$ is defined over a given class of functions if $g$ is determined for every $f$ in that class. This may sound complicated, but it is actually a mere formalization of what we are already thoroughly accustomed to doing. This is perhaps made clearer by the examples given in Table II. Observe first that although the left side of equation

| Operation | Equation (37) | Symbolic representation |
|-----------|---------------|-------------------------|
| Multiplication by 2 | $Af = 2f$ | $A = 2$ |
| Multiplication by $e^{i\phi(x)}$ | $Af = e^{i\phi}f$ | $A = e^{i\phi(x)}$ |
| Differentiation | $Af = df/dx$ | $A = d/dx$ |
| Squaring | $Af = f^2$ | $A = ?$ |
| Complex conjugation | $Af = f^*$ | $A = ?$ |

TABLE II.   Examples of operations and operators.

(37) has the appearance of a product (and indeed is a product when $A$ is a number or some given function of $x$, real or complex), there is generally nothing very much like an ordinary product involved. Observe also that operators exist, and are perfectly well defined, which have no conventional or simple symbolic representation, as indicated by the question marks in the right-hand column of Table II for the last two examples.

The operators we shall meet in quantum mechanics are *linear operators,* that is to say, operators such that

$$A(f_1 + f_2) = Af_1 + Af_2. \qquad (38)$$

All of the operators in Table II are linear except the squaring operator. That the squaring operator is non-linear follows at once from the fact that

$$A(f_1 + f_2) \equiv (f_1 + f_2)^2 = f_1^2 + 2f_1 f_2 + f_2^2,$$

which is clearly different than

$$Af_1 + Af_2 \equiv f_1^2 + f_2^2.$$

We shall frequently be concerned with sequences of operations, carried out one after the next. Thus, for example, we shall encounter situations in which an operator $B$ acts on some function and an operator $A$ then acts on the result. The net result defines a new operator, say $C$. Then we can write

$$Cf \equiv A(Bf). \tag{39}$$

It is customary to omit the parenthesis on the right and to express this relation in the form

$$Cf = ABf,$$

which then implies the purely operator statement

$$C = AB,$$

whereby $C$ is called the product of $A$ and $B$. We emphasize that the meaning of such a product is precisely that expressed by equation (39). The square of an operator is to be regarded as a special case of a product. Products of more than two operators, or higher powers of a given operator, are defined by successive application of the rule for products. Thus, for example,

$$C = A_1 A_2 \cdots A_n$$

means the result of successive operations by $A_n$ first, then $A_{n-1}$, and so on down to $A_1$.

To give some specific examples, let $A$ denote multiplication by $e^{i\phi(x)}$ and $B$ denote differentiation with respect to $x$. Then

$$ABf = e^{i\phi(x)} \frac{df}{dx},$$

$$ABAf = e^{i\phi(x)} \frac{d}{dx} \left[ e^{i\phi(x)} f(x) \right],$$

$$B^2 f = \frac{d^2 f}{dx^2},$$

and so on.

Observe that the order of the operators in a product uniquely determines the order in which these operators act. The result, in general, depends on this order, which is to say that, in general,

$$AB \neq BA$$

and the algebra of operators is *non-commutative,* in contrast to the algebra of ordinary numbers. For the particular operators $A$ and $B$ above, we see that

$$BAf = \frac{d}{dx} \left[ e^{i\phi(x)} f \right],$$

which is clearly different than $ABf$, as given above. Similarly, if $C$ is the operator of complex conjugation, then

$$ACf = e^{i\phi} f^*(x)$$

but

$$CAf = e^{-i\phi} f^*(x).$$

On the other hand, for the particular operators $B$ and $C$ defined above, the results obtained are easily seen to be indifferent to the order of operation, whence

$$BC = CB,$$

and these operators are then said to *commute* with each other, or to be *mutually commutative*.

## 6. COMMUTATION RELATIONS

An obvious aspect of the fact that dynamical variables must be regarded as operators is that dynamical quantities which are indistinguishable classically may become quite different in the quantum domain. An important example is the product $xp$ compared to $px$. Classically these are identical, but that they are not identical in quantum mechanics follows by letting each operate on some arbitrary state function. Thus, in configuration space, we have

$$xp\psi(x) = x \frac{\hbar}{i} \frac{d}{dx} \psi = \frac{\hbar}{i} x \frac{d\psi}{dx}$$

and

$$px\psi(x) = \frac{\hbar}{i} \frac{d}{dx} (x\psi) = \frac{\hbar}{i} \psi + \frac{\hbar}{i} x \frac{d\psi}{dx},$$

whence

$$(px - xp)\psi = \frac{\hbar}{i} \psi. \tag{40}$$

Since $\psi$ is arbitrary, this implies that the difference between $px$ and $xp$ is the purely numerical operator $\hbar/i$,

$$(px - xp) \equiv (p, x) = \frac{\hbar}{i}.$$

The difference between the products of two operators in the two possible orders is called the commutator, and we have introduced a special notation to represent this important quantity. Thus, if $A$ and $B$ are arbitrary operators,

$$(A, B) \equiv AB - BA = -(B, A). \tag{41}$$

In contrast to ordinary multiplication, multiplication of dynamical variables in quantum mechanics is noncommutative.

It is also instructive to examine the commutator of $p$ and $x$ in momentum space. We have

$$xp\phi(p) = -\frac{\hbar}{i}\frac{d}{dp}(p\phi) = -\frac{\hbar}{i}\phi - \frac{\hbar}{i}p\frac{d\phi}{dp}$$

and

$$px\phi(p) = p\left(-\frac{\hbar}{i}\frac{d}{dp}\right)\phi = -\frac{\hbar}{i}p\frac{d\phi}{dp},$$

whence

$$(p, x)\phi = (px - xp)\phi = \frac{\hbar}{i}\phi.$$

Since $\phi$ is arbitrary, we again conclude that

$$(p, x) = \hbar/i.$$

We thus see that although the particular form taken by $p$ and $x$ depends on the representation, their commutator does not. The commutation relation (40) can thus be regarded as *the* fundamental statement of the properties of the quantum mechanical dynamical variables representing momentum and position. Note that $\hbar$ enters as the quantitative measure of noncommutativity. In the classical limit, where $\hbar$ can be regarded as vanishingly small, the classical commutivity is seen to be recovered.

It is easy to generalize the commutation relation (40) to functions of the dynamical variables. Thus, for example, consider $[p, f(x)]$ for arbitrary $f(x)$. We have at once

$$[p, f(x)] = \frac{\hbar}{i}\frac{df(x)}{dx}, \tag{42}$$

which is most easily verified in configuration space. Similarly,

$$[f(p), x] = \frac{\hbar}{i}\frac{df(p)}{dp}, \tag{43}$$

as is easily verified in momentum space. On the other hand, of course,

$$[p, f(p)] = [x, f(x)] = 0 \tag{44}$$

whence it also follows that, if $f(x, p)$ is a well-defined operator,[5]

$$[p, f(x, p)] = \frac{\hbar}{i} \frac{\partial f(x, p)}{\partial x} \tag{45}$$

while

$$[f(x, p), x] = \frac{\hbar}{i} \frac{\partial f(x, p)}{\partial p}. \tag{46}$$

These latter are equivalent to

$$f(x, p) = f\left(x, \frac{\hbar}{i} \frac{d}{dx}\right) \tag{47}$$

in configuration space and to

$$f(x, p) = f\left(-\frac{\hbar}{i} \frac{d}{dp}, p\right) \tag{48}$$

in momentum space.

## 7. THE UNCERTAINTY PRINCIPLE

The noncommutativity of quantum mechanical operators, some special examples of which we have just been studying, has direct physical significance in terms of measurements or observations. The specification of a particular state function implies that the system in question has been prepared by a sequence of observations. Measuring the value of a given dynamical variable is equivalent to operating upon the state function with the operator representing that variable, as we shall elaborate upon later. In general, measurement produces a disturbance of the system at the quantum level. Consequently, measurement of property $A$ need not yield the same result if carried out after property $B$ is measured than if carried out before, since the disturbance produced by measuring $B$ may well cause changes in the value of $A$. If that is the case, then $A$ and $B$ do not commute. If, on the other hand, there is no interference, then $A$ and $B$ do commute.

The uncertainty principle is concerned precisely with the question of the interference produced by observation. Later we shall establish an exact general relationship between the mutual uncertainty in a pair of observables and their commutator. We now give a somewhat qualitative discussion in terms of the simultaneous determination of the particular variables position and momentum.

At the beginning of this chapter we saw that the state function $\psi_p$ of

---

[5] By a well-defined operator, we mean one for which the order of the non-commuting elements is unambiguously specified in every term of its power series development.

equation (1), which describes a particle with precisely defined momentum, contains no information at all about the location of the particle. To describe a localized particle, it was necessary to construct a wave packet. From the form of equation (3), representing such a packet, we see that the momentum is then no longer precisely fixed but is distributed over a range of values, determined by the structure of $\phi(p)$. This behavior, in which one of the variables $p$ or $x$ becomes less sharply fixed as the other becomes more so, is a significant general feature that we now briefly examine.

As a starting point, we discuss two particular wave packets as examples.

(a) **Square-Wave Packet.** Consider first a normalized square-wave packet defined by

$$\psi(x) = \frac{1}{\sqrt{2L}}\, e^{ip_0 x/\hbar}, \quad |x| \leq L$$
$$= 0 \qquad\qquad , \quad |x| > L, \tag{49}$$

which is seen to consist simply of a piece of a plane wave, corresponding to momentum $p_0$, of length $2L$ and centered at the origin. According to equation (21) we then have

$$\phi(p) = \frac{1}{\sqrt{2\pi\hbar}} \int_{-L}^{L} \frac{e^{i(p_0-p)x/\hbar}}{\sqrt{2L}}\, dx$$

or

$$\phi(p) = \sqrt{\hbar/\pi L}\; \frac{\sin\left[(p_0-p)L/\hbar\right]}{p_0 - p} \tag{50}$$

and

$$\phi^*\phi = \frac{\hbar}{\pi L}\; \frac{\sin^2\left[(p_0-p)\,L/\hbar\right]}{(p_0-p)^2}\;, \tag{51}$$

which is plotted in Figure 2. As indicated in the figure, the height of the

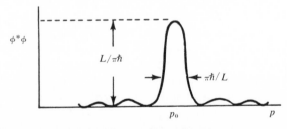

FIGURE 2.   Distribution of momenta for a square-wave packet.

main peak, centered at $p_0$, is proportional to $L$ and its width is inversely proportional to $L$.[6] The area under it is thus seen to be independent of $L$ and close to unity. In other words a wave packet which localizes a particle within a distance $\Delta x \simeq 2L$ is localized in momentum within a range $\Delta p \simeq \hbar\pi/L$. Hence $\Delta x \Delta p$ is of the order of $\hbar$ and is independent of $L$. For large $L$, the momentum becomes well defined, the space location poorly defined, and conversely for small $L$, and always in such a way that the product of the uncertainty in $x$ and that in $p$ is of the order of $\hbar$.

(b) **Gaussian Wave Packet.** As a second example, we consider a normalized Gaussian wave packet

$$\psi(x) = \sqrt{1/L\sqrt{\pi}} \, \exp\left[\frac{ip_0 x}{\hbar} - \frac{x^2}{2L^2}\right], \tag{52}$$

which describes a particle localized about the origin within a distance $L$ and with mean momentum $p_0$. In momentum space it turns out, using the techniques of Appendix I, that this packet is given by

$$\phi(p) = \sqrt{L/\hbar\sqrt{\pi}} \, \exp[-(p - p_0)^2 L^2/2\hbar^2], \tag{53}$$

which is again a Gaussian, of width inversely proportional to $L$. More precisely, it is seen that the momentum is localized within a range $\hbar/L$, centered about $p_0$. Thus once again, the product of the uncertainties is of the order of $\hbar$.

---

**Exercise 4.** Calculate $\langle x \rangle$ and $\langle p \rangle$ in both the configuration and momentum space representations for the square wave packet [equations (49) and (50)] and the Gaussian packet [equations (52) and (53)].

---

The relationship between the width of a wave packet in configuration space and its width in momentum space has been shown to be approximately the same for both square and Gaussian wave packets. We now verify that this is a general feature by examining a general wave packet, which we write in the form

$$\psi(x) = f(x) \, e^{ip_0 x/\hbar}. \tag{54}$$

We shall take $f(x)$ to be a *real, relatively smooth function* of $x$, of width

---

[6] The height of the peak is obtained by taking the limit as $(p - p_0)$ approaches zero in equation (51). The width is measured by the location of the first zero of the sine function, which occurs when its argument is $\pi$.

$L$, centered about the origin. We shall also assume $\psi$ to be normalized, which means that

$$\int_{-\infty}^{\infty} f^2(x) \, dx = 1.$$

It is easily verified that for such a wave packet, the expectation value of the momentum is $p_0$.

---

**Exercise 5.**  Verify that $\langle p \rangle = p_0$ for the wave packet of equation (54).

---

For this wave packet, the state function in momentum space is given by, according to equation (21),

$$\phi(p) = \frac{1}{\sqrt{2\pi\hbar}} \int_{-\infty}^{\infty} f(x) \, e^{i(p_0 - p)x/\hbar} \, dx.$$

By hypothesis, $f(x)$ is a relatively smooth function peaked about the origin and of width $L$. This means that the principal contribution to the integral comes from the domain $|x| \leqslant L$. Hence if

$$(p_0 - p) \, L/\hbar \ll 1,$$

the argument of the exponential factor is negligible over the effective domain of integration and $\phi(p)$ is roughly constant and proportional to the area under the function $f(x)$. As $(p_0 - p)$ increases, the exponential begins to oscillate and eventually it oscillates very rapidly over the effective domain of integration, so that the integral, and thus $\phi(p)$, becomes very small. The dividing line between these two regions of behavior comes for

$$(p_0 - p) \, L/\hbar \simeq 1,$$

which is when the oscillations begin to occur. In other words, when this condition is satisfied, $\phi(p)$ begins to significantly decrease from its maximum value at $p = p_0$. Thus the width of the wave packet in momentum space is approximately

$$\Delta p \simeq \hbar/L.$$

But the width in configuration space, $\Delta x$, is just $L$ and hence

$$\Delta p \Delta x \simeq \hbar, \tag{55}$$

provided $f(x)$ is smooth.

Suppose $f(x)$ is not smooth but contains some structure. Then $(p_0 - p)$

must become larger than before if the oscillations of the exponential are to become rapid compared to the distance characterizing this structure and $\phi(p)$ is to begin to decrease significantly. Thus the width $\Delta p$ of the wave packet in momentum space becomes correspondingly larger, and equation (55) represents the best we can do in the sense that, more generally,

$$\Delta p \Delta x \gtrsim \hbar. \tag{56}$$

Later we shall define precisely what we mean by $\Delta x$ and $\Delta p$ and shall obtain a correspondingly precise inequality. For now, we simply regard each of these quantities as some reasonable measure of the width of a wave packet in configuration and momentum space, respectively.

The physical interpretation of this mathematically trivial relation is the following. If a particle is localized in some region $\Delta x$, no matter the means, then its momentum will necessarily be localized at most within $\Delta p$ and conversely. In other words, the position and momentum of a particle cannot be simultaneously known (or determined or measured) with arbitrary precision, but only within the limits of the relation, equation (56), which is a mathematical, if not yet very accurate, version of the *uncertainty principle*, first enunciated by Heisenberg. The uncertainty principle convincingly demonstrates that classical trajectories have no precise meaning in the quantum mechanical domain.

More generally, the uncertainty relation holds between any pair of canonically conjugate, or complementary, dynamical variables. As a second example, if the energy of a system is measured within $\Delta E$, the time to which the measurement refers is uncertain within an interval $\Delta t$ such that

$$\Delta E \Delta t \gtrsim \hbar. \tag{57}$$

It takes increasingly longer to measure the energy of a system to increasing accuracy. An important aspect of this result is the information it provides about the decay of excited states. In particular, the mean lifetime and the energy width of such states are related by equation (57). For an example, see Section 3 of Chapter VII.

From our discussion, we see that the uncertainty relations are automatically built into quantum mechanics. In terms of measurement, these relations are a consequence of uncontrollable transfers of energy and momentum which necessarily take place, during the process of measurement, between the measuring apparatus and the system whose properties are being measured. As an example, consider localization of the $y$-coordinate of a particle by a slit (Figure 3). If the slit has aperture $L$,

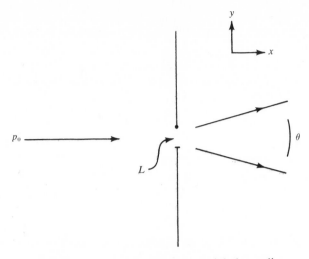

FIGURE 3.    Localization of a particle by a slit.

as indicated, so that $\Delta y \simeq L$, the diffraction pattern produced will have angular width $\theta \simeq \lambda/L = h/p_0 L = h/p_0 \Delta y$. The uncertainty of the $y$ component of momentum is $\Delta p_y = p_0 \sin\theta \simeq p_0 \theta$. Hence $\Delta p_y \Delta y \simeq h$, in agreement with the uncertainty principle. A second standard example is the use of a microscope to locate a particle. To localize a particle within a distance $\Delta x$, the wavelength of the light in which it is viewed must be of order $\lambdabar < \Delta x$. The momentum of the photon is then of order $p \gtrsim \hbar/\Delta x$. The particle is seen because of the photon it scatters or absorbs, and the momentum transferred to it in the scattering process is seen to be of the order $\hbar/\Delta x$, as the uncertainty principle demands.[7]

As implied earlier, it is important to recognize that the uncertainty principal is an integral part of quantum mechanics, as well as a natural and direct consequence of it. In a purely quantum mechanical treatment of a problem, the question of violating the uncertainty principle never arises. To the extent that quantum mechanics in its present form is correct, then the uncertainty principle necessarily follows. This is not to say, of course, that even within the framework of quantum mechanics the uncertainty principle is in any sense empty. On the contrary, it is an extremely useful guide to the understanding of the quantum mechanical properties of a system. Indeed, for systems too complicated to permit complete or exact solutions, it often provides a basis for deciding whether given effects exist or not.

As an example of where it provides very significant qualitative, and even semi-quantitative, understanding, consider a particle confined to

[7] For a detailed discussion of this and other examples, see Reference [18].

some region of radius $a$ by an attractive potential $V(r)$. For each of its coordinates, the uncertainty in position is of the order $a$, and hence $\Delta p \simeq \hbar/a$. The mean momentum being zero, this is of the order of the momentum itself and hence the mean kinetic energy $\langle T \rangle$ of the particle cannot be less than $\hbar^2/2ma^2$. The mean potential energy $\langle V \rangle$ is of the order $V(a)$. Hence the total energy $E$ is approximately given by

$$E(a) = \langle T \rangle + \langle V \rangle \simeq \frac{\hbar^2}{2ma^2} + V(a). \tag{58}$$

We observe in this expression a clear competition between kinetic and potential energies which is strictly quantum mechanical in origin. As $a$ becomes smaller, the potential energy decreases for attractive interactions, but the kinetic energy always increases. A rough estimate of the minimum possible energy of the system, or *ground state energy,* is obtained by differentiating equation (58) with respect to $a$ and setting the result equal to zero. As a specific example, consider the hydrogen atom where the confining potential is the Coulomb potential, $(-e^2/r)$. The total energy is thus approximately

$$E(a) \simeq \frac{\hbar^2}{2ma^2} - \frac{e^2}{a}.$$

This expression is easily found to assume its minimum value when $a = \hbar^2/me^2$, which is recognized to be just the Bohr radius, and the minimum energy is $(-me^4/2\hbar^2)$, which is recognized to be the correct energy of the ground state.[8] The mere existence of a ground state already completely explains why atoms do not collapse, as the classical prediction would have it. Because the energy is an absolute minimum, the atom simply cannot lose energy, by radiation or any other means, and the system is stable.

This example is a most instructive one. The uncertainty principle has been used to explain the existence of a ground state, and thus to account for the existence and stability of atoms, and it has provided us with a simple method for estimating what that energy is, at least for the hydrogen atom. With reference to the hydrogen atom, in particular, the uncertainty principle has enabled us to identify the essential characteristics of a relatively complex system, sufficiently complex that we shall not be in a position even to formulate the problem, let alone solve it, until Chapter IX.

---

[8] Our apparently casual choice of numerical factors in estimating $\langle T \rangle$ and $\langle V \rangle$, each of which is actually uncertain by a factor of two or more, was in fact designed to provide us with the correct answers for the hydrogen atom. This minor swindle in no way affects the essential nature of the results.

**Problem 1.**

(a) Consider the function $\Delta(x)$ defined as

$$\Delta(x) = \frac{1}{2\pi} \int_{-L}^{L} e^{ikx} \, dk.$$

Evaluate the integral and show that $\Delta(x)$ has the properties of $\delta(x)$ when $L \to \infty$.

(b) Consider the function

$$\Delta(x) = \frac{1}{2\pi} \int_{-\infty}^{\infty} e^{ikx - \alpha|k|} \, dk.$$

Show that $\Delta(x)$ behaves like $\delta(x)$ in the limit $\alpha \to 0$.

(c) Consider the function $\Delta(x) = Ae^{-x^2/b^2}$. Show that if $A$ is properly chosen, $\Delta(x)$ behaves like $\delta(x)$ when $b \to 0$.

(d) Show the following:

    (1) $\delta(-x) = \delta(x)$.

    (2) $a\delta(\pm ax) = \delta(x)$,      $a > 0$.

    (3) $\delta(x^2 - a^2) = \frac{1}{2} a[\delta(x - a) + \delta(x + a)]$.

    (4) $\int_{-\infty}^{\infty} f(x)\delta'(x - a) \, dx = -f'(a)$, where the prime denotes differentiation.

**Problem 2.** Find the Fourier *series* representation of the following functions in the interval $-L < x < L$:

    (a) $f(x) = x$.

    (b) $f(x) = |x|$.

    (c) $f(x) = 1$.

    (d) $f(x) = e^{-\alpha|x|}$.

For case (d), compare the behavior of the Fourier series amplitudes in the limit $\alpha L \gg 1$ with the behavior of the Fourier transform of $e^{-\alpha|x|}$.

**Problem 3.** Find the Fourier transforms of the following functions:

    (a) $f(x) = \begin{cases} x & , \quad |x| < 1 \\ 0 & , \quad |x| \geq 1. \end{cases}$

    (b) $f(x) = \begin{cases} |x| & , \quad |x| < 1 \\ 0 & , \quad |x| \geq 1. \end{cases}$

    (c) $f(x) = \begin{cases} 1 - |x|, & |x| < 1 \\ 0 & , \quad |x| \geq 1. \end{cases}$

**Problem 4.** Consider the wave packet $\psi = A \exp[-(|x|/L) + ip_0 x/\hbar]$.

    (a) Normalize $\psi$.

    (b) Calculate $\phi(p)$ and verify that it is properly normalized.

    (c) Examine the width of the wave packets in configuration space and momentum space and verify that the uncertainty relations are satisfied.

**Problem 5.** Calculate the following commutators, using whichever representation is most convenient.
 (a) $(p, x^2)$.
 (b) $(p^3, x)$.
 (c) $(p^2, x^2)$.
 (d) $[p^2, f(x)]$.

**Problem 6.** The operators $A_1, A_2, \ldots, A_6$ are defined to act as follows upon an arbitrary function $\psi(x)$: $A_1$ multiplies $\psi(x)$ by $x^2$, $A_2$ squares $\psi$, $A_3$ averages $\psi(x)$ over an interval $2L$ centered about $x$, $A_4$ replaces $x$ by $x + a$, $A_5$ replaces $x$ by $-x$, and $A_6$ differentiates $\psi$ twice.
 (a) Write out an explicit expression for $A_i\psi(x)$ in each case.
 (b) Which of the $A_i$ are linear operators?
 (c) Which pairs of the $A_i$ commute?

**Problem 7.** Use the uncertainty relation to estimate the ground state energy of the following systems:
 (a) A particle in a box of length $L$.
 (b) A harmonic oscillator of classical frequency $\omega$.
 (c) A particle sitting on a table under the influence of gravity.

**Problem 8.** Let $x_0$ and $p_0$ denote the expectation values of $x$ and $p$ for the state $\psi_0(x)$. Consider the state $\psi(x)$ defined by

$$\psi(x) = e^{-ip_0x/\hbar} \psi_0(x_0 + x).$$

Show that both $\langle x \rangle$ and $\langle p \rangle$ vanish for this state. Does this violate the uncertainty principle? Explain.

**Problem 9.** A particle of mass $m$ moving in a potential $V(x)$ is in its ground state $\psi_0(x)$. Suppose $\psi_0$ to be known, but not necessarily normalized.
 (a) Write an expression for the probability that a measurement of the particle's momentum would yield a value between $p$ and $p + dp$.
 (b) Write an expression for the probability that a measurement of the particle's kinetic energy $T$ would yield a value between $T$ and $T + dT$. Check your result by verifying that you get the correct expression for the expectation value of $T$.

**Problem 10.** Consider a state function which is real, $\psi(x) = \psi^*(x)$.
 (a) Show that $\langle p \rangle = 0$. What about $\langle p^2 \rangle$? $\langle x \rangle$?
 (b) Under what conditions on $\psi(x)$ is $\phi(p)$ real, and what then is $\langle x \rangle$?

# IV

# *Motion of a free particle*

## 1. MOTION OF A WAVE PACKET; GROUP VELOCITY

We have thus far been concerned with the state function of a system at some fixed instant. We now begin our discussion of how such states develop in time. In the present chapter we shall consider the simplest possible problem, namely that of the motion of a particle free from external influences. As a starting point, we return to equation (2) of Chapter III, which provides a general description of a free particle wave packet

$$\psi(x, t) = \frac{1}{\sqrt{2\pi\hbar}} \int_{-\infty}^{\infty} \phi(p) \exp\left[\frac{ipx}{\hbar} - i\omega(p)t\right] dp, \tag{1}$$

where $\omega(p)$ is as yet unknown. We shall assume that $\psi$ is prescribed at $t = 0$, and we shall denote its initial value by $\psi_0(x)$,

$$\psi_0(x) = \psi(x, t = 0) = \frac{1}{\sqrt{2\pi\hbar}} \int_{-\infty}^{\infty} \phi(p) \, e^{ipx/\hbar} \, dp. \tag{2}$$

Our problem then is, given an initial wave packet $\psi_0(x)$, how does it develop in time? That is, what is $\psi(x, t)$?

As a preliminary step, and as an example, we first consider *dispersionless propagation* (as in the propagation of light in free space) where $\omega$ is proportional to $p/\hbar$. The constant of proportionality has the dimensions of a velocity and we denote it by $c$. Thus, in this example,

$$\omega = cp/\hbar = 2\pi c/\lambda$$

and (1) becomes

$$\psi(x, t) = \frac{1}{\sqrt{2\pi\hbar}} \int_{\infty}^{\infty} \phi(p) \, e^{ip(x-ct)/\hbar} \, dp.$$

Comparison with equation (2) then shows that $\psi(x, t)$ is exactly the same function of $(x - ct)$ as $\psi_0$ is of $x$. In other words,

$$\psi(x, t) = \psi_0(x - ct),$$

and the wave packet simply travels to the right with velocity $c$, without distortion, whatever its initial shape.[1]

We now go back to the general case where $\omega(p)$ is unknown. We cannot, of course, calculate anything for arbitrary $\omega(p)$ and for a completely arbitrary wave packet. However, we do not need to particularize too much in order to determine the essential features. The first assumption we shall make is that $\phi(p)$ is a smooth wave packet in momentum space, centered about $p_0$, say, and of width $\Delta p$. To emphasize this behavior we rewrite $\phi(p)$ in the form

$$\phi(p) = g(p - p_0),$$

where $g$ is a smooth function which falls rapidly to zero when its argument exceeds $\Delta p$ in magnitude. This means that the main contribution to the integral in equation (1) comes from a region of width $\Delta p$ about the point $p_0$.

Secondly, we assume that $\omega(p)$ is a smooth function of $p$. If so, we can expand $\omega$ in a Taylor series about $p_0$,

$$\omega(p) = \omega(p_0) + (p - p_0) \frac{d\omega(p_0)}{dp_0} + \tfrac{1}{2}(p - p_0)^2 \frac{d^2\omega}{dp_0^2} + \cdots$$

$$\equiv \omega_0 + (p - p_0) \frac{v_g}{\hbar} + (p - p_0)^2 \alpha + \cdots,$$

where

$$\omega_0 = \omega(p_0)$$

$$v_g = \hbar \frac{d\omega}{dp_0}$$

$$\alpha = \tfrac{1}{2} \frac{d^2\omega}{dp_0^2}.$$

Introducing also the new variable $s = p - p_0$, we can now rewrite equation (1) as

$$\psi(x, t) = f(x, t) \exp\left[\frac{ip_0 x}{\hbar} - i\omega_0 t\right],$$

[1] Our treatment of this problem in optical propagation is considerably oversimplified. The case where $\omega$ and $p$ have opposite sign, $\omega = -cp/\hbar$, must also be considered. The motion of a wave packet in optics accordingly is *not* determined if only $\psi_0(x)$ is given. In addition, it turns out that $\partial\psi/\partial t(x, t = 0)$ must be prescribed. This is a consequence of the fact that the electromagnetic wave equation is of second order in the time.

where $f(x, t)$, which determines the envelope of the wave packet, is given by

$$f(x, t) = \frac{1}{\sqrt{2\pi\hbar}} \int_{-\infty}^{\infty} ds\ g(s)\ \exp\left[\frac{is}{\hbar}(x - v_g t) - i\alpha s^2 t + \cdots\right]. \qquad (3)$$

At the same time, equation (2) becomes

$$\psi_0(x) = f_0(x)\ e^{ip_0 x/\hbar}$$

where $f_0(x)$, the initial envelope function, is given by

$$f_0(x) = f(x, t = 0) = \frac{1}{\sqrt{2\pi\hbar}} \int_{-\infty}^{\infty} ds\ g(s)\ e^{isx/\hbar}. \qquad (4)$$

Now the main contribution to these integrals comes when $s$ is less than $\Delta p$ in magnitude, because of the assumed behavior of $g(s)$. Hence, if $t$ is small enough that

$$\alpha(\Delta p)^2 t \ll 1,$$

the term in the exponent of equation (3) which is quadratic in $s$ can be neglected and the envelope function is approximately

$$f(x, t) \simeq \frac{1}{\sqrt{2\pi\hbar}} \int_{-\infty}^{\infty} ds\ g(s)\ e^{is(x - v_g t)/\hbar}.$$

Comparison with equation (4) then shows that, to this approximation, $f(x, t)$ is the same function of $(x - v_g t)$ as is $f_0$ of $x$, that is to say,

$$f(x, t) = f_0(x - v_g t). \qquad (5)$$

We have thus demonstrated that for $t \ll t_0$, where

$$t_0 = \frac{1}{\alpha(\Delta p)^2} = \frac{2}{(\Delta p)^2 d^2\omega/dp_0^2} \qquad (6)$$

the wave packet travels undistorted with velocity $v_g$, where

$$v_g = \hbar d\omega/dp_0. \qquad (7)$$

The quantity $v_g$ is called the *group velocity* of the waves, since it represents the velocity of propagation of a group of waves, namely those which make up the wave packet. It should be contrasted with the phase velocity, which is the velocity with which the phase of a given pure harmonic wave advances and which is given by $v_p = \hbar\omega/p_0$. For dispersionless propagation, where $\omega$ is proportional to $p$, these two velocities are equal, but in general they are quite different.

We emphasize that our result holds only for times short compared to $t_0$, defined in equation (6). Eventually, when $t$ exceeds $t_0$, the exponential begins to oscillate very rapidly for $s$ smaller than $\Delta p$. When that happens, the effective domain of integration in $p$ is reduced in size and this produces a corresponding increase in the width of the wave packet in configuration space. This means that, in general, a wave packet initially travels undistorted with the group velocity $v_g$, but eventually begins to spread out in space. We shall give some examples later.

## 2. THE CORRESPONDENCE PRINCIPLE REQUIREMENT

We now use the following argument to deduce the form of $\omega(p)$ for quantum mechanical waves. If we have a well-defined wave packet in configuration space, we have seen that it travels with the group velocity $v_g$, at least for short enough times. In the classical limit, this limitation on the time must become unimportant, that is, $t_0$ must become very large compared to all relevant times. Assuming this to be so, we accordingly demand that

$$v_g = v_{\text{classical}} = \frac{p_0}{m}.$$

Hence, from equation (5)

$$\hbar \frac{d\omega}{dp_0} = \frac{p_0}{m}$$

and, dropping the subscript on $p_0$,

$$\hbar\omega = \frac{p^2}{2m} = E, \tag{8}$$

which is the Planck relation. In a sense, we have thus deduced the Planck relation as a consequence of our formulation and its interpretation.[2]

Note that equation (8) is arbitrary up to a constant of integration, which we have simply set equal to zero. This is related to the freedom of choice of the zero of energy and implies a similar freedom in the choice of quantum mechanical frequency. *Only differences in energy, and also therefore in frequency, are physically significant.*[3]

We next examine the time $t_0$ at which a quantum mechanical wave packet begins to spread out significantly. We have, from equations (6) and (8),

[2] Our argument demonstrates that equation (8) must hold in the classical domain. At the quantum level, one might suppose that additional terms could logically enter, provided they contribute negligibly in the classical limit. However, it is not hard to show by dimensional arguments that for free particles no such terms can exist and that equation (8) is therefore unique, up to an additive constant.

[3] We saw earlier that the absolute amplitude of quantum mechanical waves has no physical significance and we now see that the absolute frequency has none either. The contrast with classical waves is remarkable and complete.

$$t_0 = \frac{2m\hbar}{(\Delta p)^2}. \tag{9}$$

An instructive way to rewrite this is to use the uncertainty relation to express $\Delta p$ as $\hbar/\Delta x$, where $\Delta x$ is the spatial extent of the initial wave packet. Thus

$$t_0 \simeq \frac{2m(\Delta x)^2}{\hbar}.$$

For a macroscopic particle, say of mass one gram, whose position is defined to even $10^{-4}$ cms (1 micron), we have

$$t_0 = 10^{19} \text{ sec.}$$

The age of the universe is about $3 \times 10^{10}$ years or about $10^{18}$ sec. Hence, *a wave packet for a macroscopic particle holds together without spreading for a period comparable to the age of the universe.* This establishes that classical trajectories for macroscopic systems are not in conflict with quantum mechanics. On the other hand, for an electron whose position is defined to say $10^{-8}$ cm,

$$t_0 = 10^{-16} \text{ sec,}$$

and a classical description is meaningless.

It is possible to give a simple physical interpretation of the time $t_0$ along the following lines. The group velocity of propagation of de Broglie waves is $p/m$. In a time $t$, two segments of a wave packet differing in momentum by $\Delta p/2$, say, will differ in distance traveled by an amount $\Delta pt/2m$. When this distance becomes comparable with the width $\Delta x$ of the initial packet, the width of the packet will begin to increase significantly. Hence the packet begins to spread at a time $t_0$ defined by

$$\frac{\Delta p}{2m} t_0 \simeq \Delta x \simeq \frac{\hbar}{\Delta p},$$

which is equation (9). This argument implies that once wave packets start spreading they do so at a rate linear in the time. We shall shortly verify that this is indeed the case.

## 3. PROPAGATION OF A FREE PARTICLE WAVE PACKET IN CONFIGURATION SPACE

With our identification of $\omega(p)$, we can now rewrite equation (1) as

$$\psi(x, t) = \frac{1}{\sqrt{2\pi\hbar}} \int_{-\infty}^{\infty} \phi(p) \exp\left[\frac{ipx}{\hbar} - \frac{ip^2 t}{2m\hbar}\right] dp, \tag{10}$$

which is a general representation of a time-dependent state function for a free particle. If $\psi_0(x) = \psi(x, 0)$ is prescribed, $\phi(p)$ is given, according to equation (III-21), by

$$\phi(p) = \frac{1}{\sqrt{2\pi\hbar}} \int_{-\infty}^{\infty} \psi_0(x) \, e^{-ipx/\hbar} \, dx.$$

We can now express $\psi(x, t)$ in terms of its initial value by substituting this expression back into equation (10). The result is

$$\psi(x, t) = \frac{1}{2\pi\hbar} \int\int_{-\infty}^{\infty} dp \, dx' \, \psi_0(x') \exp\left[\frac{ip(x - x')}{\hbar} - \frac{ip^2 t}{2m\hbar}\right].$$

Noting that the dependence upon $p$ is explicit, we invert the order of integration and rewrite the result in the form

$$\psi(x, t) = \int_{-\infty}^{\infty} \psi(x', 0) \, K(x', x; t) \, dx', \tag{11}$$

where $K$ is given by

$$K(x', x; t) \equiv \frac{1}{2\pi\hbar} \int_{-\infty}^{\infty} dp \, \exp\left[\frac{ip(x - x')}{\hbar} - \frac{ip^2 t}{2m\hbar}\right].$$

This integral is readily evaluated by the methods of Appendix I, and we obtain, finally,

$$K(x', x; t) = \sqrt{m/(2\pi i\hbar t)} \, \exp\left[i(x - x')^2 m/2\hbar t\right]. \tag{13}$$

Equation (11) is an important result and we now analyze it in some detail. The initial state function $\psi(x', 0)$ specifies the probability amplitude that a measurement of the particle's position will reveal it to be $x'$ at $t = 0$. The state function $\psi(x, t)$ specifies the corresponding amplitude at $x$ at time $t$. Equation (11) shows how the latter is to be compounded from the former through the intermediary function $K$. This implies that $K$ can be interpreted as the probability amplitude that a particle originally at $x'$ will propagate to the point $x$ in the time interval $t$. The function $K$ is thus called the *propagator,* in this particular case, the *free particle* propagator. With this interpretation, the entire process can be described in the following way: The initial amplitude for finding the particle at $x'$ multiplied by the amplitude for propagation from $x'$ to $x$ during the time interval $t$ yields, when summed over all $x'$, the amplitude for finding the particle at $x$, just as it should. This, then, is the physical content of equation (11).

The general properties of the free particle propagator are not hard to discern. Note from either equation (12) or equation (13) that

$$K(x', x; 0) = \delta(x - x'),$$

which simply means that for infinitesimal time intervals, the amplitude for propagation to any point $x$ is negligible except from points in the immediate neighborhood of $x$. As time goes on, however, $K$ is seen to spread out more and more, and contributions come from an increasing range of values of $(x - x')$.[4]

Equations (11) and (13) provide a *complete, general* and *explicit* solution to the problem of the quantum mechanical motion of a free particle. It is precisely analogous to the solution of the corresponding classical problem,

$$x = x_0 + p_0 t / m.$$

Were our only interest the study of free particle motion, nothing more would have to be said. However, our actual aim is to pave the way for eventual generalization to the motion of a particle under external influences; we must still take the step from the quantum analog of Newton's first law to the second.

## 4. PROPAGATION OF A FREE PARTICLE WAVE PACKET IN MOMENTUM SPACE; THE ENERGY OPERATOR

We now discuss the time development of a free particle wave packet in momentum space. For this purpose, we first define time-dependent momentum space state functions $\phi(p, t)$ by writing the obvious generalization

$$\phi(p, t) = \frac{1}{\sqrt{2\pi\hbar}} \int_{-\infty}^{\infty} \psi(x, t) \, e^{-ipx/\hbar} \, dx. \tag{14}$$

Of course $t$ here simply plays the role of a parameter, which was taken to be zero for convenience in our earlier discussion. Otherwise stated, the quantity we previously denoted by $\phi(p)$ is merely $\phi(p, t = 0)$. By the Fourier integral theorem, we then have

$$\psi(x, t) = \frac{1}{\sqrt{2\pi\hbar}} \int_{-\infty}^{\infty} \phi(p, t) \, e^{ipx/\hbar} \, dp. \tag{15}$$

Comparing this expression with equation (10), we see at once that

$$\phi(p, t) = \phi(p) \, e^{-ip^2 t/2m\hbar} = \phi(p, t = 0) \, e^{-ip^2 t/2m\hbar} \tag{16}$$

while

$$\rho(p, t) = |\phi(p, t)|^2 = |\phi(p)|^2 = \rho(p, t = 0).$$

In other words, only the *phase* of an arbitrary free particle wave packet

---

[4] Note that the amplitude from remote points is finite, though small, for even very short time intervals. This reflects the non-relativistic character of our treatment. The *relativistic* propagator, on the other hand, properly vanishes outside the light cone.

in momentum space changes in time. The probability density is independent of time or is stationary and therefore the expectation value of *any* function of momentum is also independent of time. This result is no surprise. We are talking about the states of a free particle and the momentum classically is a constant of the motion. In view of our use of the correspondence principle, the momentum of a free particle must be a constant of the motion in quantum mechanics, as well, and so it has turned out.

We have emphasized that the expectation value of any function of the momentum is simple for free particles. As a particular and important example, consider the energy $E$ which is just $p^2/2m$. We have

$$\langle E \rangle = \int_{-\infty}^{\infty} \phi^*(p, t) \frac{p^2}{2m} \phi(p, t) \, dp.$$

Now, according to equation (16),

$$\frac{p^2}{2m} \phi(p, t) = -\frac{\hbar}{i} \frac{\partial \phi(p, t)}{\partial t}$$

and hence this expression can be rewritten as

$$\langle E \rangle = \int_{-\infty}^{\infty} \phi^*(p, t) \left[ -\frac{\hbar}{i} \frac{\partial \phi(p, t)}{\partial t} \right] dp.$$

Indeed, for any function $f(E)$, we have, by an obvious extension of the argument,

$$\langle f(E) \rangle = \int_{-\infty}^{\infty} \phi^*(p, t) f\left( -\frac{\hbar}{i} \frac{\partial}{\partial t} \right) \phi(p, t) \, dp. \tag{17}$$

Since $\phi(p, t)$ is an arbitrary free particle state function, we conclude that in momentum space the energy $E$ can be represented by the operation of time differentiation multiplied by $(-\hbar/i)$, that is,

$$E = -\frac{\hbar}{i} \frac{\partial}{\partial t}. \tag{18}$$

What about $E$ in configuration space? Since $t$ enters only as a parameter with respect to the transformation between momentum and configuration space, it is clear that $E$ must have exactly the same representation in both spaces. That this is correct follows upon expressing $\phi(p, t)$ and $\phi^*(p, t)$ from equation (17) in terms of $\psi$ and $\psi^*$, using equation (14), and evaluating the integral over $p$. As asserted, one obtains almost at once

$$\langle f(E) \rangle = \int_{-\infty}^{\infty} \psi^*(x, t) f\left( -\frac{\hbar}{i} \frac{\partial}{\partial t} \right) \psi(x, t) \, dx. \tag{19}$$

---

**Exercise 1.** Deduce equation (19) in the way indicated.

---

## 5. TIME DEVELOPMENT OF A GAUSSIAN WAVE PACKET

Before continuing with the general discussion, it is helpful to work out a specific example in some detail. We shall now do so for the case of a wave packet which, initially, is Gaussian in form. In particular we consider the wave packet of equation (III-52),

$$\psi(x, 0) = \psi_0(x) = \frac{1}{\sqrt{L\sqrt{\pi}}} \exp\left[\frac{ip_0 x}{\hbar} - \frac{x^2}{2L^2}\right], \qquad 20)$$

which describes a particle initially localized about the origin within a distance $L$ and whose momentum has the expectation value $p_0$. Using equations (11) and (13), we then obtain, upon evaluating the integral by the methods of Appendix I,

$$\psi(x, t) = \left[\sqrt{\pi}\left(L + \frac{i\hbar t}{mL}\right)\right]^{-1/2} \exp\left[\frac{L(-x^2/2L^2 + ip_0 x/\hbar - ip_0^2 t/2m\hbar)}{L + i\hbar t/mL}\right],$$

$$(21)$$

and the probability density is

$$\rho(x, t) = |\psi|^2 = \left[\pi\left(L^2 + \frac{\hbar^2 t^2}{m^2 L^2}\right)\right]^{-1/2} \exp\left[-\frac{(x - p_0 t/m)^2}{L^2 + \hbar^2 t^2/m^2 L^2}\right]. \quad (22)$$

---

**Exercise 2.** Consider the Gaussian initial wave packet, equation (20).
   (a)   Derive equation (21).
   (b)   Derive equation (22) from (21).
   (c)   Write out $\phi(p, t)$.
   (d)   Calculate $\langle E \rangle$ and its fluctuation $\langle (E - \langle E \rangle)^2 \rangle$.

---

Comparing equation (22) to $\rho(x, t = 0)$, which is simply

$$\rho(x, t = 0) = \frac{1}{\sqrt{\pi}L} e^{-x^2/L^2},$$

we see that $\rho(x, t)$ involves only two changes. First, the center of the wave packet moves with the group velocity $p_0/m$. Second, the width of the wave packet increases with time. Calling this width $L(t)$, we have

$$L(t) = \sqrt{L^2 + \hbar^2 t^2/m^2 L^2}.$$

The result is seen to be entirely in agreement with our earlier predictions concerning the propagation of wave packets. In particular, it is seen that the wave packet is initially undistorted and begins to spread appreciably only when $\hbar t/mL^2$ is of the order of unity, in agreement with equation (9). Further, when $t$ becomes very large, the width of the packet grows linearly at the rate

$$\frac{\hbar}{mL} = \frac{\Delta p}{m},$$

also as predicted.

The behavior of $\rho(x, t)$ for a Gaussian wave packet is sketched in Figure 1 below. The area under the curve, of course, remains the same.

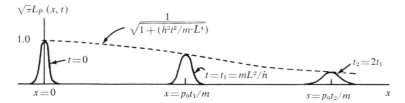

FIGURE 1.   Spreading of a Gaussian wave packet with time.

This example is particularly well-suited to a discussion of the classical limit. All quantum effects must disappear, of course, if we let $\hbar \to 0$. That this is indeed so in our example follows at once, since in that limit equation (22) becomes

$$\rho(x, t) = \frac{1}{\sqrt{\pi} L} \exp\left[-(x - p_0 t/m)^2/L^2\right],$$

which describes the classical motion of a free particle, the initial momentum of which is precisely $p_0$ but the initial position of which is distributed about the origin according to a Gaussian of width $L$. The conventional classical initial condition, in which the position and momentum are both precisely defined, is achieved by letting $L \to 0$, and we then obtain

$$\rho(x, t) = \delta(x - p_0 t/m),$$

which is the classical trajectory expressed in the language of probability densities. This last expression simply means that the probability of finding the particle anywhere except on the classical trajectory is precisely zero, as it ought to be.

There are two comments we would like to add. First, observe that two limiting processes were involved in recovering the conventional classical result and that the *order* in which they are made to occur is crucial.[5]

---

[5] The student will find it instructive to examine the result obtained if this order is reversed and *first L* and *then $\hbar$* are allowed to go to zero.

In this example, and in general, it is essential in obtaining the classical limit to let $\hbar$ vanish *before* prescribing the precise initial conditions of the classical description. Second, observe that although the limit of vanishing $\hbar$ has a meaning for the probability density, it has no meaning for the *state function*, or at least no readily discernible one at this stage of our understanding. Otherwise stated, the *amplitude* of the state function is well-defined in the classical limit; at present the *phase* is not. This should be no surprise, because only the former is a *directly* observable quantity. In Chapter VII, Section 1, we shall present a more systematic analysis of the classical limit and of its meaning for the phase, as well as the amplitude, of the state function.

## 6. THE FREE PARTICLE SCHRÖDINGER EQUATION

We return now to a discussion of the time development of an arbitrary free particle state function. We have derived several equivalent *integral* representations for the state function $\psi(x, t)$, and we now seek a *differential* description. This desire is motivated by the following consideration. An integral representation is a *global* characterization; it requires the specification of a *function*, the initial state, over all of space at some given instant. With this information, the solution at some later time can be found. We are more accustomed, however, to a *local* characterization in which information about the properties of the system in an *infinitesimal* space-time neighborhood is sufficient to define the solution. Such a local description is achieved through the intermediary of *differential equations*. More than custom is involved, of course, because differential equations provide a powerful and largely independent approach which turns out to be indispensable in treating problems more complicated than that of the free particle.

The local characterization we seek is most easily obtained from equation (10). We need merely note that, differentiating under the integral sign,

$$-\frac{\hbar}{i} \frac{\partial \psi(x, t)}{\partial t} = \frac{1}{\sqrt{2\pi\hbar}} \int_{-\infty}^{\infty} \phi(p) \frac{p^2}{2m} \exp\left[\frac{ipx}{\hbar} - \frac{ip^2 t}{2m\hbar}\right] dp,$$

and that the same result is obtained upon evaluating $-(\hbar^2/2m) \partial^2\psi/\partial x^2$, again by differentiating under the integral sign. We thus conclude that any function $\psi(x, t)$ defined by equation (10), and therefore *any* free particle state function, satisfies the partial differential equation

$$-\frac{\hbar^2}{2m} \frac{\partial^2\psi}{\partial x^2} = -\frac{\hbar}{i} \frac{\partial\psi}{\partial t}. \tag{23}$$

Otherwise stated, equation (10) is simply a way of writing the general solution of equation (23), which is *Schrödinger's equation* in configuration space for a free particle in one dimension. Note that it is explicitly complex and that it is of the first order in time, and not of second order as in optics.[6]

The interpretation of this equation is quite simple and direct if we recall that, in configuration space, the momentum operator is

$$p = \frac{\hbar}{i} \frac{\partial}{\partial x},$$

according to equation (III-30), while the energy operator is

$$E = -\frac{\hbar}{i} \frac{\partial}{\partial t},$$

according to equation (18). Hence, Schrödinger's equation is the *operator* equation

$$\frac{p^2}{2m} \psi(x, t) = E\psi(x, t) . \tag{24}$$

Classically, of course, for a free particle

$$E = p^2/2m,$$

and we see that the corresponding quantum mechanical equation requires the state function for a free particle to be such that it yields the same result when operated upon by the total energy operator $E$ as when operated upon by the kinetic energy operator $p^2/2m$. This condition ensures that the expectation value of any function of total energy is exactly the same as the expectation value of the same function of the kinetic energy,

$$\langle f(E) \rangle = \langle f(p^2/2m) \rangle,$$

which, in turn, ensures that the requirements of the correspondence principle are satisfied.

Of course, Schrödinger's equation can also be written in momentum space as

$$\frac{p^2}{2m} \phi(p, t) = E\phi(p, t) = -\frac{\hbar}{i} \frac{\partial \phi}{\partial t}, \tag{25}$$

which is actually much simpler than equation (24) because $p$ is now a purely numerical operator. Since equations (24) and (25) have exactly the same form, it is not necessary to explicitly identify the representation

---

[6] These aspects are discussed in the next section.

and we can write symbolically

$$\frac{p^2}{2m} \Psi = E\Psi, \tag{26}$$

which is intended to convey the idea that the representation is unspecified. In configuration space $\Psi = \psi(x, t)$, in momentum space $\Psi = \phi(p, t)$.

Because of its simple form, the solution to equation (25) is trivial and immediate. We can rewrite that equation in the form

$$\frac{1}{\phi} \frac{\partial \phi}{\partial t} = -\frac{ip^2}{2m\hbar},$$

whence

$$\phi(p, t) = \phi(p, t_0) \ \exp\left[-ip^2(t - t_0)/2m\hbar\right],$$

where $\phi(p, t_0)$ is arbitrary. This is recognized as an obvious generalization of our earlier result to arbitrary initial times, $t_0$. The solution of Schrödinger's equation in configuration space, where it is a partial and not an ordinary differential equation, is more complicated, and we defer this question for the moment.

## 7. CONSERVATION OF PROBABILITY

In our development of the quantum mechanical laws, we have identified $\psi^*\psi$ as a probability density and have assumed that $\psi$ is normalizable. In particular, we have assumed that, at any arbitrarily chosen instant $t$, we can choose $\psi$ in such a way that

$$\int_{-\infty}^{\infty} \psi^*(x, t)\psi(x, t) \ dx = 1,$$

and similarly in momentum space. Now the time dependence of the state functions appearing in the integrand is not at our disposal but is prescribed by Schrödinger's equation. Hence, we must verify that this condition, if imposed at one instant, will continue to be satisfied as time goes on. Given our interpretation of $\psi$ as the probability amplitude, the normalization condition is simply the statement that the probability of finding the particle somewhere is unity. We are thus seeking to verify that this probability remains unity as time goes on, which is to say that, as it must be, *probability is conserved.*

The proof of this is straightforward. Writing

$$P(t) \equiv \int_{-\infty}^{\infty} \psi^*(x, t)\psi(x, t) \ dx,$$

we want to show that $dP/dt$ is zero for any arbitrary solution of Schrödinger's equation. We have

$$\frac{dP}{dt} = \int_{-\infty}^{\infty} \frac{\partial}{\partial t} \left[\psi^*(x, t)\psi(x, t)\right] dx$$

$$= \int_{-\infty}^{\infty} \left[\frac{\partial \psi^*}{\partial t} \psi + \psi^* \frac{\partial \psi}{\partial t}\right] dx.$$

Now according to Schrödinger's equation in configuration space, equation (23),

$$\frac{\partial \psi}{\partial t} = \frac{i\hbar}{2m} \frac{\partial^2 \psi}{\partial x^2},$$

while, upon taking the complex conjugate,

$$\frac{\partial \psi^*}{\partial t} = \frac{-i\hbar}{2m} \frac{\partial^2 \psi^*}{\partial x^2}.$$

Thus,

$$\frac{dP}{dt} = \frac{i\hbar}{2m} \int_{-\infty}^{\infty} \left[\psi^* \frac{\partial^2 \psi}{\partial x^2} - \psi \frac{\partial^2 \psi^*}{\partial x^2}\right] dx$$

$$= \frac{i\hbar}{2m} \int_{-\infty}^{\infty} \frac{\partial}{\partial x} \left[\psi^* \frac{\partial \psi}{\partial x} - \psi \frac{\partial \psi^*}{\partial x}\right] dx$$

$$= \frac{i\hbar}{2m} \left[\psi^* \frac{\partial \psi}{\partial x} - \psi \frac{\partial \psi^*}{\partial x}\right] \Big|_{-\infty}^{\infty}.$$

Because the normalizability of $\psi$ requires that it vanish at $\pm\infty$, we see that the right side vanishes and $dP/dt$ is indeed zero.

This is a crucial point for our whole interpretation and we want to emphasize that *the proof would have failed had Schrödinger's equation been other than first order in the time derivative.* This is even more apparent in momentum space, where we have

$$P(t) = \int_{-\infty}^{\infty} \phi^*(p, t)\phi(p, t) \, dp$$

$$\frac{dP}{dt} = \int \left[\frac{\partial \phi^*}{\partial t} \phi + \phi^* \frac{\partial \phi}{\partial t}\right] dp$$

$$= \int \left[\frac{i\hbar}{2m} p^2 \phi^* \phi - \frac{i\hbar}{2m} p^2 \phi^* \phi\right] dp$$

$$= 0.$$

If, instead of (25), Schrödinger's equation were, for example,

$$\left(\frac{p^2}{2m}\right)^2 \phi = E^2 \phi = -\hbar^2 \frac{\partial^2 \phi}{\partial t^2},$$

the proof fails, as is readily verified.

**Exercise 3.** If $\psi^*\psi$ is interpreted as a probability density, verify that probability is not necessarily conserved for an equation of the type

$$\left(\frac{p^2}{2m}\right)^2 \Psi = E^2\,\Psi.$$

Try to isolate the specific origin of the difficulty, bearing in mind that *every* solution of Schrödinger's equation is also a solution of the above equation.

---

We have emphasized both the importance of probability conservation and the fact that it holds only because of the particular form assumed by Schrödinger's equation. When we generalize from the free particle case to that of a particle under the influence of external forces, we shall turn the argument around by *starting* from the *requirement* that the generalized Schrödinger's equation we seek be so constructed that probability conservation is *guaranteed*. As we shall see, this is a stringent condition indeed, and therefore a very useful and informative one.

From this point of view, one might well ask how we have been so lucky in our development of the free particle equations that we have arrived at just the right result. Where, in short, was the essential step taken? The answer is, as we remarked at that time, when we defined momentum states at the very beginning as *complex* exponentials and not as the real trigonometric functions appropriate to classical fields. If the arguments are traced through, it is readily seen that this definition led us first to the identification of the momentum and energy operators, and then to Schrödinger's equation as a *first-order* differential equation in the time and, therefore, as an explicitly *complex* equation.

The fact that quantum mechanics involves complex numbers in such an essential way is often regarded as a rather mysterious manifestation of the difference between the classical and quantum descriptions. Perhaps it is worth emphasizing, however, that in quantum mechanics, as in classical mechanics, the introduction of complex numbers is a matter of convenience and economy, not of necessity. Quantum mechanics can readily be reformulated, in terms of *real* functions only, in the following way. Let $\psi_1$ denote the real part of $\psi$ and $\psi_2$ denote its imaginary part; that is, write

$$\psi = \psi_1 + i\psi_2$$

where $\psi_1$ and $\psi_2$ are both real. Using this expression, Schrödinger's equation (23) then becomes, upon separation of its real and imaginary parts,

$$-\frac{\hbar^2}{2m}\frac{\partial^2 \psi_1}{\partial x^2} = -\hbar\frac{\partial \psi_2}{\partial t}$$

$$-\frac{\hbar^2}{2m}\frac{\partial^2 \psi_2}{\partial x^2} = \hbar\frac{\partial \psi_1}{\partial t},$$

which are seen to be a pair of *coupled* equations in $\psi_1$ and $\psi_2$. Evidently, every relation we have derived and used can be laboriously rewritten in this notation. Thus, we obtain as the normalization condition

$$1 = \int \left( \psi_1{}^2 + \psi_2{}^2 \right) \, dx$$

and, as the expectation value of the momentum,

$$\langle p \rangle = \hbar \int \left( \frac{\partial \psi_2}{\partial x}\psi_1 - \psi_2\frac{\partial \psi_1}{\partial x} \right) dx,$$

and so on. In short, quantum mechanics can be completely transcribed into the language of *real two-component states*.[7]

For comparison, a similar decomposition of any classical field into its real and imaginary parts yields a pair of *identical uncoupled* differential equations, one for each part. The real and imaginary parts of classical fields are thus equivalent and either can be used to *completely* characterize the field. In contrast, as we have seen, the quantum mechanical state requires two interrelated functions for its specification. This is

---

[7] Those familiar with matrix notation will have already perceived that this transcription is not quite so awkward or inconvenient as we have made it appear. Introduction of the two component column and row matrices

$$\psi \equiv \begin{pmatrix} \psi_1 \\ \psi_2 \end{pmatrix}, \qquad \psi^* \equiv (\psi_1 \psi_2)$$

and the two-by-two matrix

$$\gamma \equiv \begin{pmatrix} 0 & -1 \\ 1 & 0 \end{pmatrix}$$

permits an immediate transcription of all our results according to the simple rule: replace $i$ everywhere it appears by $\gamma$. Specifically,

$$p = -\hbar\gamma \; (\partial/\partial x)$$

$$x = \hbar\gamma \; (\partial/\partial p)$$

$$E = \hbar\gamma \; (\partial/\partial t),$$

and the operator form of Schrödinger's equation is unaltered in appearance. None of this is surprising since it is easily verified that

$$\gamma^2 = -1,$$

and we have simply provided an alternative to the conventional designation of $\sqrt{-1}$ as an imaginary number. It is instructive to compare *this* two-component theory with the two component Pauli theory of spin presented in Chapter X.

the essential content of the statement that Schrödinger's equation is intrinsically complex.

## 8. DIRAC BRACKET NOTATION

We have consistently developed our formulation in both momentum and configuration space and have attempted to present the basic laws in a form independent of the particular representation under consideration. We now introduce a notation, originated by Dirac, which will permit us to write our equations in a representation-independent way. Dirac's notation is actually far more general than we require at present, or are prepared to understand. We shall merely introduce one or two definitions and will generalize these, and add others, as the need arises.

First, we introduce a representation-independent way of writing the expectation value. Let $\Psi$ denote some arbitrary, but definite, state function, in whatever representation. Let $A$ be some arbitrary operator. The expectation value of $A$ in the state $\Psi$ is written as the *bracket expression*

$$\langle A \rangle \equiv \langle \Psi | A | \Psi \rangle .$$

In the position representation, say,

$$\langle \Psi | A | \Psi \rangle = \int \psi^*(x, t) A \psi(x, t) \ dx .$$

while in the momentum representation

$$\langle \Psi | A | \Psi \rangle = \int \phi^*(p, t) A \phi(p, t) \ dp .$$

Suppose $A$ is the numerical operator unity. We then introduce, as our second definition,

$$\langle \Psi | \Psi \rangle \equiv \langle \Psi | 1 | \Psi \rangle .$$

In configuration space, we have

$$\langle \Psi | \Psi \rangle = \int \psi^*(x, t) \psi(x, t) \ dx$$

and similarly in momentum space. In this notation, the normalization condition is simply

$$\langle \Psi | \Psi \rangle = 1 .$$

If it is desirable, or necessary, to specify the representation in this bracket notation, this is easily accomplished by explicitly giving the argument of the state function. Thus, in the position representation, we can write $\langle \psi(x, t) | A | \psi(x, t) \rangle$ and so on. One note of caution is in order. Observe that as a matter of definition $\Psi$ is written as the left factor, *not* $\Psi^*$. The operation of complex conjugation is implicit, not explicit, in

Dirac notation. Thus, in writing out the expectation value as an integral, the complex conjugate of the left-hand state function in the bracket must *always* be used.

## 9. STATIONARY STATES

We now return to the question of solving Schrödinger's equation (23) in configuration space. For this purpose we use the conventional method of *separation of variables*. Thus we seek a special solution in the form of a product

$$\psi(x, t) = \psi(x) T(t). \tag{28}$$

Substitution into equation (23) yields

$$-\frac{\hbar^2}{2m} \frac{d^2\psi}{dx^2} T = -\frac{\hbar}{i} \psi \frac{dT}{dt}$$

or, rearranging,

$$-\frac{\hbar^2}{2m} \frac{1}{\psi} \frac{d^2\psi}{dx^2} = -\frac{\hbar}{i} \frac{1}{T} \frac{dT}{dt}.$$

The left side depends only upon $x$, the right side only upon $t$, but the two must be equal to each other for all $x$ and $t$. Hence each must equal the same constant, say $\alpha$, which is called the *separation constant*,

$$-\frac{\hbar}{i} \frac{1}{T} \frac{dT}{dt} = \alpha \tag{29}$$

and, also,

$$-\frac{\hbar^2}{2m} \frac{1}{\psi} \frac{d^2\psi}{dx^2} = \alpha. \tag{30}$$

The solution to equation (29) is immediate:

$$T(t) = e^{-i\alpha t/\hbar}.$$

Hence, from equation (28),

$$\psi_\alpha(x, t) = \psi_\alpha(x)\, e^{-i\alpha t/\hbar}, \tag{31}$$

where $\psi_\alpha(x)$ must satisfy equation (30) and where the subscript denotes that we are discussing the solution corresponding to the particular separation constant $\alpha$.

The interpretation of $\alpha$ is the following. Consider the state function (31) and suppose it to be normalized. Then no matter what the particulars of $\psi_\alpha(x)$, we have

$$\langle E \rangle = \langle \psi_\alpha(x, t) | E | \psi_\alpha(x, t) \rangle = \int \psi_\alpha^* \left[ -\frac{\hbar}{i} \frac{\partial \psi_\alpha}{\partial t} \right] dx = \alpha.$$

Similarly, for any function of energy, $f(E)$, which can be expanded in a power series,

$$\langle \psi_\alpha | f(E) | \psi_\alpha \rangle = f(\alpha),$$

which means that $\psi_\alpha$ is a state in which the energy $E$ has the precise and definite numerical value $\alpha$.[8] To remind ourselves of this fact, we shall replace $\alpha$ by the *number* $E$, which is simply the numerical value of the energy for the state in question. Compare this notation with that for the states $\psi_p$ defined earlier. It will be recalled that these were states in which the linear momentum had the precise numerical value $p$. In either case this notation is somewhat unfortunate, since the symbol which is being used to denote a *numerical value* is the same as that used to denote an *operator* in other contexts. However, it is standard practice to do this, and the literature can hardly be read without keeping this ambiguity in mind. The reader cannot proceed blindly, but must distinguish which is meant from the context. Fortunately, this is relatively easy to do.

In any case, we now write our solution in the form

$$\psi_E(x, t) = \psi_E(x) \, e^{-iEt/\hbar}, \tag{32}$$

while equation (30) can be rewritten as

$$-\frac{\hbar^2}{2m} \frac{d^2 \psi_E}{dx^2} = E\psi_E(x). \tag{33}$$

Equation (33), which no longer contains the time, is called the *time independent Schrödinger's equation,* in this case, for a free particle.

The states $\psi_E(x, t)$ have one very important property. The expectation value of any operator $A$ is independent of time provided $A$ does not itself *explicitly* depend on $t$. For example, if $A = f(p, x)$, then $\langle \psi_E | A | \psi_E \rangle$ does not change with time. For this reason, the states $\psi_E$ are called *stationary states.* Stationary states are solutions of the time dependent Schrödinger's equation; they are states of definite energy. Accordingly, they are particularly simple states quantum mechanically. On the other hand, they are very complicated states, or at least singularly inappropriate ones, from the classical point of view. This follows because the expectation values of $x$ and $p$ are *independent of time* and hence bear no discernible relation to a classical state of motion. Classical states are necessarily superpositions of many stationary states.

---

[8] This is certainly a sufficient condition on $\psi_\alpha$. It can also be shown to be necessary as follows: Any statistical distribution is uniquely defined by a complete set of appropriate moments. Ours has the property $\langle E^n \rangle = \alpha^n = \langle E \rangle^n$, for all $n$. All fluctuations of $E$ about $\alpha = \langle E \rangle$ accordingly vanish and the conclusion follows.

We now complete our discussion by exhibiting the solutions to equation (31). We have at once

$$\psi_E(x) = e^{\pm i\sqrt{2mE}\,x/\hbar},$$

as is easily verified. Hence

$$\psi_E(x, t) = \exp\left[\pm \frac{i\sqrt{2mE}\,x}{\hbar} - \frac{iEt}{\hbar}\right].$$

Note that $E$ must be positive if $\psi_E$ is not to increase exponentially in one direction or the other. Now, since

$$p = \sqrt{2mE},$$

where $p$ is the numerical value of the momentum corresponding to the energy $E$, we can equally well write

$$\psi_E(x, t) = \exp\left[\pm \frac{ipx}{\hbar} - \frac{ip^2 t}{2m\hbar}\right], \qquad E = p^2/2m,$$

where the two signs in the exponent make explicit the two possible values of linear momentum corresponding to a given energy $E$. This is then a solution of Schrödinger's equation (23) for any positive value of $E$ or, equally, for any value of $p$, positive or negative. The general solution of equation (23) is a superposition of these stationary states with arbitrary amplitudes. Our original representation, equation (10), is recognized as just such a superposition. We have thus demonstrated that equation (10) is a representation of the general solution of equation (23), as claimed.

## 10. A PARTICLE IN A BOX

As an example of the use of Schrödinger's equation, we now examine the stationary states of a particle which is constrained to move in the interior of a (one-dimensional) box, but which is otherwise free. This constraint is intended to mean that outside of the box the probability of finding the particle is zero, and hence that $\psi$ must vanish. If $\psi$ is not to be discontinuous, then its behavior in the interior must be such that it is zero at the constraining walls.[9] Taking the walls of the box to be at

---

[9] We are not in a position to *prove* that $\psi$ must indeed be continuous, as assumed. The problem under consideration is not in actuality a free particle problem, since the walls of the container exert impulsive forces on the particle. In the next two chapters we discuss motion under the influence of external forces. The case of a particle in a box can be considered as the limiting case of motion in a square well potential as the potential becomes infinitely deep. When so considered, the stated continuity and boundary conditions can be verified.

$x = 0$ and $x = L$, we thus seek those solutions of the free particle Schrödinger's equation (33) which are such that

$$\psi_E(x = 0) = \psi_E(x = L) = 0. \tag{34}$$

We saw earlier that for a fixed energy $E$,

$$\psi_E = e^{\pm i \sqrt{2mE}\, x/\hbar}.$$

The most general solution for a given $E$ is thus the linear combination

$$\psi_E = A\, e^{i \sqrt{2mE}\, x/\hbar} + B\, e^{-\sqrt{2mE}\, x/\hbar}.$$

Applying the boundary conditions, equation (34), we obtain

$$A + B = 0$$

$$A\, e^{i\sqrt{2mE}\, L/\hbar} + B\, e^{-i\sqrt{2mE}\, L/\hbar} = 0,$$

or, from the first,

$$B = -A$$

and, from the second,

$$\sin \sqrt{2mE}\, L/\hbar = 0.$$

This latter condition is not satisfied for arbitrary values of $E$, but only if $E$ has particular, *discrete* values $E_n$ such that

$$\sqrt{2mE_n}\, \frac{L}{\hbar} = n\pi, \qquad n = 0, 1, 2, \ldots$$

or

$$E_n = \frac{n^2 \pi^2 \hbar^2}{2mL^2}. \tag{35}$$

The corresponding stationary state functions are, after normalization,

$$\psi_{E_n} = \sqrt{2/L}\, \sin\left(\frac{n\pi x}{L}\right), \tag{36}$$

whence it is seen that *solutions are obtained only if a half-integral number of de Broglie waves can exactly fit into the box*. Note, however, that for $n = 0$, the state function vanishes identically. Hence the state of lowest energy is that for $n = 1$,

$$\psi_{E_1} = \sqrt{2/L}\, \sin\left(\frac{\pi x}{L}\right)$$

$$E_1 = \frac{\pi^2 \hbar^2}{2mL^2} \simeq \frac{10^{-53}}{2mL^2}\ \text{ergs}.$$

In terms of $E_1$ we can conveniently express $E_n$ as

$$E_n = n^2 E_1.$$

We have thus found that, in contrast to classical physics, a particle in a box can exist only in a discrete set of states.[10] Further, we note that no state of zero energy exists, in agreement with the requirements of the uncertainty principle. A particle in a box cannot simply sit on the floor; it must always be in at least some minimal state of motion.

The spectrum of allowed energies and the corresponding wave functions are shown in Figure 2. In spite of the fact that the spacing of these energy levels increases with $n$, for a macroscopic particle this spacing is infinitesimal for all achievable energies.

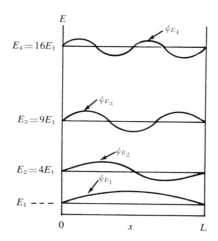

FIGURE 2.　Energy spectrum and state functions for a particle in a square well. In exhibiting the state functions, the following convention is used, here and throughout. The spatial axis $x$ for the $n$th state is drawn at a height representing the energy $E_n$. The ordinate, measured from that $x$-axis, represents the probability amplitude $\psi_n$.

For example, consider a one-gram particle in a box one cm in length. Its ground state energy is about $5 \times 10^{-54}$ ergs. For the $n \simeq 10^{30}$ state its energy will be about $10^7$ erg = one joule and for the $n \simeq 10^{33}$ state its energy will be about one million joules. The distance between adjacent states is about $10^{-24}$ ergs in the former case and $10^{-21}$ ergs in the latter. These are enormously larger than the $10^{-53}$ erg spacing in the neighbor-

---

[10] It turns out that the spectrum of states is discrete whenever the motion of a particle is *bounded,* that is, confined to a finite domain. The reason is essentially the same as in the present example, the necessity of fitting an appropriate number of de Broglie wavelengths into the domain in question.

hood of the ground state, but both are still undetectably small on the macroscopic energy scale. On the other hand, for an electron in a box two angstroms in size, $E_1 \simeq 10^{-11}$ ergs $\simeq 6$ eV, which is roughly of the order of the spacing actually observed in atoms.

We conclude with a discussion of the properties of a wave packet, or time dependent state, for a particle in a box. The most general possible such state is an arbitrary superposition of the discrete set of stationary states,

$$\psi_{E_n}(x, t) = \psi_{E_n}(x) \ e^{-iE_n t/\hbar},$$

where $E_n$ and $\psi_{E_n}$ are given by equations (35) and (36). We thus write

$$\psi(x, t) = \sum_n A_n \psi_{E_n}(x) \ e^{-iE_n t/\hbar}, \tag{37}$$

which evidently describes a state in which the energy of the particle is not precisely fixed but can take on any of the values $E_n$ with probability determined by the extent to which the $n$th state is represented in the superposition. More precisely, if $\psi(x, t)$ is normalized, we expect that a measurement of the energy will yield the value $E_n$ with a likelihood which is just $|A_n|^2$. This is an important result and we now show that our interpretation is indeed correct.

As a preliminary step, note that the stationary state functions $\psi_{E_n}$, which are simple sinusoids according to equation (36), are such that

$$\int_0^L \psi_{E_m}^*(x) \psi_{E_n}(x) \ dx = \begin{cases} 1, \ m = n \\ 0, \ m \neq n \end{cases} \equiv \delta_{mn}. \tag{38}$$

For brevity, we have introduced the *Kronecker delta* symbol $\delta_{mn}$, which is defined to be unity when $m$ is equal to $n$ and to be zero when $m$ and $n$ differ from one another. Any set of functions satisfying an equation of the type of equation (38), that is, such that each function in the set is normalized and such that different functions in the set are orthogonal to one another, is called an *orthonormal set*.

Supposing $\psi(x, t)$ to be normalized,

$$\int_0^L \psi^*(x, t) \psi(x, t) \ dx = 1,$$

we obtain from equation (37)

$$\int_0^L \left( \sum_n A_n^* \psi_{E_n}^*(x) \ e^{iE_n t/\hbar} \right) \left( \sum_m A_m \psi_{E_m}(x) \ e^{-iE_m t/\hbar} \right) dx = 1,$$

where we have used $n$ as the summation index in the expression for $\psi^*(x, t)$ and $m$ in the expression for $\psi(x, t)$. Interchanging the order of summation and integration, we now have

$$\sum_{m,n} A_n^* A_m \, e^{i(E_n - E_m)t/\hbar} \int_0^L \psi_{En}^*(x)\psi_{Em}(x) \, dx = 1.$$

Hence, using the orthonormality condition of equation (38), this becomes

$$\sum_{m,n} A_n^* A_m \, e^{i(E_n - E_m)t/\hbar} \, \delta_{mn} = 1$$

and, finally, the normalization condition is seen to require that

$$\sum_m |A_m|^2 = 1. \tag{39}$$

Next we calculate the expectation value of the energy. We have

$$\langle \psi|E|\psi \rangle = \int_0^L \psi^*(x,t) \left[ -\frac{\hbar}{i} \frac{\partial \psi(x,t)}{\partial t} \right] dx,$$

whence, using equation (37),

$$\langle \psi|E|\psi \rangle = \int_0^L \left( \sum_n A_n^* \psi_{En}^*(x) \, e^{iE_n t/\hbar} \right) \left( \sum_m A_m E_m \psi_{Em}(x) \, e^{-iE_m t/\hbar} \right) dx.$$

Thus, again using the orthonormality of the $\psi_n$, we obtain

$$\langle \psi|E|\psi \rangle = \sum_m E_m |A_m|^2. \tag{40}$$

Indeed, by the same argument, it follows that for every $s$

$$\langle \psi|E^s|\psi \rangle = \sum_m E_m^s \, |A_m|^2$$

and hence that, for any function $f(E)$,

$$\langle \psi|f(E)|\psi \rangle = \sum_m f(E_m) \, |A_m|^2. \tag{41}$$

Since $f(E)$ is arbitrary, we thus see that, as asserted, $\psi$ is a state in which a measurement of the energy yields the value $E_m$ with probability $|A_m|^2$. Otherwise stated, $A_m$ is the probability amplitude and $|A_m|^2$ the probability that the system will be found upon observation to be in its $m$th stationary state, with energy $E_m$.

Suppose now that the state function of a particle in a box is prescribed at some instant, which we take to be $t = 0$ for simplicity. Denoting this initial state by $\psi_0(x)$ we thus have

$$\psi(x, t = 0) = \psi_0(x),$$

where $\psi_0(x)$ is assumed to be known. The behavior of the system can now be readily determined, at least formally. From equation (37), setting $t = 0$, we have

$$\psi_0(x) = \sum_n A_n \psi_{E_n}(x).$$

Multiplying this expression by $\psi_{E_m}(x)$ and integrating over the box, we obtain

$$\int_0^L \psi_{E_m}^*(x)\psi_0(x)\,dx = \sum_n A_n \int\int \psi_{E_m}^*(x)\psi_{E_n}(x)\,dx.$$

By virtue of the orthonormality of the $\psi_{E_n}$, the right side reduces to the single term $A_m$ and hence we find

$$A_m = \int_0^L \psi_{E_m}^*(x)\psi_0(x)\,dx = \langle \psi_{E_m}|\psi_0\rangle. \tag{42}$$

For the given initial state $\psi_0(x)$, equation (42) then gives the probability amplitude that the system is the $m$th stationary state.

The evolution of the system in time can be studied in the following way. Substitution of equation (42) back into equation (37) yields the expression

$$\psi(x,t) = \sum_n \left(\int_0^L \psi_{E_n}^*(x')\psi_0(x')\,dx'\right) \psi_{E_n}(x)\,e^{-iE_n t/\hbar},$$

where we have used $x'$ to denote the dummy variable of integration in the expression for the amplitudes $A_n$. Rearranging this result, we thus have

$$\psi(x,t) = \int_0^L \psi_0(x')K(x',x;t)\,dx', \tag{43}$$

where the propagator $K$ for a particle in a box is given by

$$K(x',x;t) = \sum_{n=1}^\infty \psi_{E_n}^*(x')\psi_{E_n}(x)\,e^{-iE_n t/\hbar}. \tag{44}$$

These results should be compared with those for a free particle, equations (11) and (12).

The specific form of $K$ is of some interest. From equations (35) and (36), we have

$$K(x',x;t) = \frac{2}{L} \sum_{n=1}^\infty \sin\frac{n\pi x'}{L}\sin\frac{n\pi x}{L} \exp\left[-in^2\pi^2\hbar t/2mL^2\right]. \tag{45}$$

In the limit in which $L \to \infty$, the sum can be replaced by an integral and the resemblance of the result to that for a free particle is not hard to discern.

Unfortunately, it is not possible to obtain $K$ in closed form for a particle in a box. However, in the classical limit, it is possible to show that a wave packet propagates without distortion and bounces successively off the walls of the container, just as it should. The algebra involved is quite tedious, however, and we shall not work out the details.

## 11. SUMMARY

It may be helpful to briefly recapitulate our progress to this point, especially because we have been working more or less simultaneously on
two reasonably distinct conceptual levels. On the surface, our primary
concern has been with the behavior of a free particle. This we formu-
lated the equation of motion, *Schrödinger's equation,* for such a particle,
and gave its complete and general solution in terms of the *free particle
propagator.* The result was then applied to an elucidation of the relationship between the quantum and classical solutions. The technique of
*separation of variables* was introduced and applied to the motion of a
*particle in a box.* Again, a complete solution was obtained.

At a deeper level, however, our primary concern was with the under-
lying general features of the quantum mechanical description of nature,
features which recur again and again in subsequent developments. First
among these in importance were the identification of the *energy operator*
and the understanding we achieved of the essential parts played by
*probability conservation* and the *correspondence principle* in fixing the
form of the quantum mechanical equations. In the course of our analysis
we were led to the notion of a *stationary state,* the simplest possible
quantum state function. Finally, we saw that for a bound particle, the
stationary states form a *discrete* and *ortho-normal* set.

---

**Problem 1.**
   (a)   In deep water, the phase velocity of water waves of wavelength
$\lambda$ is $v_p = \sqrt{g\lambda/2\pi}$. What is the group velocity of such waves?
   (b)   The phase velocity of a typical electromagnetic wave in a wave
guide has the form

$$v_p = \frac{c}{\sqrt{1 - (\omega_0/\omega)^2}},$$

where $c$ is the velocity of light in free space and where $\omega_0$ is a certain
characteristic frequency. What is the group velocity of such waves?
Note that the phase velocity exceeds $c$. Does this violate special rela-
tivity? What about the group velocity?

**Problem 2.**   Consider a wave packet which at $t = 0$ has the form

$$\psi(x, 0) = A \, e^{ip_0x/\hbar} \, e^{-|x|/L}.$$

   (a)   Normalize $\psi(x, 0)$.
   (b)   Calculate $\phi(p, 0)$ and $\phi(p, t)$. Verify that each is normalized.
   (c)   Calculate $\langle p \rangle$ and discuss its time dependence. Do the same
for $\langle E \rangle$.
   (d)   Plot $|\phi(p, t)|^2$ against $p$, assuming that $L \gg \hbar/p_0$ and that

$L \ll \hbar/p_0$. Explain the difference in the two cases using the uncertainty principle.

(e)   Calculate $\langle x \rangle$ at $t = 0$. At any time $t > 0$. [Hint: Do this in momentum space.]

**Problem 3.**   A particle is confined in a box of length $L$.

(a)   Calculate

$$\langle \psi_{E_n}|x|\psi_{E_n} \rangle, \qquad \langle \psi_{E_n}|x^2|\psi_{E_n} \rangle, \qquad \langle \psi_{E_n}|p|\psi_{E_n} \rangle, \qquad \left\langle \psi_{E_n}\left|\frac{p^2}{2m}\right|\psi_{E_n} \right\rangle.$$

Discuss your results in each case.

(b)   The motion of a classical particle in a box is periodic with period $T = 2L/v$, where $v$ is the particle's speed. The quantum mechanical motion exhibits no such periodicity. Explain how the classical periodic motion is attained in its appropriate limit. (This is a non-trivial problem. It requires more thought than algebra for its solution.)

**Problem 4.**   A particle is in its ground state in a box of length $L$. The wall of the box at $x = L$ is suddenly moved outward to

(a)   $x = 2L$,
(b)   $x = 10L$,
(c)   $x = 100L$.

In each case, calculate the probability that the article will be found in the ground state of the expanded box. In each case, find the state of the expanded box most likely to be occupied by the particle. Explain. [Hint: What is the initial, prescribed state function of the particle in each case?]

**Problem 5.**   A particle is in its ground state in a box of length $L$. The walls of the box are suddenly dissolved so that the particle can move freely.

(a)   What is the probability that, after the walls are dissolved, the particle's momentum will be between $p$ and $p + dp$?

(b)   Plot the momentum probability density against $p$ and discuss the qualitative nature of your result. What would you expect classically? [Hint: What is the initial state function of the particle?]

(c)   Calculate $\langle x \rangle$ at $t = 0$ and at any time $t > 0$. [Hint: Do this in momentum space.]

(d)   The same assuming the particle to have been in the 100th state of the box initially.

**Problem 6.**   A particle in a box is in the state

$$\psi(x, t) = \frac{1}{\sqrt{2}} \left[ \psi_0(x, t) + \psi_1(x, t) \right],$$

where $\psi_0$ is the normalized ground state, $\psi_1$ the normalized first excited state. Calculate $\langle E \rangle$, $\langle x \rangle$, $\langle p \rangle$. Discuss the time dependence of each of these quantities.

**Problem 7.** Show that, for an arbitrary free particle wave packet,

$$\langle x \rangle_t = \langle x \rangle_{t=t_0} + \frac{\langle p \rangle}{m}\,(t - t_0)\,,$$

in agreement with the correspondence principle. Do this

    (a)   In momentum space, using equation (16).

    (b)   In configuration space, using equations (11) and (12); using equations (11) and (13).

# V

# *Schrödinger's equation*

In the preceding we have obtained the quantum mechanical equations of motion for a free particle. These results will now be generalized to the case of a particle moving under the influence of an external force. Only forces which are conservative, and thus derivable from a potential $V(x)$, will be considered. We shall show that the combined requirements of probability conservation and the correspondence principle essentially determine the result. We begin by considering the conservation of probability.

## 1. THE REQUIREMENT OF CONSERVATION OF PROBABILITY

In our analysis of free particle motion, we saw that the appearance of no higher than a first-order time derivative in Schrödinger's equation is essential if probability is to be conserved. We shall therefore assume that, in the generalized equation we seek, the time derivative again enters to the first order only. Specifically, we shall assume that Schrödinger's equation has the form, in configuration space,

$$H \psi(x, t) = -\frac{\hbar}{i} \frac{\partial \psi}{\partial t}, \tag{1}$$

where $H$ is some as yet unknown operator. Of course, if the particle is free, $H$ must reduce to the kinetic energy operator $p^2/2m$. We shall also assume that $H$ is linear, so that the superposition postulate is maintained.

Probability conservation now requires that for any physically admissible state function $\psi$, satisfying equation (1), we must have

$$\frac{dP}{dt} \equiv \frac{d}{dt} \int_{-\infty}^{\infty} \psi^*(x, t) \psi(x, t) \, dx = 0.$$

Carrying out the differentiations, we obtain

$$\frac{dP}{dt} = \int_{-\infty}^{\infty} \left( \frac{\partial \psi^*}{\partial t} \psi + \psi^* \frac{\partial \psi}{\partial t} \right) dx$$

or, using equation (1) to eliminate the time derivatives,

$$\frac{dP}{dt} = (i/\hbar) \int_{-\infty}^{\infty} [(H^* \psi^*)\psi - \psi^* H\psi] \, dx.$$

The parentheses in the first term are used to indicate that the operator $H^*$, the complex conjugate of $H$, acts only upon the function $\psi^*$ and not upon $\psi$. If this expression is to vanish then we must have

$$\int_{-\infty}^{\infty} (H^*\psi^*)\psi \, dx = \int_{-\infty}^{\infty} \psi^* H\psi \, dx$$

or, alternatively,

$$\int_{-\infty}^{\infty} (H\psi)^* \psi \, dx = \int_{-\infty}^{\infty} \psi^* H\psi \, dx. \tag{2}$$

Since this equation must hold for all times, it must equally hold for *arbitrary* admissible $\psi$ and hence is a restriction on the operator $H$. This condition is called the *Hermiticity condition* and any operator satisfying it is called an *Hermitian operator* (after the mathematician Hermite). *We have thus proved that H must be Hermitian.*

## 2. HERMITIAN OPERATORS

Before proceeding, we pause briefly to discuss the properties of Hermiticity and to give some examples. Consider some arbitrary operator $A$. If $A$ represents a physically *observable* quantity, then its expectation value must always be real. Now

$$\langle A \rangle = \int_{-\infty}^{\infty} \psi^* A\psi \, dx$$

while

$$\langle A \rangle^* = \int_{-\infty}^{\infty} \psi (A\psi)^* \, dx = \int_{-\infty}^{\infty} (A\psi)^* \psi \, dx.$$

If $\langle A \rangle$ is to be real, these two expressions must be equal, and this requirement is seen to be precisely the definition of Hermiticity.[1] We have thus

---

[1] The reader must be careful not to confuse $\langle A \rangle^*$ and $\langle A^* \rangle$. The former is written above. The latter, by definition, is

$$\langle A^* \rangle = \int \psi^* A^* \psi \, dx,$$

which generally is seen to be entirely different.

shown that *any operator which represents a physical observable must be Hermitian.*

Next we generalize our definition of Hermiticity by showing that if $A$ is Hermitian, and if $\psi_1$ and $\psi_2$ are any two admissible state functions, not necessarily the same, then

$$\int_{-\infty}^{\infty} (A\psi_1)^* \psi_2 \, dx = \int_{-\infty}^{\infty} \psi_1^* A\psi_2 \, dx. \qquad (3)$$

The proof follows by considering the state function

$$\psi = a_1\psi_1 + a_2\psi_2,$$

which is an arbitrary superposition of $\psi_1$ and $\psi_2$. By the definition of Hermiticity, we have

$$\int (A\psi)^* \psi \, dx = \int \psi^* A\psi \, dx$$

or, substituting for $\psi$ and carrying out the multiplication,

$$|a_1|^2 \int (A\psi_1)^* \psi_1 \, dx + |a_2|^2 \int (A\psi_2)^* \psi_2 \, dx + a_1^* a_2 \int (A\psi_1)^* \psi_2 \, dx$$
$$+ a_1 a_2^* \int (A\psi_2)^* \psi_1 \, dx$$
$$= |a_1|^2 \int \psi_1^* A\psi_1 \, dx + |a_2|^2 \int \psi_2^* A\psi_2 \, dx + a_1^* a_2 \int \psi_1^* A\psi_2 \, dx$$
$$+ a_1 a_2^* \int \psi_2^* A\psi_1 \, dx.$$

Since $A$ is Hermitian, the first term on the left is equal to the first term on the right, and similarly for the second terms. Hence, after rearrangement, we are left with

$$a_1^* a_2 \left[ \int (A\psi_1)^* \psi_2 \, dx - \int \psi_1^* A\psi_2 \, dx \right] =$$
$$a_1 a_2^* \left[ \int \psi_2^* A\psi_1 \, dx - \int (A\psi_2)^* \psi_1 \, dx \right].$$

Now $a_1$ and $a_2$ are arbitrary, and, in particular, their relative phase is arbitrary. But this equality must hold no matter what the relative phase and hence we conclude that *each* bracketed expression vanishes.[2] This completes the proof. Henceforth, we shall take equation (3) as the general definition of Hermiticity.

We can rewrite equation (3) in Dirac bracket notation as follows. First, we generalize the notation by writing

$$\langle \psi_1 | \psi_2 \rangle \equiv \int \psi_1^* \psi_2 \, dx.$$

Next, since $A\psi_1$ and $A\psi_2$ are themselves state functions, we can write,

---

[2] To put it another way, the right side is the complex conjugate of the left. But the phase of each side is arbitrary because $a_1$ and $a_2$ are arbitrary. The equality can thus hold if, and only if, each side is identically zero.

using the same notational definition,

$$\langle \psi_1 | A\psi_2 \rangle = \int \psi_1 {}^* A\psi_2 \, dx$$

and

$$\langle A\psi_1 | \psi_2 \rangle = \int (A\psi_1) {}^* \psi_2 \, dx.$$

Hence, if $A$ is Hermitian, from equation (3) we obtain

$$\langle A\psi_1 | \psi_2 \rangle = \langle \psi_1 | A\psi_2 \rangle \qquad (4)$$

as the definition of Hermiticity in Dirac bracket notation.

Again, it should be observed that the complex conjugate of the left-hand function is *always* understood when the Dirac bracket expressions are translated into the language of integrals.

The fact that only operators which are Hermitian have real expectation values is an important one. It means that the operator representing *any* physically observable dynamical variable must be Hermitian. It is easily verified that the linear momentum and the position of a particle each satisfy this requirement: The former is a real number in momentum space, the latter in configuration space, and real numbers are trivially Hermitian.

---

**Exercise 1.**

(a)   Prove that $p$ and $x$ are Hermitian. In each case carry your proof out in both configuration and momentum space. (Hint: Integrate by parts in the nontrivial cases.)

(b)   Prove as simply as you can that any arbitrary real function of $p$ alone is Hermitian, of $x$ alone.

(c)   Is $px$ Hermitian? Is $xp$? Is $(p, x)$? Is $(px + xp)$? Based on your results, by what quantum mechanical operator might you expect to represent the product of linear momentum and position?

---

The dynamical variables we are typically concerned with at this stage of our development are some function or other of the Hermitian operators, position and momentum. What can we say about the properties of such operator functions, that is, those constructed from operators known to be Hermitian? We approach the problem by looking at some examples. It follows trivially from the definition of Hermiticity that *the sum of Hermitian operators is Hermitian*, but if we consider the more general case of an *arbitrary* linear combination of Hermitian operators $A$ and $B$,

$$C = \alpha A + \beta B, \qquad (5)$$

then $C$ is *not* Hermitian unless $\alpha$ and $\beta$ happen to be real numbers. To see this, note that

$$\langle \psi_1 | C\psi_2 \rangle = \langle \psi_1 | (\alpha A + \beta B)\psi_2 \rangle, \tag{6}$$

while on the other hand

$$\langle C\psi_1 | \psi_2 \rangle = \langle \psi_1 | (\alpha^*A + \beta^*B)\psi_2 \rangle, \tag{7}$$

which are evidently different. This last equation follows by observing that, because $A$ is Hermitian,

$$\langle \alpha A\psi_1 | \psi_2 \rangle = \alpha^* \langle A\psi_1 | \psi_2 \rangle = \alpha^* \langle \psi_1 | A\psi_2 \rangle = \langle \psi_1 | \alpha^*A\psi_2 \rangle$$

and similarly for the term in $B$.

As a second example consider the product $AB$ of two Hermitian operators. Writing

$$D = AB \tag{8}$$

we have

$$\langle \psi_1 | D\psi_2 \rangle \equiv \langle \psi_1 | AB\psi_2 \rangle.$$

The effect of operating upon $\psi_2$ with $B$ is to produce some new function which we temporarily abbreviate as $\phi_2$. Thus we write

$$\langle \psi_1 | AB\psi_2 \rangle \equiv \langle \psi_1 | A\phi_2 \rangle = \langle A\psi_1 | \phi_2 \rangle,$$

where, in the last step, we have used the fact that $A$ is Hermitian. Re-expressing $\phi_2$ as $B\psi_2$, we then obtain

$$\langle A\psi_1 | \phi_2 \rangle = \langle A\psi_1 | B\psi_2 \rangle.$$

Thinking of $A\psi_1$ as some new function $\phi_1$, we thus write

$$\langle A\psi_1 | B\psi_2 \rangle \equiv \langle \phi_1 | B\psi_2 \rangle = \langle B\phi_1 | \psi_2 \rangle,$$

where we have now used the fact that $B$ is also Hermitian. Finally, re-expressing $\phi_1$ as $A\psi_1$ and putting all of this together, we see that

$$\langle \psi_1 | D\psi_2 \rangle \equiv \langle \psi_1 | AB\psi_2 \rangle = \langle BA\psi_1 | \psi_2 \rangle. \tag{9}$$

On the other hand, we have

$$\langle D\psi_1 | \psi_2 \rangle \equiv \langle AB\psi_1 | \psi_2 \rangle, \tag{10}$$

and hence $D$ is *not* Hermitian unless $A$ and $B$ happen to *commute*.

These apparently random results can be put into a systematic and unified framework by introducing the concept of the *adjoint*, or *Hermitian conjugate*, of an operator. Let $E$ denote some operator, not necessarily Hermitian. Its adjoint, written $E\dagger$, and called "$E$ adjoint" or "$E$ dagger," is *defined* by

$$\langle \psi_1 | E^\dagger \psi_2 \rangle \equiv \langle E\psi_1 | \psi_2 \rangle \qquad (11a)$$

or, equivalently,

$$\langle \psi_1 | E\psi_2 \rangle \equiv \langle E^\dagger \psi_1 | \psi_2 \rangle \qquad (11b)$$

for arbitrary admissible $\psi_1$ and $\psi_2$. Note that it follows from this definition that the adjoint of the adjoint of an operator is the operator itself,

$$(E^\dagger)^\dagger = E.$$

That equations (11a) and (11b) are equivalent is easily seen by going to a specific representation in either configuration or momentum space. Equation (11a) assumes the form

$$\int \psi_1^* E^\dagger \psi_2 \, dx = \int (E\psi_1)^* \psi_2 \, dx,$$

while equation (11b) states that

$$\int \psi_1^* E\psi_2 \, dx = \int (E^\dagger \psi_1)^* \psi_2 \, dx,$$

and similarly in momentum space. The second is merely the complex conjugate of the first, provided the arbitrary functions $\psi_1$ and $\psi_2$ are relabeled. To put all of this into words, we see that the adjoint of an operator acting on one function in a Dirac bracket expression is equivalent to the operator itself acting on the other function. In the language of integrals, the definition is basically a generalization of the concept of integration by parts in which the adjoint symbol attached to an operator means that the *complex conjugate* of that operator is transferred from one function to the other.

The fact that equations (11a) and (11b) must hold for *arbitrary* admissible $\psi_1$ and $\psi_2$ guarantees that these equations indeed provide at least a formally complete definition of the adjoint of an operator. However, the definition may seem so formal as to appear useless, because it contains no clear prescription for actually constructing an explicit representation of $E^\dagger$ when $E$ is regarded as known. We now show that such a prescription can, in fact, be readily given for the class of operators of interest, namely those constructed from Hermitian operators. In addition, the notion of the adjoint permits us to give a purely operator characterization of the properties of operators so constructed and of Hermiticity itself. For this latter, we see at once from equation (4) that every Hermitian operator is its own adjoint. In other words,

$$A = A^\dagger \qquad (12)$$

if $A$ is Hermitian. For this reason, Hermitian operators are often called *self-adjoint* or *self-conjugate*.

Next consider an operator such as $C$ in equation (5). Recalling that, according to our definition,

$$\langle C\psi_1|\psi_2\rangle \equiv \langle \psi_1|C\dagger\psi_2\rangle\,,$$

we see from equation (7) that

$$C\dagger = \alpha^*A + \beta^*B,$$

which is to say that

$$(\alpha A + \beta B)\dagger = \alpha^*A + \beta^*B.$$

More generally, if $A_i$ are a set of Hermitian operators and $\alpha_i$ a set of numbers,

$$(\Sigma\ \alpha_iA_i)\dagger = \Sigma\ \alpha_i^*A_i. \tag{13}$$

As a special case, we see that the adjoint of a numerical operator is just its complex conjugate,

$$\alpha\dagger = \alpha^*,$$

so that Hermitian conjugation can be regarded as the operator analog of complex conjugation for ordinary complex numbers. As a further expression of this idea, observe that $R + R\dagger$ is Hermitian and so is $(R - R\dagger)/i$. Hence an arbitrary operator $R$ can always be expressed in terms of two Hermitian operators by writing

$$R = [(R + R\dagger)/2] + i[(R - R\dagger)/2i]\,, \tag{14}$$

in analogy with the familiar way of writing complex numbers in terms of two real numbers.

Finally, considering products of operators, equation (9) shows that

$$D\dagger = BA,$$

which is to say that

$$(AB)\dagger = BA.$$

More generally, the adjoint of a product of any number of Hermitian operators is simply that product written in reverse order,

$$(ABC\cdots QR)\dagger = RQ\cdots CBA. \tag{15}$$

Equations (13) and (15) are the prescription we seek. They tell us how to construct the adjoints of arbitrary sums and products of operators and thus of arbitrary functions of operators expressible in power series form.

---

**Exercise 2.** Suppose $A, B, C, \ldots, R$ are arbitrary operators, not necessarily Hermitian.

(a)  Show that $(ABC\cdots QR)\dagger = R\dagger Q\dagger\cdots C\dagger B\dagger A\dagger$.

(b)   Show that $(\alpha A + \beta B)\dagger = \alpha^* A\dagger + \beta^* B\dagger$.

(c)   Show that if $A$ and $B$ are Hermitian, so are $i(A, B)$; $AB + BA$; $ABA$.

(d)   Show that if $A$ is Hermitian, so is $A^n$.

---

With this understanding of how to construct the adjoints of operators, equation (11b) now means that $\langle \psi_1 | A\psi_2 \rangle$ and $\langle A\dagger\psi_1 | \psi_2 \rangle$ can be regarded as two completely equivalent expressions for the *same* quantity. To express this equivalence, we now introduce the single symmetrical bracket symbol $\langle \psi_1 | A | \psi_2 \rangle$ to denote both, by writing, as a notational definition,

$$\langle \psi_1 | A | \psi_2 \rangle \equiv \langle \psi_1 | A\psi_2 \rangle = \langle A\dagger\psi_1 | \psi_2 \rangle. \tag{16}$$

This notation makes explicit the complete freedom we have gained to transfer operators from one state function to another at will, or to integrate by parts, using the rules of equations (13) and (15). In essence, then, the symbol $\langle \psi_1 | A | \psi_2 \rangle$ implies no prior commitment as to which function is to be operated upon. Of course, in actually evaluating such an expression, some choice must ultimately be made, but we are always free to do this as it suits our convenience.

## 3. THE CORRESPONDENCE PRINCIPLE REQUIREMENT

Returning now to our main task, that of determining the Hermitian operator $H$ of equation (1), we examine the correspondence principle requirements that

$$\langle \psi | p | \psi \rangle = m \frac{d}{dt} \langle \psi | x | \psi \rangle \tag{17}$$

and

$$\frac{d}{dt} \langle \psi | p | \psi \rangle = - \langle \psi \left| \frac{dV}{dx} \right| \psi \rangle, \tag{18}$$

where $\psi$ is any solution of equation (1). The first of these expresses the classical relation between momentum and velocity, the other is just Newton's second law of motion.

As a preliminary step, we examine $d/dt \langle \psi | A | \psi \rangle$ for an arbitrary operator $A(p, x, t)$. We have

$$\frac{d}{dt} \langle \psi | A | \psi \rangle = \frac{d}{dt} \int_{-\infty}^{\infty} \psi^*(x, t) A(p, x, t) \psi(x, t) \, dx$$

$$= \int_{-\infty}^{\infty} \left[ \frac{\partial \psi^*}{\partial t} A\psi + \psi^* \frac{\partial A}{\partial t} \psi + \psi^* A \frac{\partial \psi}{\partial t} \right] dx.$$

Separating out the middle term, which is recognized as the expectation value of $\partial A/\partial t$, and using equation (1) to eliminate the time derivatives in the remaining terms, we obtain

$$\frac{d}{dt} \langle \psi|A|\psi \rangle = \langle \psi \left| \frac{\partial A}{\partial t} \right| \psi \rangle + \frac{i}{\hbar} \int_{-\infty}^{\infty} [(H\psi)^* A\psi - \psi^* A H\psi] \, dx.$$

Looking now at the first term in the integral, we see that, as a consequence of the Hermiticity of $H$,

$$\int (H\psi)^* A\psi \, dx = \int \psi^* H A\psi \, dx,$$

whence

$$\frac{d}{dt} \langle \psi|A|\psi \rangle = \langle \psi \left| \frac{\partial A}{\partial t} \right| \psi \rangle + \frac{i}{\hbar} \int_{-\infty}^{\infty} (\psi^* H A\psi - \psi^* A H\psi) \, dx.$$

In this form the integral is recognized as the expectation value of the *commutator* of $H$ and $A$, and we thus obtain the very important result

$$\frac{d}{dt} \langle \psi|A|\psi \rangle = \langle \psi \left| \frac{\partial A}{\partial t} \right| \psi \rangle + \frac{i}{\hbar} \langle \psi|(H,A)|\psi \rangle. \tag{19}$$

It is worthy of note that if $A$ does not *explicitly* depend upon the time, the first term vanishes and the time rate of change is determined solely by the commutator of $H$ and $A$. The first term involves the rate at which the operator $A$ itself varies with time; the second term is generated entirely by the change of the *state function* with time.

We have just emphasized that it is the *explicit* time dependence of $A$ which is relevant in these considerations. Now we are accustomed to thinking of dynamical variables such as position and momentum as functions of the time because they indeed are in the classical domain. At the quantum level, however, the symbols $x$ and $p$ refer to *operators* which do not alter in form with time; in short, they are *time independent* operators. This means precisely and explicitly that

$$\frac{\partial x}{\partial t} = 0$$

and

$$\frac{\partial p}{\partial t} = 0,$$

and similarly for any other operators whose specification contains no reference to the time.[3]

The particular application of equation (19) we have in mind involves just these time independent position and momentum operators. Thus, considering first $x$, we obtain at once

$$\frac{d}{dt} \langle \psi | x | \psi \rangle = \frac{i}{\hbar} \langle \psi | (H, x) | \psi \rangle. \tag{20}$$

The right-hand side can be rewritten using the commutation relations derived in Chapter III. Recall that

$$(f, x) = \frac{\hbar}{i} \frac{\partial f(x, p)}{\partial p}, \tag{III-46}$$

where $f$ is any operator function of $x$ and $p$. Hence, equation (20) becomes

$$\frac{d}{dt} \langle \psi | x | \psi \rangle = \langle \psi \left| \frac{\partial H}{\partial p} \right| \psi \rangle. \tag{21}$$

In this form, all reference to $\hbar$ having disappeared, the passage to the classical limit may be taken at once. Comparison with the correspondence principle requirement expressed by equation (17) then shows that, at least in the classical limit, we must have

$$\frac{1}{m} \langle \psi | p | \psi \rangle = \langle \psi \left| \frac{\partial H}{\partial p} \right| \psi \rangle.$$

Now at any given instant, $\psi$ is a perfectly arbitrary (admissible) state function and hence this result implies the operator equation

$$\frac{p}{m} = \frac{\partial H}{\partial p}. \tag{22}$$

Next, consider the momentum operator $p$. Equation (19) at once gives

$$\frac{d}{dt} \langle \psi | p | \psi \rangle = \frac{i}{\hbar} \langle \psi | (H, p) | \psi \rangle. \tag{23}$$

This time, using the alternative commutation relation

---

[3] Our version of quantum mechanics is one in which all of the time dependence of the observables represented by such operators is carried by the state function. It was introduced by Schrödinger and is usually called the *Schrödinger representation*. An alternative, introduced by Heisenberg, is also possible. In the *Heisenberg representation*, the state function is time *independent* and all of the time dependence is carried by the dynamical operators, through the equations of motion. This representation thus bears a close resemblance to the classical description. Versions intermediate between these extremes are also possible. All of these are completely equivalent and each has its own domain of simplicity and utility. We have chosen the Schrödinger representation because it is the easiest to work with at the elementary level.

$$(p, f) = \frac{\hbar}{i} \frac{\partial f}{\partial x}, \tag{III-45}$$

we obtain

$$\frac{d}{dt} \langle \psi | p | \psi \rangle = - \langle \psi \left| \frac{\partial H}{dx} \right| \psi \rangle . \tag{24}$$

Comparison with the correspondence principle requirement expressed by equation (18) then shows that, again at least in the classical limit, we must have

$$\langle \psi \left| \frac{dV}{dx} \right| \psi \rangle = \langle \psi \left| \frac{\partial H}{\partial x} \right| \psi \rangle ,$$

which implies the operator equation

$$\frac{\partial H}{\partial x} = \frac{dV}{dx}. \tag{25}$$

Equations (22) and (25) thus express conditions on the quantity $H$ which must hold in the classical limit. It is easily verified that these conditions on $H$ are satisfied, provided that

$$H = \frac{p^2}{2m} + V(x). \tag{26}$$

The operator $H$, which is the total energy of the system expressed in terms of momentum and position variables, is recognized as the *Hamiltonian function* and equations (21) and (24) as the classical equations of motion in Hamiltonian form.[4] We thus conclude that the correspondence principle will be satisfied provided that *the quantum mechanical operator H is the same function of the quantum mechanical dynamical variables p and x as is the classical Hamiltonian of the corresponding classical dynamical variables*. The operator $H$ is seen to be Hermitian and to reduce to $p^2/2m$ for a free particle, as required.

Our argument, of course, does not rule out the possibility that $H$ could contain terms which vanish in the classical limit, because their effects would be significant only on the quantum level. It turns out, however, that no such terms occur and that equation (26), which merely gives $H$ its simplest form consistent with the correspondence principle, is correct as it stands.[5]

---

[4] See any of References [14]–[17].

[5] More precisely, equation (26) is correct for particles without spin. For particles with spin, some modifications are in fact required, as we shall eventually see.

## 4. SCHRÖDINGER'S EQUATION IN CONFIGURATION AND MOMENTUM SPACE

We have shown that the state function of a particle in one dimension must satisfy Schrödinger's equation (1), where $H$, the Hamiltonian operator, is given by equation (26). In configuration space this equation takes the form

$$-\frac{\hbar^2}{2m}\frac{\partial^2\psi}{\partial x^2} + V(x)\psi = -\frac{\hbar}{i}\frac{\partial\psi}{\partial t}. \tag{27}$$

What about in momentum space? If $V(x)$ can be expanded as a power series in $x$, then according to our general prescription, we have

$$\frac{p^2}{2m}\phi(p,t) + V\left(-\frac{\hbar}{i}\frac{\partial}{\partial p}\right)\phi(p,t) = -\frac{\hbar}{i}\frac{\partial\phi}{\partial t}. \tag{28}$$

Since, in general, $V(x)$ will contain terms of all orders in $x$, this represents a differential equation of infinite order and hence is hardly useful. A better procedure is to start with equation (27) and transform it directly to momentum space, using the convolution theorem to evaluate the transform of the product $V(x)\psi(x,t)$. In this way we obtain the *integral equation*

$$\frac{p^2}{2m}\phi(p,t) + \frac{1}{\sqrt{2\pi\hbar}}\int W(p')\phi(p-p',t)\,dp' = -\frac{\hbar}{i}\frac{\partial\phi(p,t)}{\partial t}, \tag{29}$$

where $W$ is the transform of $V(x)$,

$$W(p') = \frac{1}{\sqrt{2\pi\hbar}}\int_{-\infty}^{\infty} V(x)\,e^{-ip'x/\hbar}\,dx. \tag{30}$$

The connection between the forms of equations (28) and (29) can be established by expanding $\phi(p-p',t)$ in a power series in $p'$ and identifying the coefficient of the $n$th derivative of $\phi(p,t)$ with the $n$th term in the power series expansion of $V(x)$. In any case, unless the potential is a very special function, the momentum space equation is considerably more complicated than that in configuration space. Hence we shall concentrate mainly on the latter. The reason for the complications in momentum space, compared to the utter simplicity of the momentum space equations for a free particle, is that the momentum is no longer a constant of the motion. The presence of external forces means that the state function necessarily contains a broad mixture of pure momentum states and this mixing is explicitly represented by the integral in equation (29).

**Exercise 3.** Verify the equivalence of equations (28) and (29) in the way suggested above.

It is easy to see that the operator identification

$$E = -\frac{\hbar}{i}\frac{\partial}{\partial t}, \tag{31}$$

which we obtained earlier, retains its validity for a particle which is not necessarily free. This follows since, evidently,

$$\langle E \rangle \equiv \left\langle \frac{p^2}{2m} \right\rangle + \langle V \rangle = \langle H \rangle.$$

Hence, using equation (1),

$$\langle E \rangle = \int \psi^* H \psi \, dx = \int \psi^* \left( -\frac{\hbar}{i}\frac{\partial}{\partial t} \right) \psi \, dx,$$

where $\psi$ is any admissible solution of Schrödinger's equation, and similarly for any function of $E$, which verifies our assertion. Note, too, that the requirement that the energy be a constant of the motion (if it does not contain the time) is automatically satisfied. This follows upon setting the operator $A$ in equation (19) equal to an arbitrary function of $H$. We obtain at once

$$\frac{d\langle f(H) \rangle}{dt} = 0,$$

since $[H, f(H)] = 0$, for arbitrary $f$.

Schrödinger's equation (1) can thus be written as the operator equation

$$H\psi = E\psi, \tag{32}$$

where the Hamiltonian operator $H$ is the classical Hamiltonian function of the dynamical variables $p$ and $x$, regarded as quantum mechanical operators, and where $E$ is the operator (31). Classically, of course,

$$H = \frac{p^2}{2m} + V(x) = E,$$

and we thus see that Schrödinger's equation requires that *the state function $\psi$ be such that it yield the same result when operated upon by the Hamiltonian operator $H$ as when operated upon by the total energy operator $E$.* This then ensures that the expectation value of any function of the total energy is exactly the same as the expectation value of the same function of the Hamiltonian, as required by the correspondence principle. Explicitly,

$$\langle \psi | f(H) | \psi \rangle = \langle \psi | f(E) | \psi \rangle,$$

where $\psi$ is any solution of Schrödinger's equation.

Our formulation of quantum mechanics is now more or less completed. We must still generalize it to three dimensions and to systems of particles. Neither of these is difficult to do, although we shall postpone these generalizations for now. We must also eventually introduce the idea of spin.

The particular procedure we have followed in our development is only one of many possible schemes. At this point, it might be instructive to outline a rather more direct procedure which is a rough approximation to the historical development. Starting with the Planck relation and the de Broglie relation, the free particle Schrödinger equation (which is the equation of a de Broglie wave, $\exp [2\pi i x/\lambda]$) can be quickly written down as

$$-\frac{\hbar^2}{2m}\frac{\partial^2 \psi}{\partial x^2} = \frac{\hbar^2}{2m}\left(\frac{2\pi}{\lambda}\right)^2 \psi = \frac{p^2}{2m}\psi.$$

For a particle moving in a potential $V(x)$,

$$\frac{p^2}{2m} = E - V(x),$$

and hence it may plausibly be argued that, more generally,

$$-\frac{\hbar^2}{2m}\frac{\partial^2 \psi}{\partial x^2} + V(x)\psi = E\psi.$$

Now, according to the Planck relation, the time dependence is $e^{-iEt/\hbar}$ whence, finally, the result can be rewritten in the time dependent form

$$-\frac{\hbar^2}{2m}\frac{\partial^2 \psi}{\partial x^2} + V(x)\psi = -\frac{\hbar}{i}\frac{\partial \psi}{\partial t},$$

which is recognized as Schrödinger's equation (27) in configuration space. One can then use this result to examine the time dependence of expectation values and hence to consider the classical limit. A proof that the correspondence principle is satisfied is required under this scheme of development. Such a proof was first given by Ehrenfest and is known as *Ehrenfest's theorem*. We have more or less reversed this entire line of reasoning in our approach.

## 5. STATIONARY STATES

Just as in the free particle case, a basic set of solutions of Schrödinger's equation can be constructed by using the method of separation of variables to isolate the time dependence. In this way we obtain, as is readily verified, the set of stationary states

$$\psi_E(x, t) = \psi_E(x) \ e^{-iEt/\hbar},$$

where $\psi_E(x)$ is the solution of the *time independent Schrödinger's equation* for energy $E$,

$$H\psi_E \equiv \left[ \frac{p^2}{2m} + V(x) \right]\psi_E = -\frac{\hbar^2}{2m} \frac{d^2\psi_E}{dx^2} + V(x)\psi_E(x) = E\psi_E(x).$$
(33)

The general solution of the time dependent Schrödinger's equation is an arbitrary superposition of stationary states. Depending upon $V(x)$, these states may exist for only a discrete set of energies, for a continuous set or for a mixture. We shall nonetheless write this superposition in the form of a summation

$$\psi(x, t) = \sum_E C_E\psi_E(x) \ e^{-iEt/\hbar},$$
(34)

but it must be understood that equation (34) is symbolic. If the spectrum of energy values is continuous, the superposition sum must be replaced by an integral. If the spectrum contains both discrete and continuous states, a general superposition involves a sum over the discrete states and an integral over continuum states, as they are called. No conceptual difficulties are involved, but these aspects are best explained by considering specific examples, as we shall later do. For now, we shall simply treat the general superposition as if it were a simple summation, because it is algebraically simpler to do so.

The general state described by equation (34) is one in which the energy of the system is not precisely fixed but can take on any of the values $E$ with a likelihood determined by the amplitude, $C_E$, of the $E$th state in the superposition. More precisely, if $\psi(x, t)$ is normalized, a measurement of the energy yields the value $E$ with probability $|C_E|^2$, whence $|C_E|^2$ is the probability that a measurement of the energy of the system will reveal it to be in the $E$th state. We shall shortly give a proof of these assertions in a rather general context.

We now make the fundamental assumption that *whatever the character of the energy spectrum, it represents the totality of physically realizable energies for the system under consideration.* This assumption means that the set of functions $\psi_E$ forms a *complete set* in the sense that *any* physical realizable state function must be expressible as a superposition of stationary states.

It is convenient to choose the stationary states to be normalized, and we shall generally do so,

$$\langle \psi_E | \psi_E \rangle = \int \psi_E{}^*(x)\psi_E(x) \ dx = 1.$$

We now show that these states are also orthogonal, so that the $\psi_E$ form a complete, orthonormal set,

$$\langle \psi_{E'} | \psi_E \rangle = \int \psi_{E'}^* \psi_E \, dx = \delta_{EE'}, \tag{35}$$

where $\delta_{EE'}$ is the Kronecker $\delta$-symbol

$$\delta_{EE'} = \begin{cases} 1, & E = E' \\ 0, & E \neq E' \end{cases}.$$

The proof proceeds as follows. We have

$$H\psi_E = E\psi_E$$

and hence, multiplying by $\psi_{E'}^*$ and integrating,

$$\int \psi_{E'}^* H\psi_E \, dx = E \int \psi_{E'}^* \psi_E \, dx. \tag{36}$$

Similarly, since

$$H\psi_{E'} = E'\psi_{E'}$$

or

$$(H\psi_{E'})^* = E'\psi_{E'}^*$$

we have, upon multiplying by $\psi_E$ and integrating,

$$\int (H\psi_{E'})^* \psi_E \, dx = E' \int \psi_{E'}^* \psi_E \, dx.$$

Because $H$ is Hermitian, this can be rewritten as

$$\int \psi_{E'}^* H\psi_E \, dx = E' \int \psi_{E'}^* \psi_E \, dx.$$

Comparison with equation (36) then shows that

$$(E - E') \int \psi_{E'}^* \psi_E \, dx = 0$$

and hence the integral vanishes if $E' \neq E$, as was to be proved.

It is instructive to carry out the proof in Dirac notation, as an illustration of its use. We have at once, using equation (16) and the fact that $H$ is Hermitian,

$$\langle \psi_{E'} | H | \psi_E \rangle = \langle \psi_{E'} | H\psi_E \rangle = \langle H\psi_{E'} | \psi_E \rangle$$

whence, from Schrödinger's equation,

$$E \langle \psi_{E'} | \psi_E \rangle = E' \langle \psi_{E'} | \psi_E \rangle$$

and hence, as before,

$$\langle \psi_{E'} | \psi_E \rangle = 0, \qquad E' \neq E.$$

It may turn out, and in three dimensions it generally does, that there can be more than one stationary state corresponding to a given energy $E$.

The states are then called *degenerate*. Suppose the set of degenerate states, however many they may be, are denoted by $\psi_E^{(1)}, \psi_E^{(2)}, \ldots$.[6] Our proof, then, furnishes no information on whether the degenerate states are orthogonal to each other, and in general they are not. However, these degenerate states can always be chosen in such a way that they become orthogonal. One procedure for doing this, known as the Schmidt orthogonalization procedure, is the following. Suppose the normalized degenerate state functions $\psi_E^{(1)}, \psi_E^{(2)}, \ldots$ to be given, but suppose they are not orthogonal. Define a new set $\overline{\psi}_E^{(1)}, \overline{\psi}_E^{(2)}, \ldots$ by writing

$$\overline{\psi}_E^{(1)} = \psi_E^{(1)}$$

$$\overline{\psi}_E^{(2)} = a_1 \psi_E^{(1)} + a_2 \psi_E^{(2)}$$

$$\overline{\psi}_E^{(3)} = b_1 \psi_E^{(1)} + b_2 \psi_E^{(2)} + b_3 \psi_E^{(3)},$$
$$\vdots$$

where the coefficients are as yet unknown. These coefficients are now determined successively by requiring that the set $\overline{\psi}_E$ be orthonormal. Thus, we demand first that

$$\langle \overline{\psi}_E^{(2)} | \overline{\psi}_E^{(1)} \rangle = 0$$

$$\langle \overline{\psi}_E^{(2)} | \overline{\psi}_E^{(2)} \rangle = 1,$$

and these two equations then determine $a_1$ and $a_2$ and hence $\overline{\psi}_E^{(2)}$. Next, we demand that

$$\langle \overline{\psi}_E^{(3)} | \overline{\psi}_E^{(1)} \rangle = \langle \overline{\psi}_E^{(3)} | \overline{\psi}_E^{(2)} \rangle = 0$$

$$\langle \overline{\psi}_E^{(3)} | \overline{\psi}_E^{(3)} \rangle = 1,$$

which furnishes three equations for the determination of $b_1$, $b_2$ and $b_3$. This process is continued until all the functions are found.

Note that because the initial ordering of the degenerate states is arbitrary, the set of orthogonalized states is far from unique. Indeed, there are infinitely many possible sets, even in the simplest case. In practice, one takes advantage of this freedom to choose a set convenient to the problem at hand. In the future we shall assume that, if there are degenerate states, these have been orthogonalized in one way or another.

We next demonstrate another important property of the set of stationary states, namely that of *closure*. Closure is a mathematical statement of the completeness of the set. Let $\psi$ denote some arbitrary admissible function and expand it in the complete set $\psi_E$,

---

[6] The notation is intended to emphasize the fact that we are dealing with a group of states, each member of which has the same energy $E$. The different members of this group of degenerate states are specified by the superscripts (1), (2) and so on.

$$\psi(x) = \sum_E c_E \psi_E(x) \, . \tag{37}$$

From equation (35),

$$c_E = \int \psi_E^*(x)\psi(x) \; dx = \langle \, \psi_E | \psi \, \rangle \, . \tag{38}$$

Substituting this expression back into equation (37), we obtain, after interchanging the order of summation and integration,

$$\psi(x) = \int \psi(x') \; dx' \left( \sum_E \psi_E^* (x') \psi_E (x) \right) .$$

Since $\psi(x)$ is arbitrary, this implies that

$$\sum_E \psi_E^*(x')\psi_E(x) = \delta(x - x') \, , \tag{39}$$

which is the desired result. Any *complete* set of functions satisfies the closure condition (39). Conversely, any set of functions which satisfies the closure condition is complete.

## 6. EIGENFUNCTIONS AND EIGENVALUES OF HERMITIAN OPERATORS

Stationary states are those for which the energy has a definite, precise value $E$. It is also of great interest and importance to discuss states in which other physically observable quantities, such as linear momentum or angular momentum, have a definite, precise value. Let the observable in question be represented by the Hermitian operator $A$, and let $\psi_a$ denote the normalized state in which the observable has the precise (real) value $a$. This means that for any $n$,

$$\langle \, \psi_a | A^n | \psi_a \, \rangle = a^n$$

or that, for all $n$,

$$\int \psi_a^*(x) [A^n - a^n] \psi_a(x) \; dx = 0 \, , \tag{40}$$

which implies that operating upon $\psi_a$ with $A$ is equivalent to multiplication by $a$, that is,

$$A\psi_a = a\psi_a. \tag{41}$$

This is clearly a sufficient condition that equation (40) be satisfied; it can be shown to be a necessary condition as well.

The quantity $a$ is called an eigenvalue of the operator $A$ and $\psi_a$ is called the eigenfunction of $A$ corresponding to the eigenvalue $a$. Equation (41) is called an eigenvalue equation. In this language, the stationary state function $\psi_E(x)$ is the eigenfunction of the Hamiltonian operator corresponding to the energy eigenvalue $E$.

The eigenvalues of $A$ represent possible values of the observable represented by $A$. *Whatever the character of the eigenvalue spectrum of a given observable may be, we shall assume it to contain the totality of physically realizable values of the observable in question.* Stated mathematically, we shall assume the $\psi_a$ to form a complete set. Repeating the arguments of the last section, with $H$ replaced by $A$ and $\psi_E$ by $\psi_a$, we find at once that, corresponding to equation (35),

$$\langle \psi_{a'}|\psi_a \rangle = \delta_{aa'} \tag{42}$$

and, corresponding to equation (39),

$$\sum_a \psi_a^*(x')\psi_a(x) = \delta(x - x'). \tag{43}$$

Indeed, equations (35) and (39) are now to be regarded as special cases of equation (42) and equation (43).

According to our assumptions, any arbitrary admissible state function $\psi(x)$ can be expressed in terms of the $\psi_a$ by the general superposition

$$\psi(x) = \sum c_a\psi_a(x) \tag{44}$$

where

$$c_a = \int \psi_a^*(x)\psi(x)\ dx = \langle \psi_a|\psi \rangle. \tag{45}$$

Equations (44) and (45) are thus seen to be generalizations of Fourier series (or integrals).

The physical significance of the expansion coefficients $c_a$ can be seen as follows. Assuming $\psi(x)$ to be normalized, we have

$$1 = \langle \psi|\psi \rangle = \langle \sum_{a'} c_{a'}\psi_{a'}| \sum_a c_a\psi_a \rangle$$
$$= \sum_{a,a'} c_{a'}^*c_a \langle \psi_{a'}|\psi_a \rangle,$$

and hence, using the orthonormality condition, equation (42),

$$\sum |c_a|^2 = 1.$$

Next consider the expectation value of $A$. We have

$$\langle \psi|A|\psi \rangle = \sum_{a,a'} c_{a'}^*c_a \langle \psi_{a'}|A|\psi_a \rangle$$

whence, since $\psi_a$ is an eigenfunction of $A$ with eigenvalue $a$,

$$\langle \psi|A|\psi \rangle = \sum_{a,a'} ac_{a'}^*c_a \langle \psi_{a'}|\psi_a \rangle.$$

Once again using the orthnormality condition, we finally obtain

$$\langle \psi|A|\psi \rangle = \sum a|c_a|^2.$$

By exactly the same argument we see that, more generally,

$$\langle \psi | f(A) | \psi \rangle = \sum f(a) |c_a|^2 .$$

We thus conclude that $|c_a|^2$ is the probability that in the state $\psi$, the observable represented by the operator $A$ has the numerical value $a$. This is then the proof, and the generalization, of our assertions about the meaning of the analogous coefficients which are obtained when an arbitrary state function is expressed as a superposition of stationary states.

As a specific example, suppose that $A$ is the momentum operator,

$$A = \frac{\hbar}{i} \frac{\partial}{\partial x} .$$

Equation (41) becomes

$$\frac{\hbar}{i} \frac{d\psi_p(x)}{dx} = p\psi_p(x) ,$$

whence

$$\psi_p(x) = \frac{1}{\sqrt{2\pi\hbar}} e^{ipx/\hbar} ,$$

in agreement with our earlier discussion of state functions corresponding to a definite momentum $p$. Since these states exist for all real values of $p$, the spectrum is continuous and the superposition sum must be replaced by an integral. The momentum eigenfunctions are *not* normalizable, and the multiplicative factor $1/\sqrt{2\pi\hbar}$ has been chosen to maintain the closure relation (43), which takes the form

$$\int \psi_p{}^*(x')\psi_p(x) \, dp = \delta(x - x')$$

or

$$\frac{1}{2\pi\hbar} \int e^{ip(x-x')/\hbar} \, dp = \delta(x - x') .$$

The orthonormality condition (42) is

$$\frac{1}{2\pi\hbar} \int e^{i(p-p')x/\hbar} \, dx = \delta(p - p') ,$$

the Kronecker $\delta$ being replaced by a Dirac $\delta$-function, since $p$ is continuous.

The expansion of an arbitrary wave function as a superposition of momentum eigenfunctions is expressed as an integral, as we have said, and we rewrite equation (44) as

$$\psi(x) = \int c_p\psi_p(x) \, dp$$

where the expansion coefficient $c_p$ must now be considered as a function of the continuous variable $p$. Writing $c_p = \phi(p)$ to make this apparent, and inserting the explicit form of $\psi_p$, we obtain the familiar result

$$\psi(x) = \frac{1}{\sqrt{2\pi\hbar}} \int \phi(p) \, e^{ipx/\hbar} \, dp,$$

while equation (45) becomes the equally familiar expression

$$\phi(p) = \frac{1}{\sqrt{2\pi\hbar}} \int e^{-ipx/\hbar} \, \psi(x) \, dx.$$

Finally $|\phi(p)|^2$ is to be interpreted as the probability density that the momentum have the value $p$ for a system in the state $\psi$. Thus we have again derived the features of momentum space representations, physically speaking, or of Fourier transforms, mathematically speaking, from this general point of view.

## 7. SIMULTANEOUS OBSERVABLES AND COMPLETE SETS OF OPERATORS

We have seen that if an observation is performed on some system it will lead to a definite and precise result only if the system is in a special state, namely, an eigenstate of the observable in question. If, as before, the operator representing the observable is denoted by $A$, then its eigenstate $\psi_a$ corresponding to the eigenvalue $a$ is defined by equation (41), which we rewrite here for reference,

$$A\psi_a \equiv a\psi_a.$$

Thus far, our understanding of such an equation has mainly been something as follows. If the operator A, representing some observable, is taken as known, that is, if the effect it produces when acting upon any arbitrary admissible function is given, then equation (41) is a mathematically precise recipe for constructing the abstract states $\psi_a$ in which the observable in question has the definite value $a$. Our appreciation of the physical content of such quantum mechanical equations can be greatly enhanced, however, if we also regard these equations as *abstract representations of the physical process of measurement itself*. More specifically, operation upon some state function with an operator $A$ can be thought of as equivalent to actually measuring the observable to which $A$ corresponds. In this view, equation (41) is *directly* and simply a statement that $\psi_a$ is that state for which measurement of the observable represented by $A$ *always* yields the value $a$.

The complete specification of a quantum mechanical state, just as of a classical state, requires some number of measurements or observations, this number being determined by the degrees of freedom of the system. As an immediate generalization of our previous results, it follows that simultaneous measurement of two or more observables will lead to a definite result for each only if the system is, at one and the same time, an eigenfunction of each. This implies that such simultaneous observations are mutually independent or do not interfere with one another. If that is the case, the *order* in which the measurements are made is irrelevant and the operators representing the observables (or observations) must mutually commute. We now give a formal proof that this is so.

Consider first the situation in which there are only two observables. Denote the corresponding operators by $A$ and $B$ and let $\psi_{ab}$ be a state in which the observables represented by $A$ and $B$ have the precise values $a$ and $b$. Such a state is called a simultaneous eigenfunction of $A$ and $B$ and is defined by

$$A\psi_{ab} = a\psi_{ab}$$
$$B\psi_{ab} = b\psi_{ab}.$$

Now, evidently, we have

$$BA\psi_{ab} = ab\psi_{ab}$$
$$AB\psi_{ab} = ab\psi_{ab}$$

whence

$$(A, B)\,\psi_{ab} = 0.$$

If the set $\psi_{ab}$ is complete, then any arbitrary state $\psi$ can be expressed as a superposition of the $\psi_{ab}$. This means that $(A, B)$ gives zero when acting upon an arbitrary state and hence, as was to be proved, $(A, B) = 0$.

Note that the state $\psi_{ab}$ is degenerate with respect to $a$ or $b$ alone. For *given a* the eigenfunctions of $B$, with differing values of $b$, are all degenerate with respect to $A$, and conversely.

By an obvious extension of the argument, we see that a complete set of simultaneous eigenfunctions of a set of operators can exist only if the operators mutually commute. Bearing in mind the degeneracy associated with such simultaneous eigenfunctions, we say that a set of mutually commuting operators is *complete* if that set *uniquely* defines a complete set of states. For a structureless particle, for example, the momentum operator by itself forms such a complete set. So does the position operator. The Hamiltonian operator $H$ generally does not. Along with $H$, the complete set of operators which commute with $H$ defines the quantum mechanical constants of the motion. As we shall see, most but not all of these are analogous to classical constants of motion.

It is important to note that if, say, $A$ and $B$ commute, then our analysis in no way implies that an eigenfunction of $A$ is *necessarily* an eigenfunction of $B$ or vice versa. What we have demonstrated is that it is *possible* to define a set of simultaneous eigenfunctions of operators if they mutually commute. All this is obvious when expressed in terms of measurements. Commutivity means that independent, noninterfering observations are possible, but not, of course, that such observations have necessarily been carried out.

## 8. THE UNCERTAINTY PRINCIPLE

Observables take on definite and precise values only in their eigenstates. For more general states, observation yields values which fluctuate about the average or expectation value from one measurement to the next. For some given admissible state $\psi$ and some given observable $A$, we now introduce the *uncertainty* $\Delta A$ as a quantitative measure of these fluctuations. The uncertainty is defined as the root mean square deviation of the observed values of $A$ from the expected value. Thus we write[7]

$$(\Delta A)^2 = \langle (A - \langle A \rangle)^2 \rangle = \langle A^2 \rangle - \langle A \rangle^2, \tag{48}$$

where all expectation values are taken with respect to the state $\psi$, which we shall assume to be normalized as a matter of convenience.

Consider now a pair of observables represented by the Hermitian operators $A$ and $B$. If $A$ and $B$ commute, and thus are noninterfering observations, we have seen that states exist for which each has a definite value. However, if $A$ and $B$ do not commute, so that measurement of one introduces a disturbance into the measurement of the other, then both cannot be simultaneously known with arbitrary precision. This mutual uncertainty is not a matter of experimental technique but is a question of principle, the *uncertainty principle*. As a precise statement of that principle, we now prove that for any admissible state $\psi$,

$$(\Delta A)^2 (\Delta B)^2 \geq \tfrac{1}{4} |\langle (A, B) \rangle|^2, \tag{49}$$

---

[7] The equivalence of the two expressions given in equation (48) is established by squaring out the first. Thus

$$\langle (A - \langle A \rangle)^2 \rangle = \langle (A^2 - 2A \langle A \rangle + \langle A \rangle^2) \rangle.$$

Now $\langle A \rangle$ is some ordinary number and the expectation value of any number is just the number itself. Hence, for all $n$,

$$\langle ( \langle A \rangle^n ) \rangle = \langle A \rangle^n.$$

Proceeding term by term,

$$\langle (A - \langle A \rangle)^2 \rangle = \langle A^2 \rangle - 2\langle A \rangle^2 + \langle A \rangle^2$$
$$= \langle A^2 \rangle - \langle A \rangle^2,$$

as was to be proved.

where again all expectation values are taken with respect to $\psi$. We shall show further that the equality sign in equation (49), which indicates the *minimum possible uncertainty*, holds only for states such that, with $c$ some constant,

$$[A - \langle A \rangle]\psi = c[B - \langle B \rangle]\psi \qquad (50)$$

and such that, at the same time,

$$\langle [A - \langle A \rangle][B - \langle B \rangle] + [B - \langle B \rangle][A - \langle A \rangle] \rangle = 0. \qquad (51)$$

The starting point of the proof is the observation that for any arbitrary pair of functions $f$ and $g$,

$$\int_{-\infty}^{\infty} \left| f\left( \int_{-\infty}^{\infty} gg^* \, dx \right) - g\left( \int_{-\infty}^{\infty} fg^* \, dx \right) \right|^2 dx \geq 0. \qquad (52)$$

Because the integrand can never be negative, the equality sign holds only if the integrand is identically zero, that is, if

$$f = cg, \qquad (53)$$

where $c$ is an arbitrary constant.

Squaring out the integrand of equation (52) and combining terms, we obtain, after some algebra,

$$\int ff^* \, dx \int gg^* \, dx \geq \int fg^* \, dx \int f^*g \, dx, \qquad (54)$$

which is the famous *Schwartz inequality.* Now replace $f$ by $F\psi$ and $g$ by $G\psi$, where

$$F = A - \langle \psi|A|\psi \rangle$$
$$G = B - \langle \psi|B|\psi \rangle. \qquad (55)$$

Since $A$ and $B$ are Hermitian, and thus have real expectation values, $F$ and $G$ are also Hermitian. Consequently, referring back to equation (48), equation (54) becomes

$$(\Delta A)^2(\Delta B)^2 \geq \int F\psi G^*\psi^* \, dx \int F^*\psi^*G\psi \, dx$$
$$= \int \psi F^*G^*\psi^* \, dx \int \psi^*FG\psi \, dx$$

or

$$(\Delta A)^2(\Delta B)^2 \geq \left| \int \psi^*FG\psi \, dx \right|^2 = |\langle \psi|FG|\psi \rangle|^2.$$

Now the quantity $\langle \psi|FG|\psi \rangle$ is complex, because $FG$ is not Hermitian. To separate it into its real and imaginary parts, we introduce the identity [8]

$$FG = \frac{1}{2}(FG + GF) + \frac{i}{2}\frac{(F, G)}{i}$$

---

[8] See Section 2, particularly equation (14), and also Exercise 2, part (c), page 91.

whence we obtain

$$(\Delta A)^2 \ (\Delta B)^2 \geqslant \tfrac{1}{4} |\langle \psi | FG + GF | \psi \rangle + i \langle \psi \left| \frac{(F, G)}{i} \right| \psi \rangle|^2. \quad (56)$$

Because the first term under the absolute value squared sign in equation (56) is real and the second imaginary, the right side of that equation is just the sum of the squares, and we thus obtain

$$(\Delta A)^2(\Delta B)^2 \geqslant \tfrac{1}{4} |\langle \psi | FG + GF | \psi \rangle|^2 + \tfrac{1}{4} |\langle \psi | [F, G] | \psi \rangle|^2. \quad (57)$$

The first term involves the expectation value of the quantum analog of the classical dynamical variable which is the product of $A$ and $B$. This expectation value is certainly state dependent and at any instant can be made to vanish. The right side, involving the commutator, often may have a state *independent* character because, as we have seen, the commutator of operators representing dynamical variables is, in at least some cases, a pure number. Hence the essential content of equation (57) is that there is a fundamental limitation on the simultaneous determination of $A$ and $B$ which is beyond our control because, no matter what, we must always have

$$(\Delta A)^2(\Delta B)^2 \geqslant \tfrac{1}{4} |\langle \psi | (F, G) | \psi \rangle|^2 \qquad (58)$$

or, using equation (55),

$$(\Delta A)^2(\Delta B)^2 \geqslant \tfrac{1}{4} |\langle \psi | (A, B) | \psi \rangle|^2,$$

which is equation (49). Further, the equality holds if, and only if, the first term of equation (57) vanishes, which is recognized as equation (51), and if, in addition, equation (53) is simultaneously satisfied. Recalling that $f$ and $g$ are $F \psi$ and $G \psi$, respectively, this condition is simply

$$F \psi = cG \psi,$$

which is recognized as equation (50).

As a particular example, consider the position and momentum,

$$A = p, \qquad B = x.$$

Equation (49) then states that

$$(\Delta p)^2 (\Delta x)^2 \geqslant \frac{\hbar^2}{4},$$

which is the precise inequality promised in our earlier discussion in Section 7 of Chapter III. Further, $(\Delta p)^2 (\Delta x)^2$ takes on its minimum possible value, $\hbar^2/4$, only if the state $\psi$ is such that

$$(p - p_0) \psi = c(x - x_0) \psi \qquad (59)$$

and

$$\langle \psi | (p - p_0)(x - x_0) + (x - x_0)(p - p_0) | \psi \rangle = 0, \qquad (60)$$

where we have used the abbreviations

$$\langle \psi | p | \psi \rangle = p_0$$

$$\langle \psi | x | \psi \rangle = x_0.$$

The state function defined by equations (59) and (60) is called the *wave packet of minimum uncertainty*. We now determine this state.

Equation (59) is equivalent to

$$\frac{\hbar}{i} \frac{d\psi}{dx} = [p_0 + c(x - x_0)]\psi$$

whence

$$\psi = A \exp\left[\frac{ip_0 x}{\hbar} + \frac{ic(x - x_0)^2}{2\hbar}\right].$$

Next consider equation (60). Perhaps the simplest procedure is the following: The commutation relation between $p$ and $x$ permits us to write

$$(p - p_0)(x - x_0) = \frac{\hbar}{i} + (x - x_0)(p - p_0),$$

whence equation (60) becomes

$$\frac{\hbar}{i} \langle \psi | \psi \rangle + 2\langle \psi | (x - x_0)(p - p_0) | \psi \rangle = 0.$$

Using equation (59), this expression can be rewritten in the form

$$\frac{\hbar}{i} \langle \psi | \psi \rangle + 2c \langle \psi | (x - x_0)^2 | \psi \rangle = 0,$$

whence

$$c = \frac{i\hbar}{2} \frac{\langle \psi | \psi \rangle}{\langle \psi | (x - x_0)^2 | \psi \rangle}.$$

This shows $c$ to be a positive, purely imaginary number. Writing therefore

$$c \equiv i\hbar/L^2,$$

we finally obtain, upon putting all of this together,

$$L^2 = 2 \frac{\int (x - x_0)^2 \, e^{-(x - x_0)^2/L^2} \, dx}{\int e^{-(x - x_0)^2/L^2} \, dx},$$

which is identically satisfied for all $L$. The minimum wave packet is therefore simply a general Gaussian

$$\psi = A \exp\left[\frac{ip_0 x}{\hbar} - \frac{(x - x_0)^2}{2L^2}\right],$$

which can be normalized by proper choice of $A$.

## 9. WAVE PACKETS AND THEIR MOTION

We have assumed that the eigenfunctions of any Hermitian operator, say $A$, form a complete set and have seen that these functions can always be chosen to be orthonormal. This means that any arbitrary state function, $\psi(x, t)$ can be expressed as the superposition

$$\psi(x, t) = \sum_a c_a(t)\psi_a(x),$$

where the $\psi_a$ are eigenfunctions of $A$ with eigenvalue $a$, and where $|c_a(t)|^2$ is the probability that the particle will be found in the state $a$ at time $t$. If $\psi(x, t)$ is prescribed at some initial instant, say $t_0$, then in virtue of the orthonormality of the $\psi_a$, we have

$$c_a(t_0) = \int \psi_a^*(x)\psi(x, t_0)\ dx$$

and the $c_a(t)$ can thus be presumed known initially. What about the subsequent development of the $c_a$ with time? This development is determined by Schrödinger's equation, and no useful general prescription can be given. Only if $A = H$ and $a = E$ is the description simple. In that case,

$$c_E(t) = c_E\ e^{-iEt/\hbar}$$

and

$$c_E(t_0) = c_E\ e^{-iEt_0/\hbar} = \int \psi_E^*(x)\ \psi(x, t_0)\ dx,$$

whence

$$c_E(t) = e^{-iE(t-t_0)/\hbar} \int \psi_E^*(x)\psi(x, t_0)\ dx.$$

Substitution into the superposition

$$\psi(x, t) = \sum_E c_E(t)\psi_E(x)$$

then gives, as the generalization of our earlier results for a free particle,

$$\psi(x, t) = \int \psi(x', t_0)K(x', x;\ t - t_0)\ dx', \tag{61}$$

where the propagator $K$ is given by

$$K(x', x; t - t_0) = \sum_E \psi_E^*(x')\psi_E(x)\ e^{-iE(t-t_0)/\hbar}. \tag{62}$$

Of course, as before,

$$K(x', x; 0) = \delta(x - x'),$$

so that $K$ is initially sharply localized, but as time goes on, $K$ spreads out over a broader and broader region.

Equations (61) and (62) provide a complete formal solution to the problem of a particle's motion under the influence of external forces, a solution expressed in terms of the complete set of eigenfunctions of the Hamiltonian. We emphasize that the solution is a formal one because, even if these eigenfunctions are explicitly known, the infinite summation in equation (62) cannot be evaluated in closed form except in special cases. Thus we were able earlier to find the free particle propagator, equation (IV-13), and we shall eventually obtain $K$ for a uniformly accelerated particle and for the harmonic oscillator.[9] These special cases together furnish at least a reasonable qualitative guide to the general characteristics of propagators. We remark, finally, that, as might be expected, $K$ can generally be constructed, and is very useful, in the classical limit.

## 10. SUMMARY: THE POSTULATES OF QUANTUM MECHANICS

We now give a brief summary in the form of a list of basic postulates. These are:

(1)  Every physically realizable state of motion of a system is described by a state function $\psi$. Physically admissible state functions are normalizable and single-valued. Any superposition of state functions is also a state function.

(2)  Dynamical variables are represented by Hermitian operators which act on state functions. The spectrum of eigenvalues of a given operator comprises the totality of physically realizable values for the observable in question. The simultaneous eigenfunctions of any complete set of mutually commuting operators is a uniquely determined complete set of state functions. For a structureless particle in one dimension, the position variable is itself such a complete set. So is the linear momentum variable. These two operators are characterized by the commutation relation $(p, x) = \hbar/i$.

(3)  The time development of a state function is given by Schrödinger's equation, $H\psi = E\psi$, where $H = p^2/2m + V(x)$ is the Hamiltonian operator and where

$$E = -\frac{\hbar}{i} \frac{\partial}{\partial t}$$

---

[9] For the former, see Problem VI-8; the latter is derived in Chapter VI, Section 6.

is the energy operator.

(4) The expectation value of any dynamical variable $A$ when the system is in a state $\psi$ is $\langle \psi|A|\psi \rangle$. The expectation value is to be interpreted as the average of a series of measurements of $A$ performed upon an ensemble of identical systems each of which is in the same state $\psi$.

---

**Problem 1.** Consider an operator $A$ and its eigenfunctions $\psi_a$, defined over the interval $0 \leqslant x \leqslant L$ and satisfying the boundary conditions $\psi_a(x=0) = \psi_a(x=L)$. [These are called periodic boundary conditions.] Determine the eigenvalues and eigenfunctions of $A$, if any, for the following cases:

(a)  $A = \dfrac{d}{dx}$.

(b)  $A = i\dfrac{d}{dx} + k$, $k$ is real, positive and fixed.

(c)  $A =$ the integral operator defined by $A\psi \equiv i \int_0^x \psi(y)\, dy$.

In each case, prove whether or not $A$ is Hermitian. Is the set of $\psi_a$ complete in each case?

**Problem 2.** Consider the eigenvalue equation

$$A\psi_a(x) = a\psi_a(x)$$

defined over the interval $-L \leqslant x \leqslant L$ and subject to the boundary conditions

$$\psi_a(-L) = \psi_a(L) = 0.$$

(a)  If $A = (d/dx)^n$, for what values of $n$, if any, is $A$ Hermitian? Explain.

(b)  Find the eigenfunctions of $A$ corresponding to $a = 0$ for each of the cases $n = 3, 4, 5$. If there are any degeneracies for a given $n$, use the Schmidt procedure to orthogonalize the degenerate states.

**Problem 3.** Give the *time dependence* of the state function for

(a)  a system which is in an eigenstate of its Hamiltonian with eigenvalue $\beta$.

(b)  a system which is an unspecified mixture of an $n$-fold degenerate set of eigenstates of its Hamiltonian with eigenvalue $\beta$.

**Problem 4.** Find the smallest interval in $x$ over which $\sin \pi x/x_0$ and $\cos \pi x/x_0$ are orthogonal. Over this same interval are $\sin \pi x/x_0$ and $\sin 2\pi x/x_0$ orthogonal? $\sin \pi x/x_0$ and $\exp[i\pi x/x_0]$? $\sin \pi x/x_0$ and $\sin \pi x/2x_0$?

**Problem 5.** Let $\phi_n$ denote the ortho-normal stationary states of a system corresponding to energy $E_n$. At time $t = 0$, the normalized state function of the system is $\psi = \Sigma\, c_n \phi_n$. Assuming the $\phi_n$ and $c_n$ to be given,

(a)  write the state function of the system for $t > 0$.

(b)  what is the probability that a measurement of the energy at time $t$ will yield the value $E_n$?

(c)  what is the expectation value of the energy at any time $t$?

**Problem 6.** $A$ and $B$ are two arbitrary Hermitian operators. Work out which, if any, of the following operators (i) $AB$, (ii) $A^2$, (iii) $AB - BA$, (iv) $AB + BA$, (v) $ABA$

(a)  are Hermitian.

(b)  have real non-negative expectation values.

(c)  have pure imaginary expectation values.

(d)  are purely numerical operators.

**Problem 7.**

(a)  $A$ and $B$ are Hermitian operators and are such that $A^2 = B^2 = 2$. Deduce the eigenvalues of each.

(b)  Let $\psi_b$ denote the eigenfunctions of $B$ corresponding to the eigenvalues $b$. Suppose $A$ to be such that $\langle \psi_b | A | \psi_b \rangle = 0$. From the definition of uncertainty, calculate the uncertainty in $A$ for a state of definite $B$.

(c)  Use your result and the uncertainty principle to deduce the value of $\langle \psi_b | (A, B) | \psi_b \rangle$. Verify your answer by direct evaluation of the expectation value of the commutator.

**Problem 8.**

(a)  If $A$, $B$ and $C$ are arbitrary operators, prove that

$$(A, BC) = (A, B)C + B(A, C).$$

(b)  Use your result to show that

$$\frac{d}{dt}\langle AB \rangle = \langle \frac{\partial A}{\partial t}\, B \rangle + \langle A\, \frac{\partial B}{\partial t} \rangle + \frac{i}{\hbar}\langle (H, A)B \rangle + \frac{i}{\hbar}\langle A(H, B) \rangle.$$

**Problem 9.** For a system described by the Hamiltonian $H = p^2/2m + V(x)$, obtain an expression for $d\,\langle p^2/2m \rangle/dt$. Discuss the relation of your result to the classical work-energy theorem.

**Problem 10.** We have asserted that the commutation relation $(p, x) = \hbar/i$ is fundamental. In configuration space, where $x$ is a numerical operator, we have represented $p$ as the differential operator

$$p = \frac{\hbar}{i}\frac{d}{dx},$$

which is consistent with the commutation relation. More generally, one *could* equally well write

$$p = \frac{\hbar}{i} \frac{\partial}{\partial x} + f(x),$$

where $f(x)$ is arbitrary, without violating this commutation relation. Suppose this is done.

(a)   Find the new momentum eigenfunctions $\psi_p$.

(b)   Express an arbitrary state function $\psi(x)$ as a superposition of these momentum eigenfunctions.

(c)   Verify that no physically observable properties of the system depend on $f(x)$ and hence that the most convenient choice, $f(x) = 0$ generally, is always permissible.

**Problem 11.** Consider the propagator $K$ of equation (62).

(a)   Show that $K$ satisfies the integral equation

$$K(x', x; t - t_0) = \int K(x', x'', t - t_1) K(x'', x; t_1 - t_0) \, dx''.$$

(b)   Interpret this equation in terms of the time development of a state function first from $t_0$ to $t_1$ and then from $t_1$ to $t$.

(c)   Equation (61) was derived from the time dependent Schrödinger's equation and is entirely equivalent to it. Verify this by deriving Schrödinger's equation from equation (61). Assume the $\psi_E(x)$ are known to be solutions of $H\psi_E(x) = E\psi_E(x)$ with $H$ given.

**Problem 12.** Let $\psi_n(x)$ denote the orthonormal stationary states of a system corresponding to energy $E_n$. Let $\psi(x, 0)$ denote the normalized state function of the system at time $t = 0$. Suppose $\psi(x, 0)$ to be such that a measurement of the energy of the system would yield the value $E_1$ with probability $\frac{1}{2}$, $E_2$ with probability $\frac{1}{4}$, and $E_3$ with probability $\frac{1}{4}$.

(a)   Write the most general expression you can for $\psi(x, 0)$ in terms of the $\psi_n(x)$, consistent with the given data.

(b)   Write an expression for the state function $\psi(x, t)$ at time $t > 0$.

(c)   Which of the following quantities have expectation values which are *independent* of the time for this system when it is in the state $\psi(x, t)$ of part (b)?

(i) position, $\langle x \rangle$.          (iv) Hamiltonian, $\langle H \rangle$.

(ii) kinetic energy, $\langle p^2/2m \rangle$.

(iii) potential energy, $V(x)$.          (v) impulse, $\left\langle -t \dfrac{dV}{dx} \right\rangle$.

(d)   The same as (c) when the system is in one of its stationary states $\psi_n$.

**Problem 13.** Let $C$ denote the operator which changes a function into its complex conjugate,

$$C\psi \equiv \psi^*.$$

(a)   Determine whether or not $C$ is Hermitian.

(b)   Find the eigenvalues of $C$.

(c)   Do the eigenfunctions of $C$ form a complete set? Are they orthogonal? Briefly explain your answers.

**Problem 14.**  Show that if the potential energy $V(x)$ is changed by a constant amount everywhere, the stationary state wave functions are unaltered. What about the energy eigenvalues?

**Problem 15.**  Show that the expectation value of the square of an Hermitian operator can never be negative.

**Problem 16.**

(a)   Using the result given in problem 8, part (b), find an expression for $d/dt \langle \psi | px | \psi \rangle$ and then, by considering a stationary state $\psi_E$, prove the so-called virial theorem

$$\langle \psi_E | T | \psi_E \rangle = \tfrac{1}{2} \left\langle \psi_E \left| x \frac{dV}{dx} \right| \psi_E \right\rangle.$$

(b)   Find the corresponding virial theorem for *classical* mechanics by translating your proof into classical language.

**Problem 17.**

(a)   Find an expression for $(\Delta x)^2 (\Delta T)^2$ where $T = p^2/2m$ is the kinetic energy operator.

(b)   Find an expression for $(\Delta p)^2 (\Delta H)^2$; for $(\Delta x)^2 (\Delta H)^2$.

(c)   Why are the results so much less useful and interesting than those obtained in the text for the mutual uncertainties in $x$ and $p$?

# VI

# *States of a particle in one dimension*

## 1. GENERAL FEATURES

We now consider the stationary states $\psi_E(x)$ of a particle moving in a potential $V(x)$. Such states are solutions of the time independent Schrödinger equation $H\psi_E = E\psi_E$, which, when written out in configuration space, is the ordinary second-order differential equation

$$-\frac{\hbar^2}{2m}\frac{d^2\psi_E}{dx^2} + V(x)\psi_E = E\psi_E. \qquad (1)$$

Before going on to consider the solutions of this equation for particular potentials, we discuss the character of its solutions for some more or less general potential, which we suppose to have the form sketched in Figure 1. In drawing the figure, the zero of energy has been chosen in such a

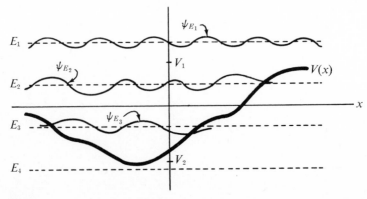

FIGURE 1.   States of motion in a potential $V(x)$ for various energies.

way that $V(x)$ vanishes for $x \rightarrow -\infty$. As $x$ increases, we have taken $V(x)$ to become increasingly negative until it attains some minimum value $V_2$, after which it is taken to increase and eventually to attain the value $V_1$ when $x \rightarrow +\infty$. A potential such as this is quite general enough for our purposes.

The character of the states of motion, both classically and quantum mechanically, is completely determined by the energy. We distinguish four regions of energy as follows:

(1)  $E > V_1$, for example $E_1$ in the figure.
(2)  $0 < E < V_1$, for example $E_2$ in the figure.
(3)  $V_2 < E < 0$, for example $E_3$ in the figure.
(4)  $E < V_2$, for example $E_4$ in the figure.

We now take these up in order.

(1)  $E > V_1$. In this region the kinetic energy $E - V(x)$ is everywhere positive. Classically there are two independent states of motion, one in which the particle moves toward the right, the other in which it moves toward the left. Quantum mechanically there are also two independent, if more complicated, states of motion which are just the two independent solutions of the second-order differential equation (1). Since the kinetic energy is everywhere positive, we see from equation (1) that $d^2\psi_E/dx^2$ and $\psi_E$ have opposite sign everywhere. Hence $\psi_E$ is always concave toward the $x$-axis or, in other words, oscillatory. Consequently $\psi_E$ is bounded but extends to infinity in both directions.[1] These properties it shares with the eigenfunctions of momentum $\psi_p$, and, like them, it is not strictly physically admissible. Also like the $\psi_p$, these states represent very useful idealizations from which the true physical states, wave packets, can be constructed. The two independent quantum mechanical states do not correspond to motion solely to the left or right; a particle incident from the left, for example, does not necessarily continue indefinitely; it is sometimes reflected. The states are thus rather complicated. Nonetheless, these states occur for any energy $E$ which exceeds $V_1$, and we thus conclude that *the energy spectrum is continuous and doubly degenerate in this region.*

(2)  $0 < E < V_1$. In this region the kinetic energy is positive to the left of the intersection of $E$ and $V(x)$, negative to the right. The point of intersection, where $E = V(x)$ and the kinetic energy vanishes, is the classical turning point of the motion. Classically, a particle moving to the right is reflected at the turning point and returns to the left; this is the only general type of motion. Quantum mechanically there are still two independent solutions of Schrödinger's equation, but only one is admissible (even as an idealized state), as we now show. To the left of the classical turning point, the solutions of equation (1) are oscillatory.

[1] The solution labeled $\psi_{E_1}$ in Figure 1 illustrates this behavior.

To the right, however, where $d^2\psi_E/dx^2$ has the same sign as $\psi_E$, the solutions are convex with respect to the $x$-axis and hence either increase without limit or decrease strongly to zero as $x$ increases. The general solution to equation (1) contains an arbitrary superposition of these two types of terms, but only the particular solution which decreases to zero is permitted by the requirements of physical admissibility. One such solution, labeled $\psi_{E_2}$, is illustrated in Figure 1. We thus conclude that *the spectrum is continuous and nondegenerate in this region.* The solutions resemble the classical ones in the sense that a particle is always reflected.

(3)   $V_2 < E < 0$.  In this region there are two classical turning points and the kinetic energy is positive between them, negative elsewhere. Classically, the motion is *periodic,* with the period uniquely determined by the energy. There is thus only one type of motion for a given energy. Quantum mechanically, no motion is possible at all unless the energy has exactly the right value, one of a set of proper values which comprises a discrete, nondegenerate spectrum. The argument is the following. The general solution of equation (1) contains both increasing and decreasing terms in the region to the right of the right-hand turning point. The same behavior occurs in the region to the left of the left-hand turning point. A physically admissible solution, however, must decrease to zero in *both* these domains. Suppose we choose that solution which has the desired property on the right. Call this particular solution, which is uniquely specified, $\psi_{E, \text{right}}$. Now suppose we choose instead that solution which decreases to zero on the left. Call it $\psi_{E, \text{left}}$. It is also unique. Both $\psi_{E, \text{right}}$ and $\psi_{E, \text{left}}$ oscillate in the region between turning points, and in general they will not join smoothly together in their common domain of existence. As $E$ is varied, however, the curvature of these functions, and therefore their rate of oscillation, is altered and hence, if $E$ has *exactly* the right value, $\psi_{E, \text{left}}$ and $\psi_{E, \text{right}}$ can be made to join smoothly together. In a rough sense, this corresponds to choosing $E$ in such a way that a definite member of de Broglie wavelengths will fit between the turning points. (See the solution labeled $\psi_{E_3}$, Figure 1.) The particular discrete values of $E$ for which this happens are the *allowed* energies. Because the particle is confined to a finite domain, these states are called bound states. We thus conclude that *the bound state spectrum in one dimension is discrete and nondegenerate.*

(4)   $E < V_2$.  In this region the kinetic energy is everywhere negative and no classical motion is possible. No quantum mechanical states exist either, because the solutions of equation (1) are everywhere convex with respect to the $x$-axis and hence must increase without limit in one direction or the other.

## 2. CLASSIFICATION BY SYMMETRY; THE PARITY OPERATOR

Most of the potentials of interest in the microscopic domain describe the interactions between pairs of particles. Such potentials are usually symmetric functions of position and this feature permits a considerable simplification of the analysis. We thus consider now the special case in which $V(x)$ is symmetric,

$$V(x) = V(-x). \tag{2}$$

The simplification we refer to is a consequence of the fact that when $V$ is symmetric, so is the Hamiltonian $H$. Hence, when $H$ operates upon any function it does not change the symmetry properties of that function. This means that Schrödinger's equation can be separated into two independent equations, one involving only symmetric state functions, the other only antisymmetric state functions, as we now proceed to demonstrate.

For this purpose we introduce a new operator $P$, called the *parity operator*, which is defined by

$$Pf(x) = f(-x), \tag{3}$$

where $f(x)$ is arbitrary. Thus, *the parity operator simply changes the sign of the space coordinates in any function upon which it operates.*

Note that $P$ is Hermitian since we have, for arbitrary admissible $\psi_1$ and $\psi_2$,

$$\int_{-\infty}^{\infty} \psi_1{}^*(x) P\psi_2(x)\ dx = \int_{-\infty}^{\infty} \psi_1{}^*(x)\psi_2(-x)\ dx$$

$$= \int_{-\infty}^{\infty} \psi_1{}^*(-x)\psi_2(x)\ dx = \int_{-\infty}^{\infty} (P\psi_1)^*\psi_2\ dx,$$

where we have changed integration variables from $x$ to $-x$ in going to the second line.

Let us now find the eigenvalues and eigenfunctions of this Hermitian operator. We seek, that is, the solutions of

$$P\psi_\alpha = \alpha\psi_\alpha.$$

Now, operating on this equation with $P$, we obtain

$$P^2\psi_\alpha = \alpha P\psi_\alpha = \alpha^2\psi_\alpha$$

but, for *any* function $f(x)$,

$$P^2 f(x) \equiv P[Pf(x)] = Pf(-x) = f(x).$$

Hence upon comparison we see that $\alpha^2 = 1$ and *the eigenvalues of the parity operator are plus or minus unity.* Denoting the corresponding eigenstates by $\psi_+$ and $\psi_-$, we then have

$$P\psi_+ = \psi_+, \qquad P\psi_- = -\psi_-, \tag{4}$$

which is to say that

$$\psi_+(-x) = \psi_+(x)$$
$$\psi_-(-x) = -\psi_-(x).$$

Accordingly, $\psi_+$ is *any* even function, $\psi_-$ is *any* odd function. Note that $\psi_+$ and $\psi_-$ are properly orthogonal. We thus have the curious spectacle of an Hermitian operator with only two eigenvalues, each of which is infinitely degenerate.

The eigenfunctions of $P$ are obviously complete, that is to say, any function can always be expressed as a sum of its symmetric and anti-symmetric parts. Specifically, we write

$$f(x) = f_+(x) + f_-(x),$$

where $f_+$ is defined as

$$f_+(x) \equiv \frac{f(x) + f(-x)}{2},$$

and is obviously symmetric, while $f_-$ is defined as

$$f_-(x) \equiv \frac{f(x) - f(-x)}{2},$$

and is obviously antisymmetric.

An interesting way of rewriting these expressions is to replace $f(-x)$ by $Pf(x)$. We then obtain

$$f_+(x) = \frac{1+P}{2} f(x)$$

$$f_-(x) = \frac{1-P}{2} f(x).$$

The quantities

$$P_\pm \equiv \frac{1 \pm P}{2} \tag{5}$$

are called projection operators; $P_+$ projects out of a general state its symmetric or even part, $P_-$ its antisymmetric or odd part. Note that

$$P_\pm^2 = P_\pm$$
$$P_+P_- = P_-P_+ = 0 \tag{6}$$
$$P_+ + P_- = 1,$$

which exemplify the general properties of such projection operators. In any case, we have, succinctly,

$$P_{\pm} f(x) = f_{\pm}(x) . \tag{7}$$

---

**Exercise 1.** Verify the expressions in equation (6) by substitution of equation (5).

---

To proceed, we need only remark that $P$ commutes with the Hamiltonian if $V(x)$ is symmetric. This is so because

$$P(H\psi) = H(-x)\psi(-x) = H(x)\psi(-x) = HP\psi .$$

Of course, the $P_{\pm}$ also commute with $H$ and hence, operating on Schrödinger's equation with $P_{\pm}$, we finally obtain the *independent* pair of equations

$$H\psi_{E,+} = E\psi_{E,+}$$
$$H\psi_{E,-} = E\psi_{E,-}. \tag{8}$$

This then tells us that *the stationary states in a symmetric potential can always be classified according to their parity,* that is, they can always be chosen to have definite symmetry, without loss of generality. More particularly, since *bound* states in one dimension are non-degenerate, *each bound state in a symmetric potential must either be even or odd.*

## 3. BOUND STATES IN A SQUARE WELL

As a specific example, we now consider bound states in one of the simplest possible potentials, the square well. This potential is defined by

$$V(x) = \begin{cases} -V_0, & -a < x < a \\ 0, & |x| > a, \end{cases} \tag{9}$$

and is sketched in Figure 2. Note that $V(x)$ is symmetric, so that the bound states must have a definite symmetry, which greatly simplifies the algebra, as we shall see.

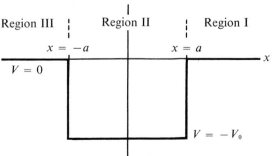

FIGURE 2. The square well potential.

Since we are seeking bound states, the energy $E$ must be negative. It is therefore convenient to introduce the *binding energy* $\epsilon$, defined as

$$\epsilon = -E, \tag{10}$$

whence Schrödinger's equation becomes

$$-\frac{\hbar^2}{2m} \frac{d^2\psi}{dx^2} + V(x)\psi_E = -\epsilon\psi_E. \tag{11}$$

In the region to the right of $x = a$, which we call region I (see Figure 2), $V(x)$ vanishes and equation (11) reduces to

$$\frac{d^2\psi_E}{dx^2} - \frac{2m\epsilon}{\hbar^2} \psi_E = 0, \tag{12}$$

which has the general solution $\psi_E \sim e^{\pm\sqrt{2m\epsilon}\,x/\hbar}$. However, only the negative exponential is permitted if $\psi_E$ is not to increase without limit as $x$ approaches infinity. Hence, in this region

$$\psi_E = C_I\, e^{-\sqrt{2m\epsilon}\,x/\hbar}. \tag{13}$$

Similarly, in the region to the left of $x = -a$, region III, $\psi_E$ also satisfies equation (12), and hence has the form

$$\psi_E = C_{III}\, e^{\sqrt{2m\epsilon}\,x/\hbar}, \tag{14}$$

where now only the positive exponential appears since otherwise $\psi$ would increase without limit as $x$ approaches minus infinity.

Finally, in the central region between $x = -a$ and $x = a$, region II, where $V(x) = -V_0$, equation (11) becomes

$$\frac{d^2\psi}{dx^2} + \frac{2m}{\hbar^2} (V_0 - \epsilon)\psi_E = 0.$$

Hence $\psi_E$ is oscillatory in this region, and we express its general solution in the form

$$\psi_E = C_{II}^{(+)} \cos\left(\sqrt{2m(V_0 - \epsilon)}x/\hbar\right) + C_{II}^{(-)}\sin\left(\sqrt{2m(V_0 - \epsilon)}x/\hbar\right). \tag{15}$$

We thus see that the state function $\psi_E$ has a different form in each of the three regions. Since $V(x)$ is discontinuous in crossing from one region to the next, equation (11) tells us that the second derivative of $\psi_E$ must also be discontinuous, but that $\psi_E$ itself and its slope are continuous. Hence we have two continuity conditions which must be satisfied at each boundary between regions, and these yield four linear and homogeneous equations in the coefficients $C_I$, $C_{II}^{(+)}$, $C_{II}^{(-)}$ and $C_{III}$. If these equations are to have a solution, then the determinant of the coefficients must vanish. The condition that this be so then determines the allowed value or values of $\epsilon$.

To reduce the algebra, we now make explicit use of our prior knowledge that the bound states in a symmetric potential have a definite symmetry; they must be either even or odd. We take up these two possibilities in order.

**Even States:** $\psi_E(x) = \psi_E(-x)$. For the even states, we have at once that $C_{\mathrm{III}} = C_{\mathrm{I}}$ and that, in equation (15), $C_{\mathrm{II}}^{(-)} = 0$. Further, any continuity condition imposed at $x = a$ is automatically satisfied at $x = -a$, by symmetry. We thus need apply these conditions only at, say, $x = a$. Doing so, we obtain:

*continuity of* $\psi$

$$C_{\mathrm{I}}\, e^{-\sqrt{2m\epsilon}\, a/\hbar} = C_{\mathrm{II}}^{(+)} \cos\ (\sqrt{2m(V_0 - \epsilon)}\ a/\hbar)$$

*continuity of* $d\psi/dx$

$$\frac{-\sqrt{2m\epsilon}}{\hbar}\, C_{\mathrm{I}}\, e^{-\sqrt{2m\epsilon}\, a/\hbar} = -\frac{\sqrt{2m(V_0 - \epsilon)}}{\hbar}\, C_{\mathrm{II}}^{(+)} \sin\ (\sqrt{2m(V_0 - \epsilon)}\ a/\hbar).$$

Taking the ratio of the second equation to the first, we see that

$$\sqrt{\epsilon} = \sqrt{V_0 - \epsilon}\ \tan\ (\sqrt{2m(V_0 - \epsilon)}\ a/\hbar)\ , \qquad (16)$$

and this is the equation which determines the energy spectrum of the even states. It is a transcendental equation and its roots cannot be determined algebraically. A straightforward graphical procedure, which gives a complete characterization, is to simply plot the left-hand and right-hand sides of equation (16) as a function of $\epsilon$; the roots are then the points of intersection of the two curves. This is illustrated in Figure 3. Since the left side becomes imaginary for negative $\epsilon$, while the right side is real, *roots exist only for positive* $\epsilon$. This is entirely consistent with our expectations, of course, since negative $\epsilon$, or negative binding energy, means that the state is, in fact, not bound. On the other hand, the right side becomes and remains negative for $\epsilon > V_0$, as shown in the figure,

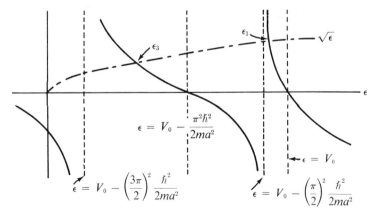

FIGURE 3.  Graphical solution of equation (16) for the allowed energies of the even states of a particle in a square well. The solid curve is the right side of equation (16), the dot-dash curve is the left side. The intersections give the allowed energies. For the particular $V_0$ chosen, there are two such energies. The energy of the ground state is labeled $\epsilon_1$, that of the next higher even state $\epsilon_3$.

and hence there are no states with binding energy which exceeds the depth $V_0$ of the well, also in agreement with expectation. Taken together, we see that these two limitations mean that the even bound state energies are confined to the interval $0 \leq \epsilon < V_0$. The number of roots and, therefore, the number of even bound states, are seen from the figure to depend only upon the number of zeros of the right side of equation (16) for $\epsilon \leq 0$. As indicated, the zeros of the right side occur when

$$\epsilon = V_0 - (n\pi)^2 \hbar^2 / 2ma^2, \qquad n = 0, 1, 2, \ldots,$$

that is, when $\epsilon = V_0$, $\epsilon = V_0 - \pi^2 \hbar^2 / 2ma^2$, and so on. In the illustration there are two zeros and therefore two such states. The energy of the most tightly bound even state is labeled $\epsilon_1$, the energy of the other state is labeled $\epsilon_3$. The reason for this choice of numerical subscripts will shortly become clear. As $V_0$ is decreased, the curve representing the right side of equation (16) is translated to the left, and when $V_0 < \pi^2 \hbar^2 / 2ma^2$, there is only one zero and one even bound state. Similarly, when $V_0$ increases, the curve is translated to the right and when $V_0 \geq (2\pi)^2 \hbar^2 / 2ma^2$, there are three zeros and three even bound states, and so on. To summarize, when

$$(n\pi)^2 \hbar^2 / 2ma^2 \leq V_0 < [(n+1)\pi]^2 \hbar^2 / 2ma^2, \qquad (17)$$

there are $n + 1$ even bound states. Note that so long as $V_0$ is positive, there is always at least one such state.

**Odd States:** $\psi_E(x) = -\psi_E(-x)$. For the odd states, $C_{\text{III}} = -C_{\text{I}}$ while $C_{\text{II}}^{(+)}$ in equation (15) must vanish. Again, any continuity condition imposed at $x = a$ is automatically satisfied at $x = -a$ by symmetry. Proceeding in the same way as before, we find that for the odd states, $\epsilon$ is determined by the equation

$$\sqrt{\epsilon} = -\sqrt{V_0 - \epsilon} \ \text{ctn} \ (\sqrt{2m(V_0 - \epsilon)} \ a/\hbar),$$

which we again solve graphically, as illustrated in Figure 4. We have used the same $V_0$ as in Figure 3, and it turns out that there are again two allowed states. We have labeled their energies $\epsilon_2$ and $\epsilon_4$. Comparison with Figure 3 shows that $\epsilon_1 > \epsilon_2 > \epsilon_3 > \epsilon_4$. Turning to general values of $V_0$, we see, by the same arguments as before, that the number of odd bound states equals the number of zeros of the right side of equation (18) for $\epsilon \geq 0$ and that, more specifically, there are exactly $(n + 1)$ odd bound states for

$$[(n + \tfrac{1}{2})\pi]^2 \hbar^2 / 2ma^2 \leq V_0 < [(n + \tfrac{3}{2})\pi]^2 \hbar^2 / 2ma^2. \qquad (19)$$

Note, however, that there are *no* bound odd states at all for

$$V_0 < \left(\frac{\pi}{2}\right)^2 \frac{\hbar^2}{2ma^2}. \qquad (20)$$

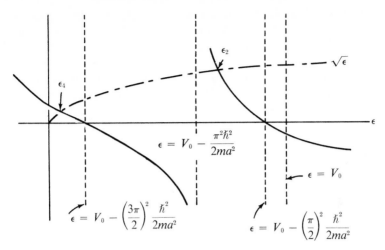

FIGURE 4.   Graphical solution of equation (18) for the allowed energies of the odd states of a particle in a square well. The solid curve is the right side of equation (18), the dot-dash curve is the left side. The intersections give the allowed energies. The energy of the lowest odd state is labeled $\epsilon_2$, that of the next highest odd state $\epsilon_4$.

This is quite an important result, as we shall see when we come to the discussion of three-dimensional problems.

We can sum up our results in the following way. Suppose we consider what happens as we gradually increase $V_0$ from zero. At first, we have only one bound state, an even state. As soon as the point is reached at which

$$V_0 = \left(\frac{\pi}{2}\right)^2 \frac{\hbar^2}{2ma^2},$$

the first odd state appears. Then, when

$$V_0 = \frac{\pi^2 \hbar^2}{2ma^2},$$

the second even state appears. When

$$V_0 = \left(\frac{3\pi}{2}\right)^2 \frac{\hbar^2}{2ma^2},$$

the second odd state appears, and so on. For any given $V_0$, the spectrum consists of an *interlacing of even and odd states,* the lowest or ground state always being even, the next odd, and so on, with the total number of states depending on the magnitude of $V_0$. In Figure 5 we sketch the spectrum and state functions for the particular case illustrated in Figures 3 and 4, where $V_0$ is such that there are two even and two odd states.

From this figure we can sense that as $V_0$ is decreased, that is, as the well is made gradually shallower, the highest state, labeled $\epsilon_4$, is pushed out, then the next state, and so on, just as our previous analysis indicated. Note, however, that equations (17), (19) and (20) are really restrictions on the quantity $V_0\,a^2$, and not on $V_0$ alone, as we have assumed for simplicity in our discussion. Thus a decrease in $a^2$ is indistinguishable in its effect on the energy (but not on the state functions) from a decrease in $V_0$. If $a^2$ is gradually decreased, thereby making the well narrower, we again successively squeeze out the higher states.

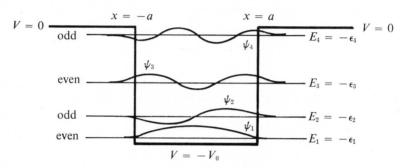

FIGURE 5.  Bound state energies and wave functions in a square well. Note that each higher energy state has one more node than the preceding one.

Note also that the state functions extend into the classically forbidden region beyond the walls of the potential well. According to equations (13) and (14), the wave functions fall off exponentially in this region, as $e^{-\sqrt{2m\epsilon}|x|/\hbar}$. Hence the distance extended, say $b$, is roughly given by

$$b \simeq \frac{\hbar}{\sqrt{2m\epsilon}},$$

which is the distance over which the exponential decreases by $1/e$. For the lowest states, where $\epsilon$ is large, this distance is very small and the state function is correspondingly small at the boundary. In the limit of an infinitely deep potential, these states actually vanish at the boundaries, and their energies, measured with respect to the bottom of the well, can easily be shown to increase as $n^2$, in agreement with our previous analysis of the states of a free particle in a box.

On the other hand, when $\epsilon$ is small, as for example $\epsilon_4$ in Figure 5, the distance $b$ becomes very large and the particle is by no means confined to the interior of the potential, as would be required classically.

## 4. THE HARMONIC OSCILLATOR

We next consider the most important single problem in quantum mechanics, the harmonic oscillator. Not unrelated to its importance is its simplicity; it is one of the two or three non-trivial examples which can be explicitly solved in complete generality. We shall devote to it the full attention it thereby merits.

The potential energy of an harmonic oscillator, of classical frequency $\omega$, is most conveniently written in the form

$$V(x) = \tfrac{1}{2} m\omega^2 x^2,$$

whence the time independent Schrödinger's equation is

$$-\frac{\hbar^2}{2m} \frac{d^2\psi}{dx^2} + \frac{1}{2} m\omega^2 x^2 \psi = E\psi. \tag{21}$$

From the general discussion of Section 1, we thus see that the states of the harmonic oscillator have positive energy and that they are discrete and nondegenerate. We now proceed to find these states by two quite different methods. The first is the power series method, the second is the method of factorization.

**The Method of Power Series.** We introduce the dimensionless variable $y$ by writing

$$y = \sqrt{m\omega/\hbar}\; x, \tag{22}$$

in terms of which Schrödinger's equation becomes

$$\frac{d^2\psi_E(y)}{dy^2} + \left(\frac{2E}{\hbar\omega} - y^2\right)\psi_E(y) = 0.$$

For $y$ approaching infinity, the term in $E$ is negligible compared to the term in $y^2$ and it is easy to verify that $\psi_E(y)$ behaves like $e^{\pm y^2/2}$, multiplied by some algebraic factor. The physically admissible solution must contain only the minus sign and we thus write, without loss of generality,

$$\psi_E(y) = u(y)\; e^{-y^2/2}. \tag{23}$$

Substitution into Schrödinger's equation then gives, after some manipulation,

$$\frac{d^2u}{dy^2} - 2y \frac{du}{dy} + \left(\frac{2E}{\hbar\omega} - 1\right) u = 0. \tag{24}$$

We shall solve this equation by expanding $u$ in a power series in $y$. Since the harmonic oscillator potential is symmetrical, we know in advance that the states must have definite parity. We thus consider separately the even and odd states.

*Even States:* $u(y) = u(-y)$. We seek a solution in the form of a power series. Since $u$ is symmetrical, only even powers of $y$ enter and we therefore write

$$u(y) = \sum_{0}^{\infty} a_s y^{2s}. \qquad (25)$$

Substitution into equation (24) then gives

$$\sum_{0}^{\infty} 2s(2s-1)\, a_s y^{2(s-1)} + \sum_{0}^{\infty} \left( \frac{2E}{\hbar\omega} - 1 - 4s \right) a_s y^{2s} = 0.$$

Replacing $s$ by $s+1$ in the first summation, this can be rewritten as

$$\sum_{0}^{\infty} \left[ 2(s+1)(2s+1)a_{s+1} + \left( \frac{2E}{\hbar\omega} - 1 - 4s \right) a_s \right] y^{2s} = 0.$$

The coefficient of each power of $y$ must separately vanish and hence we obtain the *recursion relation*,

$$a_{s+1} = \frac{4s + 1 - (2E/\hbar\omega)}{2(s+1)(2s+1)} a_s. \qquad (26)$$

Given $a_0$, all of the expansion coefficients can be successively determined.[2] Thus,

$$a_1 = \frac{1}{2}\left( 1 - \frac{2E}{\hbar\omega} \right) a_0$$

$$a_2 = \frac{1}{12}\left( 5 - \frac{2E}{\hbar\omega} \right) a_1 = \frac{1}{24}\left( 5 - \frac{2E}{\hbar\omega} \right)\left( 1 - \frac{2E}{\hbar\omega} \right) a_0$$

$$\vdots$$

and we have now obtained a series representation of the general symmetrical solution of equation (24). Let us next examine the form of this solution for large $y$. Its behavior is determined by the character of $a_s$ for large $s$. From equation (26), we see that

$$\frac{a_{s+1}}{a_s} \simeq \frac{1}{s}, \qquad s \to \infty.$$

This ratio is exactly the same as that of the coefficients in the power series expansion of the exponential function. We thus conclude that, for $y^2$ approaching infinity, $u(y)$ diverges like $e^{y^2}$, whence, from equation (23), $\psi_E(y)$ diverges like $e^{y^2/2}$. This is no surprise, since we have already argued that the general solution of Schrödinger's equation behaves like $e^{\pm y^2/2}$ for large $y$. Because the general solution must contain

---

[2] The essential role of the transformation of equation (23) can now be made clear. The reader will readily verify that if $\psi_E$, rather than $u$, is expanded in a power series, the recursion formula obtained connects *three* expansion coefficients and not two, as in equation (26). Such three-term recursion relations are, in general, extremely difficult to solve.

terms of both signs, the positive exponential necessarily dominates, and so we have found. This catastrophe can be avoided only if the series representation *terminates* after, say, $r$ terms. From equation (26), we see that this will be so if, and only if, $E$ has one of the discrete values

$$E = \frac{\hbar\omega}{2}\,(4r+1) = (2r + \tfrac{1}{2})\,\hbar\omega, \qquad r = 0, 1, 2, \ldots,$$

since $a_{r+1}$, and therefore all subsequent $a_s$, then vanish. We have thus, finally, obtained an *infinite set* of solutions, one for each value of the integer $r$. These symmetrical states, still unnormalized, are given explicitly by

$$r = 0, \qquad E = \hbar\omega/2, \qquad \psi_E = a_0\, e^{-y^2/2}$$

$$r = 1, \qquad E = 5\hbar\omega/2, \qquad \psi_E = a_0(1 - 2y^2)\, e^{-y^2/2} \qquad (27)$$

$$r = 2, \qquad E = 9\hbar\omega/2, \qquad \psi_E = a_0(1 - 4y^2 + \tfrac{4}{3}\, y^4)\, e^{-y^2/2}.$$
$$\vdots$$

Before discussing these solutions further, we go on to consider the odd states.

*Odd States:* $u(y) = -u(-y)$. For this case we express $u$ as a series in odd powers of $y$,

$$u(y) = \sum_0^\infty a_s y^{2s+1}.$$

Substitution into equation (24) then yields, after the same kind of manipulation as before, the recursion relation

$$a_{s+1} = \frac{4s + 3 - 2E/\hbar\omega}{2(s+1)(2s+3)}\, a_s.$$

Again the solution diverges unless the series terminates after $r$ terms, say, and hence $E$ can only take on the discrete values

$$E = (2r + \tfrac{3}{2})\, \hbar\omega; \qquad r = 0, 1, 2, \ldots,$$

and the corresponding unnormalized solutions are

$$r = 0, \qquad E = 3\hbar\omega/2, \qquad \psi_E = a_0 y\, e^{-y^2/2}$$

$$r = 1, \qquad E = 7\hbar\omega/2, \qquad \psi_E = a_0 y(1 - \tfrac{2}{3}\, y^2)\, e^{-y^2/2}. \qquad (28)$$
$$\vdots$$

Putting this all together, we see that the lowest state is even and its energy is $\hbar\omega/2$, the next state is odd with energy $3\hbar\omega/2$, the next even with energy $5\hbar\omega/2$, and so on. The complete spectrum is thus expressible as

$$E_n = (n + \tfrac{1}{2})\, \hbar\omega; \qquad n = 0, 1, 2 \ldots, \qquad (29)$$

and the states are even or odd according to whether $n$ is even or odd. This spectrum is just that adduced by Planck, except for the additive constant, $\hbar\omega/2$. The necessity for such a term in the spectrum, the famous *zero point energy*, is quite clear from the uncertainty principle.

---

**Exercise 2.**   Use the uncertainty principle to deduce the order of magnitude of the zero point energy.

---

The most convenient and customary way to write the normalized state functions[3] is the following. Letting $\psi_n$ denote the state with energy $E_n$, we write

$$\psi_n(x) \equiv \frac{c_n}{2^{n/2}} h_n(y) \, e^{-y^2/2}, \qquad y = \sqrt{m\omega/\hbar} \, x$$

$$c_n = \left( \frac{\sqrt{m\omega/\hbar\pi}}{n!} \right)^{1/2}. \tag{30}$$

The $h_n(y)$ are polynomials of degree $n$, in even or odd powers of $y$ according to whether $n$ is even or odd. These polynomials are called *Hermite polynomials*, and the first few are[4]

$$h_0(y) = 1 \qquad\qquad h_3 = 8y^3 - 12y$$

$$h_1(y) = 2y \qquad\qquad h_4 = 16y^4 - 48y^2 + 12 \tag{31}$$

$$h_2(y) = 4y^2 - 2 \qquad h_5 = 32y^5 - 160y^3 + 120y.$$

The first four states of the harmonic oscillator are sketched in Figure 6. Note that, as usual, the number of nodes in the wave function increases by one as we go from one energy level to the next higher one.

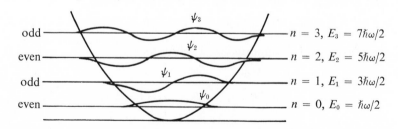

FIGURE 6.   Energy states and wave functions for the harmonic oscillator.

---

[3] The normalization of the state functions will be discussed in Sections 4 and 5.

[4] See Reference [7] for a more extensive list.

One curious feature of the harmonic oscillator is its complete symmetry with respect to configuration and momentum space. Since the Hamiltonian operator is simply

$$H = \frac{p^2}{2m} + \frac{1}{2} m\omega^2 x^2, \tag{32}$$

we see that Schrödinger's equation in momentum space is identical to that in configuration space, provided $m$ is everywhere replaced by $1/m\omega^2$. It is then easy to verify that the normalized state functions in *momentum space* are given by

$$\phi_n(p) = \frac{d_n}{2^{n/2}} h_n(q) \, e^{-q^2/2}, \qquad q = \frac{p}{\sqrt{m\omega\hbar}}$$

$$d_n = \left( \frac{1}{\sqrt{m\omega\hbar}\,\pi\,n!} \right)^{1/2}. \tag{33}$$

A plot of these functions, superposed on a plot of the kinetic energy $p^2/2m$, has exactly the same appearance as Figure 6.

**The Method of Factorization.** The second method, which we now present, is a *purely operator method* for obtaining the eigenstates of the harmonic oscillator. This scheme, introduced by Dirac, is called the method of factorization.

We introduce two new operators, which we define in the following way. The first, denoted by $a$, is expressed in the form

$$a \equiv \sqrt{m\omega/2\hbar}\; x + i \sqrt{1/2m\omega\hbar}\; p, \tag{34}$$

while the second is defined as the *adjoint* of the first. It is thus the operator analog of the complex conjugate. According to equation (V-13), it is given by

$$a\dagger = \sqrt{m\omega/2\hbar}\; x - i \sqrt{1/2m\omega\hbar}\; p, \tag{35}$$

where, as in Chapter V, we have used the dagger to denote the adjoint.

For future reference we note that in configuration space $a$ and $a\dagger$ are simply expressed, in terms of the dimensionless variable $y$ of equation (30), by

$$a = \frac{1}{\sqrt{2}} \left( y + \frac{d}{dy} \right)$$

$$a\dagger = \frac{1}{\sqrt{2}} \left( y - \frac{d}{dy} \right), \tag{36}$$

and similarly, in momentum space, in terms of the variable $q$ of equation (33), by

$$a = \frac{i}{\sqrt{2}}\left(q + \frac{d}{dq}\right)$$

$$a\dagger = -\frac{i}{\sqrt{2}}\left(q - \frac{d}{dq}\right). \tag{37}$$

Next, we observe that

$$aa\dagger = \frac{1}{\hbar\omega}\left(\frac{p^2}{2m} + \frac{1}{2}m\omega^2 x^2\right) + \frac{i}{2\hbar}(p, x),$$

as is easily verified by using equations (34) and (35) to express the left side in terms of $x$ and $p$ and carrying out the indicated multiplication. Recognizing the term in parentheses as the Hamiltonian and evaluating the commutator, we then obtain

$$aa\dagger = \frac{1}{\hbar\omega}H + \frac{1}{2}, \tag{38}$$

where $H$ is the Hamiltonian operator.[5] In exactly the same way we also find

$$a\dagger a = \frac{1}{\hbar\omega}H - \frac{1}{2}. \tag{39}$$

Subtracting equation (39) from equation (38), we see that $a$ and $a\dagger$ satisfy the commutation rule

$$(a, a\dagger) = 1, \tag{40}$$

while adding these two equations we see that

$$aa\dagger + a\dagger a = \frac{2}{\hbar\omega}H. \tag{41}$$

Schrödinger's equation can thus be written in three different but *completely equivalent* ways, corresponding to equations (38), (39) and (41):

$$\left(aa\dagger - \frac{1}{2}\right)\psi_E = \frac{E}{\hbar\omega}\psi_E \tag{42a}$$

$$\left(a\dagger a + \frac{1}{2}\right)\psi_E = \frac{E}{\hbar\omega}\psi_E \tag{42b}$$

$$(aa\dagger + a\dagger a)\psi_E = \frac{2E}{\hbar\omega}\psi_E. \tag{42c}$$

[5] Giving the method its name is the fact that, except for an additive constant, the Hamiltonian has now been *factored* into a product of the new operators $a$ and $a\dagger$.

As the essential step in actually solving Schrödinger's equation, we now show that if we know $\psi_E$ for some given $E$, we can successively construct an infinite set of solutions from it. To do this, we operate on equation (42a) from the left with $a\dagger$, that is, we write

$$a\dagger \left( aa\dagger - \frac{1}{2} \right) \psi_E = \frac{E}{\hbar\omega} \, a\dagger\psi_E$$

or

$$\left( a\dagger a - \frac{1}{2} \right) a\dagger\psi_E = \frac{E}{\hbar\omega} \, a\dagger\psi_E.$$

Next, adding $a\dagger\psi_E$ to each side, we obtain

$$\left( a\dagger a + \frac{1}{2} \right) a\dagger\psi_E = \frac{E + \hbar\omega}{\hbar\omega} \, a\dagger\psi_E.$$

Comparison with equation (42b) then shows that $a\dagger\psi_E$ is also a solution of Schrödinger's equation, but with energy eigenvalue $E + \hbar\omega$, that is,

$$a\dagger\psi_E = C_+\psi_{E+\hbar\omega}, \tag{43}$$

where $C_+$ is a constant. Repeating this procedure, we thus generate from the given state an infinite ladder of equally spaced states of energies $E + \hbar\omega$, $E + 2\hbar\omega$, $E + 3\hbar\omega$, and so on. Since operating with $a\dagger$ raises the energy by one step, $a\dagger$ is called the raising operator, or more commonly, the *creation operator*.

We next show that from the given $\psi_E$ we can also successively construct states of lower energy. To do this, we operate on Schrödinger's equation in the form equation (42b) with $a$. We then find, in exactly the same way as before,

$$\left( aa\dagger - \frac{1}{2} \right) a\psi_E = \frac{E - \hbar\omega}{\hbar\omega} \, a\psi_E, \tag{44}$$

whence we can write

$$a\psi_E = C_-\psi_{E-\hbar\omega}. \tag{45}$$

For obvious reasons the operator $a$ is called the lowering operator, or the *annihilation operator*.

Repeated operation with $a$ thus extends the ladder downward from $E$, again in steps of $\hbar\omega$. However, the energy of an harmonic oscillator state can never be negative,[6] and hence the ladder must have a bottom

---

[6] This is clear from our discussion in Section 1. It also follows from the fact that the harmonic oscillator Hamiltonian is positive definite and hence so must be its expectation value in *any* state.

rung. Call the energy of this lowest state $E_0$, the corresponding eigen-function $\psi_0$.

For $E = E_0$, equation (44) becomes

$$\left( aa\dagger - \frac{1}{2} \right) a\psi_0 = \frac{E_0 - \hbar\omega}{\hbar\omega} \, a\psi_0 \,.$$

Unless $a\psi_0$ vanishes, this equation states that $a\psi_0$ is an eigenstate with eigenvalue $E_0 - \hbar\omega$, in contradiction to the fact that $E_0$ is the lowest state. We thus conclude that $\psi_0$ must be such that

$$a\psi_0 = 0, \tag{46}$$

whence, from equation (42b), we see that $E_0$ is uniquely determined to be

$$E_0 = \hbar\omega/2 \,.$$

Assuming $\psi_0$ known for the moment, we also see that the remaining states of the harmonic oscillator can be constructed by successive appli-cation of $a\dagger$ to $\psi_0$, the energy increasing at each step by $\hbar\omega$. For the $n$th state, we thus have

$$\psi_n = \frac{c_n}{c_0} (a\dagger)^n \, \psi_0, \tag{47}$$

where $c_n$ is a normalizing constant, and where, as before,

$$E_n = (n + \tfrac{1}{2}) \, \hbar\omega \,.$$

To explicitly construct $\psi_0$ in configuration space, we use equation (36) to rewrite equation (46) in the form

$$\frac{1}{\sqrt{2}} \left( y + \frac{d}{dy} \right) \psi_0 = 0 \,,$$

which is easily solved to give

$$\psi_0 = c_0 \, e^{-y^2/2} \,,$$

in agreement with our earlier result. Again using equation (36), equa-tion (47) becomes

$$\psi_n = \frac{c_n}{2^{n/2}} \left( y - \frac{y}{dy} \right)^n e^{-y^2/2} \,. \tag{48}$$

It turns out, as we show later, that $c_n$ is the same normalizing constant as in equation (30). Comparison of our result with that equation then shows that

$$h_n(y) = e^{y^2/2} \left( y - \frac{d}{dy} \right)^n e^{-y^2/2} \,, \tag{49}$$

which is a very compact and convenient representation of the Hermite polynomials. We shall shortly give an even simpler and more convenient representation.

Equations (43) and (45) are very useful in a number of applications, and we now evaluate the proportionality constants contained in these relations. From equation (48) we have

$$a^\dagger \psi_n = \frac{c_n}{c_0} (a^\dagger)^{n+1} \psi_0 = \frac{c_n}{c_{n+1}} \psi_{n+1}, \tag{50}$$

whence, from the explicit expression for the $c_n$ given in equation (30),

$$a^\dagger \psi_n = \sqrt{n+1}\ \psi_{n+1}. \tag{51}$$

Next consider the more difficult case of operation upon $\psi_n$ with the annihilation operator $a$. We shall obtain the result in two different ways, just to illustrate how manipulations can be carried out entirely in the language of these operators. The first is straightforward. It starts with equation (47), from which we obtain, upon operating with $a$,

$$a\psi_n = \frac{c_n}{c_0} aa^{\dagger n} \psi_0.$$

It follows from the fundamental commutation relation, equation (40), that

$$aa^{\dagger n} = a^{\dagger n}a + na^{\dagger n-1}. \tag{52}$$

---

**Exercise 3.** Prove this result using the commutation relation, equation (40). (We remark that the proof can also be carried out, and trivially, by the methods of the next section.)

---

Because $a$ annihilates $\psi_0$, the first term gives no contribution and we thus obtain

$$a\psi_n = n \frac{c_n}{c_0} a^{\dagger n-1} \psi_0 = n \frac{c_n}{c_{n-1}} \psi_{n-1},$$

whence, finally, again using equation (30),

$$a\psi_n = \sqrt{n}\ \psi_{n-1}. \tag{53}$$

The second method is less direct. Consider equation (42a) for the $(n-1)$st state, that is, for $E = (n - \frac{1}{2}) \hbar\omega$. We have, rearranging terms,

$$aa^\dagger \psi_{n-1} = n\psi_{n-1}.$$

However, from equation (51),

$$a^\dagger \psi_{n-1} = \sqrt{n}\ \psi_n,$$

and the result follows at once.

---

**Exercise 4.** Verify the commutation relation, equation (40), by operating with the commutator on $\psi_n$ and using equations (51) and (53) to evaluate the result.

---

The reader may be tempted, at this stage, to regard the algebraic operator techniques of the factorization approach as merely a device for circumventing the inelegant power series method of solution. It should be recognized, however, that it has also furnished us with a *representation independent* solution of the harmonic oscillator problem which is *complete* in every way. Not only is it complete, it is, in fact, generally much more convenient than the explicit configuration space representation involving, as it does, the unwieldly Hermite polynomials. The essential feature here is that the state functions are very simple expressions in terms of $a$ and $a^\dagger$, and so are the dynamical variables, according to equations (34) and (35). Hence expectation values can be readily computed in the language of these operators. As an example, we briefly discuss the normalization of the harmonic oscillator eigenfunctions.

Consider then, for arbitrary $n$,

$$\langle \psi_n | \psi_n \rangle = \left| \frac{c_n}{c_0} \right|^2 \langle a^{\dagger n}\psi_0 | a^{\dagger n}\psi_0 \rangle .$$

Transferring $a^{\dagger n}$ from the second factor to the first, we obtain, by the definition of the adjoint,

$$\langle \psi_n | \psi_n \rangle = \left| \frac{c_n}{c_0} \right|^2 \langle \psi_0 | a^n a^{\dagger n} | \psi_0 \rangle .$$

To evaluate this expression, we observe that, according to equation (52), and because $a\psi_0 = 0$,

$$aa^{\dagger n}\psi_0 = na^{\dagger n-1}\psi_0 .$$

Similarly, using this result, we find

$$a^2 a^{\dagger n}\psi_0 = naa^{\dagger n-1}\psi_0 = n(n-1)a^{\dagger n-2}\psi_0 .$$

Repeating this process $n$ times, we then obtain

$$a^n a^{\dagger n}\psi_0 = n!\psi_0 ,$$

whence

$$\langle \psi_n | \psi_n \rangle = \left| \frac{c_n}{c_0} \right| \, n! \, \langle \psi_0 | \psi_0 \rangle .$$

Thus, if $\psi_0$ is normalized, so will the $\psi_n$ be normalized, provided that

$$|c_n|^2 = \frac{|c_0|^2}{n!},$$

in agreement with equation (30).

As a second example, consider the expectation value of the kinetic energy for the $n$th state. Inverting equations (34) and (35) we have

$$p = \frac{\sqrt{2m\omega\hbar}}{2i} \, (a - a\dagger),$$

whence

$$\left\langle \psi_n \left| \frac{p^2}{2m} \right| \psi_n \right\rangle = -\frac{\hbar\omega}{4} \, \langle \psi_n | (a - a\dagger)^2 | \psi_n \rangle$$

$$= -\frac{\hbar\omega}{4} \, \langle \psi_n | a^2 - aa\dagger - a\dagger a + a\dagger^2 | \psi_n \rangle .$$

Now $a^2$ operating on $\psi_n$ yields the state $\psi_{n-2}$, whence that term vanishes because the states form an orthonormal set. Similarly, $a\dagger^2$ operating on $\psi_n$ yields $\psi_{n+2}$, and again there is no contribution. We thus obtain, retaining only the contributing terms,

$$\left\langle \psi_n \left| \frac{p^2}{2m} \right| \psi_n \right\rangle = \frac{\hbar\omega}{4} \, \langle \psi_n | aa\dagger + a\dagger a | \psi_n \rangle ,$$

which is one-half the expectation value of the Hamiltonian according to equation (42c). Thus we find

$$\left\langle \psi_n \left| \frac{p^2}{2m} \right| \psi_n \right\rangle = \frac{E_n}{2} = \frac{1}{2} \left( n + \frac{1}{2} \right) \hbar\omega ,$$

in agreement with the classical relation between the mean kinetic energy and the total energy of an oscillator.

Even if one fails to recognize the particular combination of operators above as the Hamiltonian, the calculation can easily be worked out directly using equations (51) and (53). Thus we have

$$a\dagger a \psi_n = \sqrt{n} \, a\dagger \psi_{n-1} = n \psi_n$$

and

$$aa\dagger \psi_n = \sqrt{n+1} \, a \psi_{n+1} = (n+1) \psi_n ,$$

whence the result follows at once.[7]

These examples illustrate the kind of manipulations of the creation and annihilation operators which must be carried out when quantities of physical interest are to be calculated. In the next section we develop a new representation in which the algebra becomes entirely trivial.

## 5. THE CREATION OPERATOR REPRESENTATION

In the preceding we introduced a transformation from the dynamical variables $p$ and $x$ to a new set of dynamical variables $a$ and $a\dagger$, satisfying the commutation relation (40). We now introduce a new representation, the creation operator representation, in which $a\dagger$ is a purely *numerical* operator and $a$ is the *operator of differentiation* in $a\dagger$ space.[8] Of course, $a$ and $a\dagger$ are somewhat peculiar dynamical variables classically, since each is complex. Although some complications are encountered as a result, these turn out not to be very serious.

To develop the representation we seek, we first obtain the eigenfunctions $\psi_{a\dagger}$ of the creation operator in configuration space. In terms of the dimensionless variable $y$, using equation (36), we seek the solutions of

$$\frac{1}{\sqrt{2}}\left(y - \frac{d}{dy}\right)\psi_{a\dagger}(y) = a\dagger\psi_{a\dagger}(y), \qquad (54)$$

where $a\dagger$ is a (complex) number. The states $\psi_{a\dagger}$ are thus states in which the creation operator has the *definite numerical value* $a\dagger$. Such states are nonphysical because $a\dagger$ is not a physical observable, but we can nonetheless use them to construct physical states, as we shall see. In any case, the solution of equation (54) is easily seen to be

$$\psi_{a\dagger}(y) \sim e^{y^2/2 - \sqrt{2}\,ya\dagger}.$$

Note that $\psi_{a\dagger}$ is a strongly divergent function of $y$, so that these states are clearly unnormalizable. We are, of course, free to multiply this solution by an arbitrary factor, independent of $y$, and it turns out to be convenient to do so. Specifically, we introduce the factor $e^{a\dagger^2/2}$, in which case the solution to equation (54) is written as

$$\psi_{a\dagger}(y) = \exp\left[y^2/2 - \sqrt{2}\,ya\dagger + a\dagger^2/2\right].$$

---

[7] It may help the reader to appreciate the remarkable simplicity of the operator methods if he attempts to work out either of these examples in the configuration space representation of equation (49), or of equations (30) and (31).

[8] We are motivated here by the observation that the commutator of $a$ and $a\dagger$ is the same as that of $p$ and $x$ up to the factor $\hbar/i$. That this factor is imaginary has important consequences, as we shall see. However, our procedure is quite analogous to that followed in treating momentum and configuration space representations.

With this choice, the annihilation operator $a$ assumes its simplest form in $a\dagger$ space consistent with the commutation relation (40), namely

$$a = \frac{d}{da\dagger}, \tag{55}$$

as we shall shortly verify.[9]

We now want to represent an arbitrary state function $\psi(x)$ as a superposition of the eigenfunctions of $a\dagger$, just as we earlier represented arbitrary state functions as superpositions of momentum eigenfunctions or of energy eigenfunctions (stationary states). Since $a\dagger$ is complex, we are free to specify a convenient path of integration in the complex $a\dagger$-plane when making this superposition. We choose it to lie along the *axis of imaginaries* and thus write

$$\psi(y) = \frac{1}{i\sqrt{2\pi}} \int_{-i\infty}^{i\infty} f(a\dagger) \, \exp\left[ y^2/2 - \sqrt{2}\, ya\dagger + a\dagger^2/2 \right] da\dagger. \tag{56}$$

To make the meaning of this expression clear, we introduce a new *real* variable $\alpha$ by writing

$$a\dagger = i\alpha,$$

whence $\psi(y)$ assumes the more comfortable form

$$\psi(y) = \frac{1}{\sqrt{2\pi}} \int_{-\infty}^{\infty} f(\alpha) \, \exp\left[ y^2/2 - i\sqrt{2}\, y\alpha - \alpha^2/2 \right] d\alpha$$

or

$$\psi(y) \, e^{-y^2/2} = \frac{1}{\sqrt{2\pi}} \int_{-\infty}^{\infty} \left[ f(\alpha) \, e^{-\alpha^2/2} \right] e^{-i\sqrt{2}\, y\alpha} \, d\alpha, \tag{57}$$

which is simply a Fourier integral representation of $\psi(y) \, e^{-y^2/2}$ in terms of $f(\alpha) \, e^{-\alpha^2/2}$. This relation can be inverted to give

$$f(\alpha) \, e^{-\alpha^2/2} = \frac{1}{\sqrt{\pi}} \int_{-\infty}^{\infty} \psi(y) \, e^{-y^2/2} \, e^{i\sqrt{2}\, y\alpha} \, dy \tag{58}$$

or, reintroducing $a\dagger$,

$$f(a\dagger) = \frac{1}{\sqrt{\pi}} \int_{-\infty}^{\infty} \psi(y) \, \exp\left[ -y^2/2 + \sqrt{2}\, ya\dagger - a\dagger^2/2 \right] dy. \tag{59}$$

These expressions demonstrate the completeness of the eigenfunctions ı of $a\dagger$ in view of the known properties of Fourier integral representations.

The pair of functions $\psi(y)$ and $f(a\dagger)$ provide *completely equivalent* descriptions of any state of the harmonic oscillator system, the former,

---

[9] See Problem 10, Chapter V, for a discussion of the same question with respect to the representation of the momenum operator in configuration space.

of course, in configuration space and the latter in creation operator space. According to equations (46) and (59) each is uniquely defined if the other is given, and each therefore contains precisely the same information. We can thus speak of $f(a\dagger)$ as the *creation space representation* of a state function. Note, however, that normalization in configuration space does *not* lead to normalization, in the usual sense, in creation space, as may be seen from equations (57) and (58) with the aid of the convolution theorem, equation (III-19). Indeed, a direct meaning cannot even be assigned to normalization integrals of the standard form in creation space. This feature, which is a consequence of the unphysical character of that space, causes no difficulty provided that all questions of normalization are simply referred to configuration space, and that is what we shall do.[10]

We next verify that in creation space the annihilation operator is just the operator of differentiation, as given by equation (55). To do so, recall that, according to equation (36),

$$a\psi(y) = \frac{1}{\sqrt{2}}\left(y + \frac{\mathrm{d}}{dy}\right)\psi(y) .$$

Hence, from equation (56), differentiating under the integral sign,

$$a\psi(y) = \frac{1}{i\sqrt{2\pi}}\int_{-i\infty}^{i\infty} f(a\dagger)\,(\sqrt{2}\,y - a\dagger)\,\exp\,[y^2/2 - \sqrt{2}\,ya + a\dagger^2/2]\,da\dagger$$

$$= \frac{1}{i\sqrt{2\pi}}\int_{-i\infty}^{i\infty} f(a\dagger)\left(\frac{-d}{da\dagger}\right)\exp\,[y^2/2 - \sqrt{2}\,ya\dagger + a\dagger^2/2]\,da\dagger .$$

Integrating by parts, we then obtain, because the integrated part vanishes,

$$a\psi(y) = \frac{1}{i\sqrt{2\pi}}\int_{-i\infty}^{i\infty}\left[\frac{d}{da\dagger}f(a\dagger)\right]\exp\,[y^2/2 - \sqrt{2}\,ya\dagger + a\dagger^2/2]\,da\dagger .$$

In other words, this states that if $f(a\dagger)$ is equivalent in creation space to $\psi(y)$ in configuration space, then $df/da\dagger$ is equivalent to $a\psi$. In short, the creation space representation of $a\psi(y)$ is seen to be $df(a\dagger)/da\dagger$. In the same way, it can be seen that, for an arbitrary function $g$, the creation state representation of $g(a)\psi(y)$ is $g(d/da\dagger)f(a\dagger)$. Hence it follows that, in creation space, the operator $a$ is indeed represented by $d/da\dagger$, as was to be proved.

We are now in a position to trivially construct the harmonic oscillator eigenfunctions in $a\dagger$ space. Recall that, according to equation (47),

[10] Normalization integrals can be defined by introducing a second set of functions, $\psi_a$, the eigenfunctions of the annihilation operator $a$. It can be shown that the $\psi_{a\dagger}$ and $\psi_a$ form what is called a *bi-orthogonal* set of functions. The mathematical properties of bi-orthogonal sets are well known, and, in fact, are only a little more complicated than the familiar properties of ordinary orthonormal sets.

these are expressible in terms of the ground state $\psi_0(y)$ by

$$\psi_n(y) = \frac{c_n}{c_0}\,(a^\dagger)^n\psi_0(y)\,,$$

where the $c_n$ are normalizing constants and where $\psi_0(y)$ is such that

$$a\psi_0(y) = 0\,.$$

Hence in $a^\dagger$ space,

$$f_n(a^\dagger) = \frac{c_n a^{\dagger n} f_0(a^\dagger)}{c_0}\,,$$

where $f_n(a^\dagger)$ is the creation space representation of $\psi_n(y)$ and where, therefore, $f_0(a^\dagger)$ is such that

$$af_0(a^\dagger) = 0\,.$$

Because $a$ is the operator of differentiation with respect to $a^\dagger$, this tells us that $f_0$ is a constant, namely $c_0$, and the harmonic oscillator state function in creation space, corresponding to the energy

$$E_n = (n + \tfrac{1}{2})\,\hbar\omega\,,$$

is the *monomial*

$$f_n(a^\dagger) = c_n a^{\dagger n} \tag{60}$$

which is incomparably simpler than the Gaussian multiplied by Hermite polynomials which we found in configuration (or momentum) space.

Using equation (56), we can now write a very useful and elegant representation of the harmonic oscillator eigenfunctions. Specifically, this representation is

$$\psi_n(y) = \frac{c_n}{i\sqrt{2\pi}} \int_{-i\infty}^{i\infty} a^{\dagger n}\exp\left[y^2/2 - \sqrt{2}\,ya^\dagger + a^{\dagger 2}/2\right]\,da^\dagger. \tag{61}$$

As a first demonstration of the utility of these results, we again calculate the normalization constant $c_n$. As an exercise, we also verify that, as they must be, the $\psi_n$ are orthogonal to each other. In the light of our earlier discussion we must normalize in configuration space, and we consider therefore

$$\int \psi_m{}^*(y)\psi_n(y)\,dy = \frac{c_m{}^*c_n}{|c_0|^2}\int (a^{\dagger m}\psi_0)^* a^{\dagger n}\psi_0\,dy.$$

Recalling that $a^{\dagger n}$ operating under the complex conjugation sign on the first function in such an integral can be replaced by $a^n$ operating on the second function, according to the definition of the adjoint, we then obtain

$$\int \psi_m{}^*(y)\psi_n(y)\ dy = \frac{c_m{}^*c_n}{|c_0|^2}\int \psi_0{}^*a^m a^{\dagger n}\psi_0\ dy.$$

Next we express the factor $a^m a^{\dagger n}\psi_0(y)$ in the integrand in terms of its creation space representation. Using equation (55) and equation (61) we have

$$a^m a^{\dagger n}\psi_0(y) = \frac{c_0}{i\sqrt{2\pi}}\int_{-i\infty'}^{i\infty}\left[\left(\frac{d}{da\dagger}\right)^m a^{\dagger n}\right]\exp\left[y^2/2 - \sqrt{2}\,ya\dagger + a\dagger^2/2\right]\ da\dagger$$

which clearly vanishes for $m > n$. On the other hand, for $m < n$, we obtain the same result by carrying out the procedure in the opposite sense, transferring $a\dagger$, operating on the second function in the integrand, to $a$ operating on the first. We thus see that the integrand vanishes for $m \neq n$ and the $\psi_n$ are indeed orthogonal. Now finally, for $m = n$, we have, performing the differentiations,

$$a^n a^{\dagger n}\psi_0(y) = \frac{n!}{i\sqrt{2\pi}}c_0\int_{-i\infty}^{i\infty}\exp\left[y^2/2 - \sqrt{2}\,ya\dagger + a\dagger^2/2\right]\ da\dagger$$

$$= n!\psi_0(y)$$

and thus,

$$\int \psi_n{}^*\psi_n\ dy = \frac{|c_n|^2}{|c_0|^2}n!\int_{-\infty}^{\infty}|\psi_0|^2\ dy.$$

Hence if the $\psi_n$ are to be normalized, we find as before,

$$|c_n|^2 = \frac{|c_0|^2}{n!}.$$

We must still determine $c_0$, but this is easily done. We require normalization in $x$-space,

$$\int \psi_0{}^*(x)\psi_0(x)\ dx = 1.$$

Recalling that $x = \sqrt{\hbar/m\omega}\ y$, we thus obtain the normalization requirement in the form

$$\int |\psi_0(y)|^2\ dy = \sqrt{m\omega/\hbar}.$$

According to equation (48), we had, with $n = 0$,

$$\psi_0 = c_0\,e^{-y^2/2},$$

so that, upon evaluating the integral, we find at once,

$$c_0 = (m\omega/\hbar\pi)^{1/4}$$

and, finally, in agreement with equation (30),

$$c_n = \left( \frac{\sqrt{m\omega/\hbar\pi}}{n!} \right)^{1/2}.$$

As a second example of the utility of the integral representation, equation (61), we now construct the so-called *generating function* for the Hermite polynomials $h_n(y)$. From equation (30), we have

$$h_n(y) = 2^{n/2} \, e^{y^2/2} \frac{\psi_n(y)}{c_n},$$

whence, from equation (61), we obtain the integral representation

$$h_n(y) = \frac{1}{i\sqrt{2\pi}} \int_{-i\infty}^{i\infty} (\sqrt{2} \, a\dagger)^n \exp [y^2 - \sqrt{2} \, a\dagger y + a\dagger^2/2] \, da\dagger. \quad (62)$$

Multiplying this expression by $s^n/n!$ and summing over $n$, we obtain

$$\sum_{n=0}^{\infty} \frac{h_n(y)}{n!} \, s^n = \frac{1}{i\sqrt{2\pi}} \sum_{n=0}^{\infty} \int_{-i\infty}^{i\infty} \frac{(\sqrt{2} \, a\dagger s)^n}{n!} \exp [y^2 - \sqrt{2} \, a\dagger y + a\dagger^2/2] \, da\dagger.$$

Interchanging the order of summation and integration on the right, the summation is easily carried out, and it then follows that

$$\sum_{n=0}^{\infty} \frac{h_n(y)}{n!} \, s^n = \frac{1}{i\sqrt{2\pi}} \int_{-i\infty}^{i\infty} \exp [y^2 + \sqrt{2} \, a\dagger (s - y) + a\dagger^2/2] \, da\dagger.$$

To evaluate the integral, write $a\dagger = i\alpha$, whence we obtain, after completing the square in the exponent,

$$\sum_{n=0}^{\infty} \frac{h_n(y)}{n!} \, s^n = \frac{1}{\sqrt{2\pi}} e^{-s^2+2sy} \int_{-\infty}^{\infty} \exp [-\tfrac{1}{2}[\alpha - i\sqrt{2}(s - y)]^2] \, d\alpha$$

or, finally, upon evaluating the Gaussian integral,

$$e^{-s^2+2sy} = \sum_{n=0}^{\infty} \frac{h_n(y)}{n!} \, s^n. \quad (63)$$

The quantity on the left is called the generating function for the Hermite polynomials because these polynomials are generated as the coefficients in an expansion of that function in powers of $s$, that is,

$$h_n(y) = \left[ \frac{d^n}{ds^n} (e^{-s^2+2sy}) \right]_{s=0}.$$

---

**Exercise 5.** For $n = 0, 1, 2, 3$, verify that the $h_n(y)$ given in equation (31) are actually obtained from equation (62); from equation (63); from equation (49).

---

## 6. MOTION OF A WAVE PACKET IN THE HARMONIC OSCILLATOR POTENTIAL

We now consider the motion of an arbitrary wave packet in an harmonic oscillator potential. To obtain the general solution we must evaluate the propagator of equation (V-62),

$$K(x', x; t - t_0) = \sum_E \psi_E^*(x')\psi_E(x) \ e^{-iE(t-t_0)/\hbar},$$

in terms of which

$$\psi(x, t) = \int_{-\infty}^{\infty} \psi(x', t_0) K(x', x; t - t_0) \ dx'. \tag{64}$$

For the harmonic oscillator specifically, we must thus find

$$K(x', x; \tau) = \sum_n \psi_n^*(x')\psi_n(x) \ e^{-i(n+1/2)\omega\tau}, \tag{65}$$

where for brevity we have introduced the elapsed time $\tau$, given by

$$\tau = t - t_0. \tag{66}$$

Before going on, we note one remarkable feature, peculiar to the harmonic oscillator alone among quantum mechanical systems. Because the energy states are equally spaced, the motion of a wave packet is periodic. The period is seen to be twice the classical period for $\psi$, and exactly the classical period for $|\psi|^2$. Thus, no matter the initial conditions, wave packets do not spread indefinitely, but alternately spread and shrink. We shall exhibit this behavior in an example shortly.

The expression, equation (65), for the propagator is not too difficult to evaluate in closed form using the creation state representation. Working in $y$-space, we have from equation (61)

$$K(y', y; \tau) = \frac{\sqrt{m\omega/\hbar\pi}}{2} e^{(y^2+y'^2-i\omega\tau)/2}$$

$$\times \sum_n \int_{-i\infty}^{i\infty} da\dagger \int_{i\infty}^{-i\infty} da\dagger' \frac{(a\dagger a\dagger')^n \ e^{-in\omega\tau}}{n!}$$

$$\times \exp\left[-\sqrt{2}(ya\dagger + y'a\dagger') + (a\dagger^2 + a\dagger'^2)/2\right].$$

Recognizing that

$$(a\dagger a\dagger')^n \ e^{-in\omega\tau} = [a\dagger a\dagger' \ e^{-i\omega\tau}]^n$$

and performing the summation over $n$, we then obtain

$$K(y', y; \tau) = \frac{\sqrt{m\omega/\hbar\pi}}{2\pi} e^{(y^2+y'^2-i\omega\tau)/2} \int_{-i\infty}^{i\infty} da\dagger \int_{i\infty}^{-i\infty} da\dagger'$$

$$\times \exp\left[-\sqrt{2} (ya\dagger + y'a\dagger') + (a\dagger^2 + a\dagger'^2)/2 + a\dagger a\dagger' \ e^{-i\omega\tau}\right]. \tag{67}$$

Replacing $a\dagger$ by a new variable $i\alpha$, and replacing $a\dagger'$ by $-i\beta$, the integrals are seen to be of the standard Gaussian type. After the usual manipulations (see Appendix I) we eventually find, re-expressing $y$ and $y'$ in terms of $x$ and $x'$,

$$K(x',x;\tau) = \sqrt{m\omega/(\pi\hbar\ 2i\sin\omega\tau)}\ \exp\left[\frac{im\omega}{\hbar}\ \frac{(x^2+x'^2)\cos\omega\tau - 2xx'}{2\sin\omega\tau}\right],$$
(68)

which is the desired closed-form result.[11]

In this expression, the multiplicative factor $(2i\sin\omega\tau)^{-1/2}$ is to be interpreted as

$$(2i\sin\omega\tau)^{-1/2} = e^{-i\omega\tau/2}\ (1 - e^{-2i\omega\tau})^{-1/2},$$

so that $K$ is seen to exhibit the predicted periodicity. Specifically, we observe that, as was already obvious from equation (65),

$$K\left(x',x;\tau+\frac{2m\pi}{\omega}\right) = (-1)^m\ K(x',x;\tau).$$
(69)

Additionally, it also follows from equation (65) that

$$K\left(x',x;\tau+\frac{(2m+1)\pi}{\omega}\right) = (-i)^{2m+1}\ K(x',-x;\tau).$$
(70)

That this last follows equally from equation (65) may be seen directly. We have

$$K(x',x;\tau+(2m+1)\pi/\omega)$$

$$= \sum_n \psi_n{}^*(x')\psi_n(x)\ \exp\left[-i(n+\tfrac{1}{2})[\omega\tau+(2m+1)\pi]\right]$$

$$= (-i)^{2m+1}\sum_n \psi_n{}^*(x')\psi_n(x)\,(-1)^n\ e^{-i(n+1/2)\omega\tau}.$$

Recalling that $\psi_n(x)$ is even for even $n$ and odd for odd $n$, and hence that

$$\psi_n(-x) = (-1)^n\psi_n(x),$$

we thus can write

$$K(x',x;\tau+(2m+1)\pi/\omega) = (-i)^{2m+1}\sum_n \psi_n{}^*(x')\psi_n(-x)\ e^{-i(n+1/2)\omega\tau}$$

$$= (-i)^{2m+1}\ K(x',-x;\tau),$$

---

[11] Observing that, in the limit $\omega \to 0$, the harmonic oscillator becomes a free particle, we can check our algebra by examining the form assumed by the propagator in that limit. We obtain at once

$$K(x',x;\tau) = \sqrt{\frac{m}{2\pi i\hbar\tau}}\ \exp\left[im(x-x')^2/2\hbar\tau\right],$$

in precise agreement with equation (IV-13).

in agreement with equation (70).

The meaning of equations (69) and (70) can be understood as follows. Using equation (64), we see that equation (69) gives

$$\psi(x, t_0 + 2m\pi/\omega) = (-1)^m \psi(x, t_0),\qquad(71)$$

while equation (70) gives

$$\psi(x, t_0 + (2m + 1)\pi/\omega) = (-1)^{2m+1} \psi(-x, t_0).\qquad(72)$$

Suppose now that $\psi(x, t_0)$ describes a particle centered at $x_0$ and moving with mean momentum $p_0$. An example of such a wave packet is the expression

$$\psi(x, t_0) = e^{ip_0x/\hbar} f(x - x_0),$$

where $f$ is real and attains its maximum when its argument is zero. After a time interval $\pi/\omega$, equal to one-half of the classical period, equation (72) tells us that

$$\psi(x, t_0 + \pi/\omega) = -i\ e^{-ip_0x/\hbar} f(-x - x_0),$$

so that the wave packet is centered at $x = -x_0$, is unaltered in shape and is moving with mean momentum $-p_0$. After one period, $2\pi/\omega$, equation (71) shows that

$$\psi(x, t_0 + 2\pi/\omega) = -e^{ip_0x/\hbar} f(x - x_0),$$

so that the wave packet has returned to its original position, unaltered in shape, and with its original mean momentum $p_0$. Thus the motion proceeds indefinitely, and in this respect the wave packet moves exactly as would a classical particle.[12]

As a particular example, we consider a Gaussian wave packet initially centered at the origin and with initial mean momentum $p_0$,

$$\psi(x, t_0) = \frac{1}{\sqrt{L\sqrt{\pi}}} \exp\left[-x^2/2L^2 + ip_0x/\hbar\right].$$

Substituting this expression into equation (64), and evaluating the usual Gaussian integrals, we eventually find

$$|\psi(x, t)|^2 = \frac{1}{\sqrt{\pi}L(\tau)} \exp\left[-[x - (p_0/m\omega)\sin\omega\tau]^2/L^2(\tau)\right],\qquad(73)$$

where $\tau = t - t_0$, and

$$L(\tau) = \sqrt{L^2 \cos^2\omega\tau + (\hbar/m\omega L)^2 \sin^2\omega\tau}.\qquad(74)$$

---

[12] These conclusions do *not* depend on the particular form chosen for $\psi(x, t_0)$; they apply to an arbitrary wave packet.

In agreement with our prediction, the particle is seen to oscillate precisely as would a classical particle which starts from the origin with initial momentum $p_0$. The width of the wave packet $L(\tau)$ also oscillates, with twice the classical frequency, $\omega$, between its two extreme values $L$, at the origin, and $\hbar/m\omega L$, at the classical turning point. Note that if the initial width $L$ is larger than $\sqrt{\hbar/m\omega}$, the wave packet actually *diminishes* in size as it moves away from its initial position. No violation of the uncertainty principle is involved, of course, since the width of the wave packet in momentum space always changes in such a way as to exactly compensate for the change in width of the wave packet in configuration space. The wave packet in this example is always Gaussian and therefore always a minimum uncertainty wave packet.

Even more remarkable, perhaps, is the fact that if the initial Gaussian packet is such that

$$L = \sqrt{\hbar/m\omega}$$

then, according to equation (74),

$$L(\tau) = L = \sqrt{\hbar/m\omega}$$

and is constant. In other words, the shape of such a wave packet is independent of time! Note that with this choice of $L$, the shape of the wave packet is exactly that of the ground state eigenfunction. In effect then, if a particle in the ground state of an harmonic oscillator could be given some initial momentum $p_0$, it would oscillate indefinitely with the classical frequency and amplitude, and the *shape* of its state function would be unaltered with time.

---

**Exercise 6.**
   (a)   Obtain equation (68) from equation (67).
   (b)   Derive equation (73).

---

## 7. CONTINUUM STATES OF A SQUARE WELL POTENTIAL

So far we have considered only bound states, first in a square well, and then in the harmonic oscillator. In the harmonic oscillator, bound states constitute the entire spectrum. However, for the square well potential, there also exist a continuous set of states of positive energy, which we now examine.

Using the same notation as in Section 3, and referring to Figure 2, we seek solutions of

$$-\frac{\hbar^2}{2m}\frac{d^2\psi_E}{dx^2} + V(x)\psi_E = E\psi_E, \qquad E > 0, \tag{75}$$

where

$$V(x) = \begin{cases} -V_0, & -a < x < a \\ 0, & |x| > a. \end{cases}$$

The square well potential has bound states only for $V_0 > 0$ (attractive potentials), but of course continuum states exist for $V_0 < 0$ (repulsive potentials) as well, and we shall consider both cases.

There is no difficulty in constructing the solutions we seek, only a difficulty in interpreting them once we have them. The nature of this difficulty is the following. For $|x| > a$, that is, in regions I and III of Figure 2, equation (75) takes the form

$$\frac{d^2\psi_E}{dx^2} + \frac{2mE}{\hbar^2}\,\psi_E = 0, \tag{76}$$

and hence the solutions are oscillatory. They can be expressed as linear combinations of the de Broglie waves $\exp[\pm i\sqrt{2mE}\,x/\hbar]$. These solutions thus extend to infinity and are not normalizable. Accordingly, the stationary states are not physically admissible. To obtain a true physical state one must construct a wave packet as a superposition of such states. As an example, let us consider a wave packet centered at some large negative value of $x$ and traveling toward the potential with mean momentum $p = \sqrt{2mE}$. Such a wave packet will travel very nearly as a free-particle wave packet until it reaches the region of the potential. As a result of its interaction with the potential, some fraction of the wave packet will be transmitted through the potential well and some fraction will be reflected. In other words, a particle incident upon the potential will sometimes be transmitted and somtimes reflected.[13] Note that in region III, $x < -a$, there appears in the course of time a wave packet traveling to the left, in addition to the incident wave packet traveling to the right. On the other hand, in region I, $x > a$, only a wave packet traveling to the right is encountered.

Now suppose we let the initial wave packet become broader and broader. As it does so, its width in momentum space, and therefore also its energy spread, becomes narrower and narrower. Indeed, by making the wave packet broad enough, the energy spread can be made infinitesimal. Further, the time required for the incident wave packet to complete its interaction with the potential becomes longer and longer. Hence, in the limit of an enormously broad wave packet, the description reduces to

[13] See Section 10, particularly Figures 8-13, for a description of the numerical solution to the problem of a wave packet incident upon a square well potential.

that of a stationary state in which incident, reflected and transmitted waves *coexist in time*. Specifically, in the example under discussion, we have

Region III, $x < a$: $\quad \psi_E(x) = A\ e^{i\sqrt{2mE}\ x/\hbar} + B\ e^{-i\sqrt{2mE}\ x/\hbar}$

Region I, $x > a$: $\quad \psi_E \quad = C\ e^{i\sqrt{2mE}\ x/\hbar}$,
$$\tag{77}$$

where $A$, $B$ and $C$ are the amplitudes of the incident, reflected and transmitted waves respectively. Writing

$$R = |B/A|^2 \tag{78}$$
$$T = |C/A|^2,$$

the reflection coefficient $R$ is the probability an incident particle will be reflected, and the transmission coefficient $T$ is the probability that it will be transmitted.[14]

The calculation of the stationary state wave function, a tedious calculation indeed, now proceeds as follows. In region II, $-a < x < a$, which is the region in which the potential differs from zero, Schrödinger's equation (75) becomes

$$\frac{d^2\psi_E}{dx^2} + \frac{2m}{\hbar^2}\ (E + V_0)\psi_E = 0,$$

and the general solution in this region is thus

$$\psi_E = b_1 \exp\left[i\sqrt{2m(E + V_0)}\ x/\hbar\right] + b_2 \exp\left[-i\sqrt{2m(E + V_0)}\ x/\hbar\right]. \tag{79}$$

Just as in the bound state problem, $\psi_E$ and its slope must be continuous at $x = a$ and at $x = -a$. Hence for $x = a$ we obtain

$$C\ e^{ika} = b_1\ e^{iKa} + b_2\ e^{-iKa}$$
$$ikC\ e^{ika} = iK(b_1\ e^{iKa} - b_2\ e^{-iKa}) \tag{80}$$

and, for $x = -a$,

$$A\ e^{-ika} + B\ e^{ika} = b_1\ e^{-iKa} + b_2\ e^{iKa}$$
$$ik(A\ e^{-ika} - B\ e^{ika}) = iK(b_1\ e^{-iKa} - b_2\ e^{+iKa}), \tag{81}$$

where we have introduced the two wave numbers

$$k = \frac{\sqrt{2mE}}{\hbar}$$
$$K = \frac{\sqrt{2m(E + V_0)}}{\hbar}. \tag{82}$$

[14] An *ensemble* of incident wave packets is equivalent to a *beam* of incident particles. Accordingly, the following language is commonly used to describe the state characterized by equation (77): A beam of relative intensity or flux $|A|^2$ is incident upon the potential. The reflected beam has intensity $|B|^2 = R|A|^2$ and the transmitted beam intensity $|C|^2 = T|A|^2$.

From equation (80), we can express $b_1$ and $b_2$ in terms of $C$,

$$b_1 = \frac{1}{2}\left(1 + \frac{k}{K}\right) C \, e^{i(k-K)a}$$

$$b_2 = \frac{1}{2}\left(1 - \frac{k}{K}\right) C \, e^{i(k+K)a}.$$

Equation (81) then becomes

$$A \, e^{-ika} + B \, e^{ika} = \left[\frac{1}{2}\left(1 + \frac{k}{K}\right) e^{i(k-2K)a} + \frac{1}{2}\left(1 - \frac{k}{K}\right) e^{i(k+2K)a}\right] C$$

$$k(A \, e^{-ika} - B \, e^{ika} = K\left[\frac{1}{2}\left(1 + \frac{k}{K}\right) e^{i(k-2K)a} - \frac{1}{2}\left(1 - \frac{k}{K}\right) e^{i(k+2K)a}\right] C.$$

Introducing the amplitude transmission coefficient $\tau = C/A$, we then obtain, after some elementary algebra,

$$\tau = \frac{C}{A} = \frac{2 e^{-ika}}{\left(1 + \frac{k}{2K} + \frac{K}{2k}\right) e^{i(k-2K)a} + \left(1 - \frac{k}{2K} - \frac{K}{2k}\right) e^{i(k+2K)a}}$$

$$= \frac{e^{-2ika}}{\cos 2Ka - \frac{i}{2}\left(\frac{k}{K} + \frac{K}{k}\right) \sin 2Ka}$$

or, finally,

$$\tau = \frac{e^{-2i\sqrt{2mE}\, a/\hbar}}{\cos\left(2\sqrt{2m(E + V_0)}\, a/\hbar\right) - \frac{i}{2}\left(\frac{\sqrt{E}}{\sqrt{E + V_0}} + \frac{\sqrt{E + V_0}}{\sqrt{E}}\right) \sin\left(2\sqrt{2m(E + V_0)}\, a/\hbar\right)}. \tag{83}$$

Introducing also the amplitude reflection coefficient $\rho = B/A$, we obtain, after some more algebra,

$$\rho = \frac{\frac{i}{2} e^{-2i\sqrt{2mE}\, a/\hbar} \left(\sqrt{\frac{E + V_0}{E}} - \sqrt{\frac{E}{E + V_0}}\right) \sin\left(2\sqrt{2m(E + V_0)}\, a/\hbar\right)}{\cos\left(2\sqrt{2m(E + V_0)}\, a/\hbar\right) - \frac{i}{2}\left(\sqrt{\frac{E}{E + V_0}} + \sqrt{\frac{E + V_0}{E}}\right) \sin\left(2\sqrt{2m(E + V_0)}\, a/\hbar\right)}. \tag{84}$$

Consider first the case of attractive potentials, $V_0 > 0$. In that instance, we obtain

$$R = |\rho|^2 = \frac{[V_0^2/4E(E + V_0)] \sin^2\left(2\sqrt{2m(E + V_0)}\, a/\hbar\right)}{1 + [V_0^2/4E(E + V_0)] \sin^2\left(2\sqrt{2m(E + V_0)}\, a/\hbar\right)} \tag{85}$$

and

$$T = |\tau|^2 = \frac{1}{1 + [V_0^2/4E(E + V_0)] \sin^2\left(2\sqrt{2m(E + V_0)}\, a/\hbar\right)}, \tag{86}$$

where we have used the fact that

$$\left(\sqrt{E/(E+V_0)} + \sqrt{(E+V_0)/E}\right)^2 = 4 + \frac{V_0{}^2}{E(E+V_0)}$$

to simplify these expressions for the reflection and transmission coefficient. Observe that, as it must according to our interpretation,

$$R + T = 1. \tag{87}$$

Also observe that transmission resonances occur, in which $T = 1$ and $R = 0$, whenever

$$2\sqrt{2m(E+V_0)}\, a/\hbar = n\pi.$$

Now the momentum of the particle in the interior of the potential is given by

$$\bar{p} = \sqrt{2m(E+V_0)}.$$

Hence, these resonances occur when the width $2a$ of the potential is a half-integral number times the de Broglie wave length $h/\bar{p}$ in the interior of the potential, just as we claimed intuitively in our discussion in Chapter I.

From equations (85) and (86), it can be seen that when $E$ is sufficiently large compared to $V_0$, the reflection coefficient approaches zero and the transmission coefficient unity, this being the expected result in the classical limit. Actually, the approach to this limit is surprisingly slow. Even if $E$ is as large as $V_0$ in magnitude, the probability that an incident particle will be reflected can exceed 10 per cent. It must be mentioned, however, that for more realistic potentials than a square well, with its unphysical discontinuities, the approach to the classical result is much more rapid.

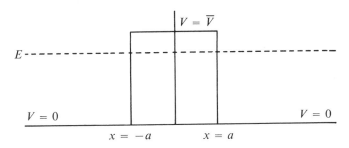

FIGURE 7. The tunnel effect or penetration through a barrier.

We next turn to the more interesting case of repulsive potentials, $V_0 < 0$. As long as $E + V_0 > 0$, the results are as given above and equations (85) and (86) are both applicable as they stand. Suppose, however, that $E + V_0 < 0$. Write $V_0 = -\bar{V}$, so that $\bar{V}$ represents the *height* of the potential barrier. The case under consideration is thus that for which $E < \bar{V}$, as illustrated in Figure 7. From equations (83) and (84), we then find

$$R = |\rho|^2 = \frac{[\bar{V}^2/4E(\bar{V} - E)] \sinh^2 (2\sqrt{2m(\bar{V} - E)} \, a/\hbar)}{1 + [\bar{V}^2/4E(\bar{V} - E)] \sinh^2 (2\sqrt{2m(\bar{V} - E)} \, a/\hbar)} \quad (88)$$

and

$$T = |\tau|^2 = \frac{1}{1 + [\bar{V}^2/4E(\bar{V} - E)] \sinh^2 (2\sqrt{2m(\bar{V} - E)} \, a/\hbar)}, \quad (89)$$

and, of course, equation (87) is still satisfied, as it must be. Here we have an explicit example of the tunnel effect, which was also discussed in Chapter I. There is a finite probability that a particle will be transmitted through a classically forbidden region, that is, one in which its kinetic energy is negative. Note that if the barrier is wide enough, or the kinetic energy is negative enough, that

$$2\sqrt{2m(\bar{V} - E)} \, a/\hbar \gg 1.$$

Then, to good approximation,

$$\sinh (2\sqrt{2m(\bar{V} - E)} \, a/\hbar) \simeq \frac{1}{2} \exp [2\sqrt{2m(\bar{V} - E)} \, a/\hbar].$$

In that case the transmission coefficient is the exponentially small quantity

$$T = \frac{16E(\bar{V} - E)}{\bar{V}^2} \exp [-4\sqrt{2m(\bar{V} - E)} \, a/\hbar], \quad (90)$$

so that the deviation from the classical result is negligible. Otherwise stated, only if the potential barrier is very thin, and the potential energy barely negative, will there be significant tunneling through a barrier.

One final remark must be added. We have derived a solution of Schrödinger's equation which corresponds to a particle incident on the potential well from the left. We could equally well have chosen to describe a particle incident from the right, of course. This freedom is simply a reflection of the existence of two linearly independent solutions to the second-order Schrödinger's equation, in agreement with our earlier prediction of the twofold degeneracy of continuum states of the type we have just been discussing.

## 8. CONTINUUM STATES IN GENERAL; THE PROBABILITY FLUX

The most important potentials on the microscopic level describe the interaction between particles and thus have the property that

$$V(x) \to 0, \qquad |x| \to \infty.$$

It is accordingly of interest to discuss the continuum states of such potentials. For simplicity, we restrict our attention to the case in which $V(x)$ vanishes more rapidly than $x^{-1}$ as $|x|$ tends to infinity.[15] Under these circumstances, it is not hard to see that, for sufficiently large $|x|$, the state function is expressible as a linear combination of the de Broglie waves $\exp[\pm i\sqrt{2mE}\, x/\hbar]$, just as for the square well case. Indeed, our interpretation of the solutions can be taken over unaltered, and we write at once

$$x \to -\infty, \; \psi_E(x) = A \; (e^{i\sqrt{2mE}\, x/\hbar} + \rho \; e^{-i\sqrt{2mE}\, x/\hbar}), \qquad (91)$$

and

$$x \to +\infty, \; \psi_E(x) = A\tau \; e^{i\sqrt{2mE}\, x/\hbar}, \qquad (92)$$

where we have explicitly introduced the amplitude reflection and transmission coefficients $\rho$ and $\tau$. Of course, the precise form of $\rho$ and $\tau$ depends on the detailed nature of the potential, but in any case, $R = |\rho|^2$ is the probability that a particle incident from infinity will be reflected, and $T = |\tau|^2$ is the probability that it will be transmitted. As before, we must demand that equation (87), $R + T = 1$, be satisfied, no matter what the detailed form of the potential, if our interpretation is to be correct. We now verify that this is indeed the case. To do so we must temporarily revert to the wave packet description, of which the stationary state solutions are the limiting case. Denote the state function by $\psi(x, t)$, as usual, and consider the probability that the particle is to be found at some instant $t$ in the fixed but arbitrary region of space between $x = x_1$ and $x = x_2$, where for definiteness we take $x_1 < x_2$. This probability is

$$P(x_1, x_2; t) = \int_{x_1}^{x_2} \psi^*(x, t)\psi(x, t) \; dx \qquad (93)$$

and, as we have indicated, it is a function of time. Specifically, as the wave packet sweeps across the region of space under examination, we expect $P(x_1, x_2; t)$ to increase from an initially negligible value to something close to unity and then to decrease practically to zero again. In any case, we have

$$\frac{\partial P(x_1, x_2; t)}{\partial t} = \int_{x_1}^{x_2} \left( \frac{\partial \psi^*}{\partial t} \psi + \psi^* \frac{\partial \psi}{\partial t} \right) dx$$

or, since $\psi$ is a solution of Schrödinger's equation,

---

[15] The case in which $V(x)$ falls off exactly as $1/x$ is of great importance, but requires special treatment which is, at present, beyond our capabilities.

$$-\frac{\hbar^2}{2m}\frac{\partial^2\psi}{\partial x^2}+V(x)\psi=-\frac{\hbar}{i}\frac{\partial\psi}{\partial t},$$

we see that

$$\frac{\partial P(x_1, x_2; t)}{\partial t}$$

$$=\frac{i}{\hbar}\int_{x_1}^{x_2}\left(\left[-\frac{\hbar^2}{2m}\frac{\partial^2\psi^*}{\partial x^2}+V(x)\psi^*\right]\psi-\psi^*\left[-\frac{\hbar^2}{2m}\frac{\partial^2\psi}{\partial x^2}+V(x)\psi\right]\right)dx$$

$$=\frac{i\hbar}{2m}\int_{x_1}^{x_2}\left(\psi^*\frac{\partial^2\psi}{\partial x^2}-\psi\frac{\partial^2\psi^*}{\partial x^2}\right)dx$$

$$=\frac{i\hbar}{2m}\int_{x_1}^{x_2}\frac{\partial}{\partial x}\left(\psi^*\frac{\partial\psi}{\partial x}-\psi\frac{\partial\psi^*}{\partial x}\right)dx.$$

The right side is thus a perfect differential and we obtain finally

$$\frac{\partial P(x_1, x_2; t)}{\partial t}=j(x_1, t)-j(x_2, t),\tag{94}$$

where

$$j(x, t)=\frac{\hbar}{2mi}\left(\psi^*\frac{\partial\psi}{\partial x}-\psi\frac{\partial\psi^*}{\partial x}\right).\tag{95}$$

The interpretation of the result is the following. The change per unit time of the probability of finding the particle in the interval between $x_1$ and $x_2$, $\partial P/\partial t$, is the difference between the probability per unit time that the particle entered at $x_1$, $j(x_1, t)$ and that it emerged at $x_2$, $j(x_2, t)$. Thus $j(x, t)$ has the significance of a probability flux or current, directed toward positive $x$.[16]

Equation (94) is simply a general statement of probability conservation, applied to a finite domain rather than to all of space. The particular application we have in mind is to the stationary state limit of a wave packet. In that limit, we have

$$\psi(x, t)=\psi_E(x, t)=\psi_E(x)\,e^{-iEt/\hbar},$$

and hence it follows from its definition, equation (93), that $P(x_1, x_2; t)$

---

[16] It may be helpful to compare this result and its interpretation to that expressing conservation of charge. Suppose $Q(x_1, x_2; t)$ represents the net charge between the points $x_1$ and $x_2$ on a long wire, and suppose that $I(x, t)$ is the electric current flowing in the wire at a given point $x$, the positive direction of flow being toward positive $x$. Then in a time interval $\delta t$, the net charge flowing into the wire at $x_1$ is $I(x_1, t)\,\delta t$, and that flowing out at $x_2$ is $I(x_2, t)\,\delta t$. Hence the net increase in charge, $\delta Q$, is $[I(x_1, t)-I(x_2, t)]\,\delta t$, or

$$\frac{\partial Q(x_1, x_2; t)}{\partial t}=I(x_1, t)-I(x_2, t),$$

which has the same form as equation (94) and thus supports our interpretation of $j$ as a (probability) current.

is independent of time. The left side of equation (94) therefore vanishes, and we conclude that, for arbitrary $x_1$ and $x_2$, $j(x_1) = j(x_2)$, which means that $j(x)$ *is a constant*, say $j_0$. For $|x| \to \infty$, $\psi_E(x)$ is given by (91) or (92) and it is easily verified that

$$x \to -\infty, j_0 = |A|^2 v(1 - |\rho|^2) = |A|^2 v(1 - R)$$

$$x \to +\infty, j_0 = |A|^2 v |\tau|^2 = |A|^2 v T$$

where $v = \sqrt{2mE/h}$ is the classical velocity of the particle. Since these must be equal, we thus see that $R + T = 1$, as required.

The form of $j_0$ in each region is not unexpected, as the following argument shows. Consider first the region $x \to \infty$, where $\psi_E$ is given by equation (92). We have

$$\rho(x) = \psi_E^* \psi_E = |A|^2 \, |\tau|^2 = |A|^2 T,$$

where $\rho(x)$ is the probability density. Thus the probability flux $j$ is simply $\rho v$, just as in hydrodynamics. Similarly for $x \to -\infty$, the net probability flux is the difference between the flux to the right $|A|^2 v$ and the independent flux to the left $|A|^2 vR$. Note that the interference terms between incident and reflected waves have completely cancelled out, as they must (why?).

## 9. PASSAGE OF A WAVE PACKET THROUGH A POTENTIAL

We have considered nonphysical stationary state solutions as the limiting case of very broad wave packets. Our arguments concerning this limiting process were crucial to our interpretation of the solutions, but they are essentially irrelevant as far as the mathematical properties of these solutions are concerned. Quite independently of this interpretation, we are at liberty to view the stationary states simply as idealized states from which physical states can be constructed by superposition. We now show how to do this. Since the $\psi_E$ form a complete set, we can express an arbitrary solution of the time dependent Schrödinger's equation in the form

$$\psi(x, t) = \int \psi_E(x) \, e^{-iEt/\hbar} f(E) \, dE, \tag{96}$$

where the integral extends over the continuum states and must also be regarded as including the discrete states, if any. Now suppose we want to construct a wave packet which at time $t = 0$ is centered at large negative $x$, say at $x = -x_0$, and is moving to the right with mean momentum, $p_0 = \sqrt{2mE_0}$. For $x$ large and negative, $\psi_E$ is given by (91). Hence, absorbing the arbitrary amplitude $A$ into $f(E)$, we have, as $x \to -\infty$,

$$\psi(x, t) = \int f(E)[e^{i\sqrt{2mE} \, x/\hbar} + \rho(E) \, e^{-i\sqrt{2mE} \, x/\hbar}] \, e^{-iEt/\hbar} \, dE. \tag{97}$$

If we thus choose $f(E)$ to have the form

$$f(E) = g(E - E_0) \, e^{i\sqrt{2mE}\, x_0/\hbar}, \qquad (98)$$

where $g$ is peaked about $E = E_0$, we will have constructed such a wave packet, as is readily seen upon recognizing that these expressions are equivalent to the by now familiar momentum space representations of a wave packet. The first term in equation (97) yields the initial wave packet, starting at $x = -x_0$ and moving to the right. The second term is initially negligible, but eventually, after sufficient time has elapsed for the packet to reach the potential and be reflected back, that is, after a time of the order of $(x_0 + x_1) \, m/p_0$, it will begin to build up into the reflected wave packet at $x = x_1$. This can be directly verified, but in any case it is guaranteed by the correspondence principle. In a moment we show how this actually works out for the transmitted wave packet. First note, however, that neglecting the second term of equation (97), we have explicitly

$$\psi(x, 0) \simeq \int f(E) \, e^{i\sqrt{2mE}\, x/\hbar} \, dE,$$

or

$$\psi(x, 0) = \int g(E - E_0) \, e^{i\sqrt{2mE}\,(x+x_0)/\hbar} \, dE$$
$$= h(x + x_0) \, e^{ip_0(x+x_0)/\hbar},$$

where $h(x + x_0)$ is the envelope of the initial wave packet, centered about $x = -x_0$. This can be simplified as follows. Since $g(E - E_0)$ is peaked about $E \simeq E_0$, introduce a new variable $\eta$ by writing

$$E = E_0 + \eta.$$

We then have, expanding in a Taylor series,

$$\sqrt{2mE} = \sqrt{2m(E_0 + \eta)} \simeq \sqrt{2mE_0} + \tfrac{1}{2} \sqrt{(2m/E_0)} \, \eta + \cdots$$

$$\simeq p_0 + \frac{m}{p_0} \eta + \cdots \, .$$

Hence, neglecting higher-order terms,

$$h(x + x_0) \simeq \int g(\eta) \, \exp\left[i(m/p_0)\,\eta \, (x + x_0)/\hbar\right] \, d\eta. \qquad (99)$$

To obtain the transmitted packet as $x \to +\infty$, we have, using equation (92),

$$\psi(x, t) = \int f(E)\tau(E) \, \exp\left[i\sqrt{2mE}\, x/\hbar - iEt/\hbar\right] \, dE$$
$$\simeq \exp\left[i[p_0(x + x_0) - E_0 t]/\hbar\right] \int g(\eta)\tau(E_0 + \eta)$$
$$\times \exp\left[i[m/p_0 \, (x + x_0) - t]\,\eta/\hbar\right] \, d\eta.$$

Recalling that $|\tau|^2 = T$, we write

$$\tau(E_0 + \eta) = \sqrt{T(E_0 + \eta)}\ e^{i\delta(E_0 + \eta)},$$

and we now assume that $T(E)$ is a rather slowly varying function of $E$, but that $\delta(E)$ need not be. In other words, we make the plausible assumption that the magnitude of the transmission coefficient does not change much when $E$ is changed a little, but that its phase may change significantly. With this assumption,

$$\tau(E_0 + \eta) \simeq \sqrt{T(E_0)}\ \exp\left[i\delta(E_0) + i\eta\ d\delta/dE_0\right] = \tau(E_0)\ e^{i\eta\ d\delta/dE}$$

and

$$\psi(x, t) \simeq \tau(E_0) \exp\left[i[p_0(x + x_0) - E_0 t]/\hbar\right] \int g(n)$$
$$\times \exp\left[i[m/p_0(x + x_0) - t + \hbar d\delta/dE_0]\eta/\hbar\right]\ d\eta.$$

Comparison with equation (91) then shows that, for $x \rightarrow +\infty$,

$$\psi(x, t) \simeq \tau(E_0)\ h\left[x + x_0 - \frac{p_0}{m}(t - \hbar\ d\delta/dE_0)\right]$$
$$\times \exp\left[i[p_0(x + x_0) - E_0 t]/\hbar\right]. \qquad (100)$$

Thus, to this approximation, the transmitted packet appears undistorted but reduced in amplitude by the transmission coefficient $\tau(E_0)$. The packet appears at the position $x$ at a time $t$ given by

$$t = \frac{m}{p_0}(x + x_0) + \hbar\ \frac{d\delta}{dE_0}.$$

The first term is recognized as the time for a free particle with momentum $p_0$ to travel from $-x_0$ to $x$. The second term is the increment of time introduced by the forces which act on the particle during its passage through the potential. Calling this increment $\Delta t$, we thus have

$$\Delta t \simeq \hbar\ \frac{d\delta}{dE}. \qquad (101)$$

Perhaps it should be emphasized that these last results are only approximate in nature. Specifically, two rather different assumptions were made in deriving equations (100) and (101). The first was that in the various Taylor series expansions about $E_0$ the quadratic (and higher) terms in $\eta$ could be neglected. These terms lead to a *spreading* of the wave packet, in just about the same way as for a free-particle wave packet. Reference to the discussion of Chapter IV shows that this spreading is negligible if the wave packet is sufficiently broad. The second assumption was that the *magnitude* of the transmission coefficient $\tau$ does not change significantly over the energy width of the wave packet. The neglected terms in this case lead to a *distortion* of the wave packet, and again this is negligible for sufficiently broad wave packets. Even if some spreading and distortion occur, however, the qualitative features of the results are not strikingly affected. A transmitted wave packet ultimately *does* appear and its time delay *is* expressed through the dependence of $\delta$ on energy. These features comprise the essential contents of equations (100) and (101).

## 10. NUMERICAL SOLUTION OF SCHRÖDINGER'S EQUATION

Our remarks thus far have been exclusively directed to the solutions of Schrödinger's equation by analytic methods. However, the availability of modern high-speed computers has made it a relatively simple and routine matter to obtain such solutions numerically. In particular, the *ordinary* differential equations which define the stationary states are readily integrated numerically, in a matter of seconds or fractions of a second, for both bound states and continuum states, and for quite complicated potential energy functions. On the other hand, the *partial differential time dependent* Schrödinger's equation is a great deal more complicated, and its solution, while feasible, is a significant task for even the fastest and largest computers. As a result, the time dependent problem of the motion of wave packets has not yet been very extensively studied by numerical methods and our knowledge of the details of their behavior is largely semi-quantitative. The discussion in the last section is rather typical in this respect.

This situation has recently been improved to a considerable degree by the construction of exact numerical solutions for a Gaussian wave packet incident upon a square well potential.[17] The results are shown for a number of different circumstances in Figures 8 through 13.[18] In Figures 8, 9 and 10, a wave packet is incident upon an attractive potential of fixed width and depth, and its behavior is studied as the mean energy of the wave packet state is varied. The figures are largely self-explanatory, consisting of a sequence of "snapshots" which show the packet approaching the well and then interacting with it to generate reflected and transmitted packets. In Figure 8 the mean energy of the

[17] A. Goldberg, H. M. Schey, and J. L. Schwartz, "Computer-Generated Motion Pictures of One-Dimensional Quantum Mechanical Transmission and Reflection Phenomena," American Journal of Physics, *35*, 177 (1967).

[18] Note that in these figures the envelope of the wave packets is displayed on an arbitrary ordinate scale of probability density, while the potential well is displayed on an independently arbitrary energy scale. The height of the packet relative to the magnitude of the potential has no significance whatsoever. Thus the well could have been drawn one-tenth as large or ten times as large, without affecting in any way the depicted behavior of the wave packet.

packet state is one-half the depth of the attractive potential, in Figure 9 the mean energy is equal to the well depth and in Figure 10 it is twice the well depth. In agreement with expectations, the reflected packet decreases rapidly in magnitude with increasing energy, and at the highest energy the classical limit of no discernible reflection is seen to be achieved.

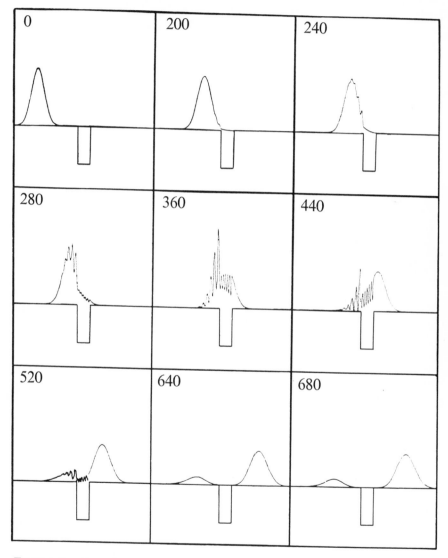

FIGURE 8.    Gaussian wave-packet scattering from a square well. The average energy is one-half the well depth. Numbers denote the time of each configuration in arbitrary units.

Figure 11, 12 and 13 show the behavior of packets for this same set of energies, but incident upon a repulsive potential, or barrier. Here one goes from practically total reflection at low energies to practically total transmission at high, both in agreement with the classical behavior. The intermediate energy case, in which both reflected and transmitted packets are

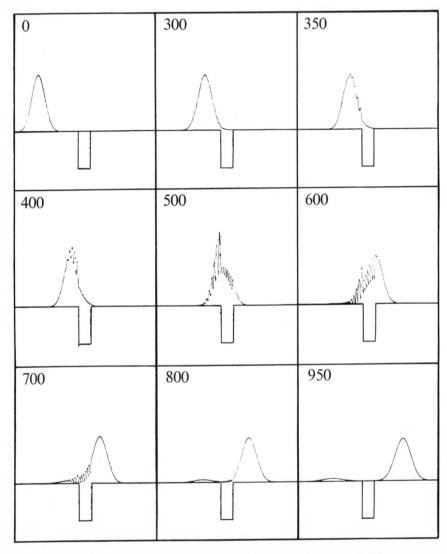

FIGURE 9. Gaussian wave-packet scattering from a square well. The average energy is equal to the well depth. Numbers denote the time of each configuration in arbitrary units.

formed, is particularly interesting, especially in view of the rather unexpected trapping of a portion of the wave packets within the potential, as shown in Figure 12(b).

These exact results demonstrate very clearly and explicitly our earlier assertions about the qualitative behavior of wave packets. The gradual formation of the reflected and transmitted packets is clearly visible, as

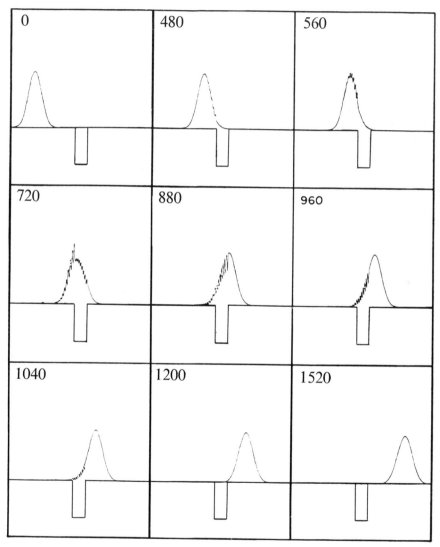

FIGURE 10.    Gaussian wave-packet scattering from a square well. The average energy is twice the well depth. Numbers denote the time of each configuration in arbitrary units.

is the gradual spreading of the wave packet with the passage of time. As expected, this spreading is noticeably less for higher than for lower energy packets (why?). Note, however, that the wave packets suffer negligible distortion, at least visually. This is rather remarkable, considering the highly complicated and easily visible structure which develops during the actual period of interaction between the packet and the poten-

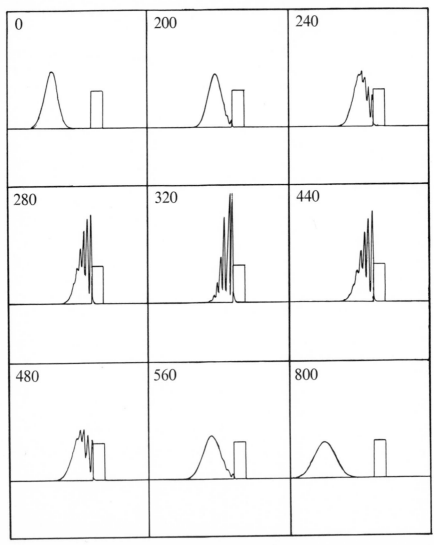

FIGURE 11. Gaussian wave-packet scattering from a square barrier. The average energy is one-half the barrier height. Numbers denote the time of each configuration in arbitrary units.

tial. The time increment associated with the passage of the packet through the potential, although certainly present, unfortunately is too small to be apparent on the time scale in the figures.

We conclude with some brief remarks on the methods commonly used to numerically integrate differential equations, such as Schrödinger's.[19] Consider first the matter of stationary states. Here we seek to solve the time independent Schrödinger's equation, which we write in the form

$$\ddot{\psi} + f(x)\psi = 0 \tag{102}$$

where each dot over a symbol denotes differentiation with respect to $x$ and where

$$f(x) = \frac{2m}{\hbar^2}(E - V). \tag{103}$$

We assume $f(x)$ to be an exactly known function.

*By a numerical solution to equation (102) we mean the set of values, to some assigned level of accuracy, of both $\psi$ and $\dot{\psi}$ at a discrete set of points, $x_n$, called mesh points.*[20] We shall take these points to be equidistant from each other, and of spacing $\epsilon$, whence we write

$$x_n = n\epsilon, \qquad n = 0, \pm 1, \pm 2, \ldots \tag{104}$$

Since we are seeking the solution of a second-order differential equation, two integration constants must be fixed to uniquely specify the solution; we shall take these to be the values of $\psi$ and $\dot{\psi}$ at some given point, which, for convenience, we choose to be the origin. Our problem may now be stated as follows: Given $\psi(0)$ and $\dot{\psi}(0)$, how do we construct $\psi(x_n)$ and $\dot{\psi}(x_n)$ over whatever domain of mesh points may be of interest?[21] We now answer this question by constructing a pair of simple difference equations for these quantities.

Consider first

---

[19] See Reference [7], Chapter 13, and Reference [9], Chapter 10, for a general discussion. For the application to Schrödinger's equation for bound states, see R. S. Caswell, "Improved Fortran Program for Single Particle Energy Levels and Wavefunctions in Nuclear Structure Calculations," National Bureau of Standards Technical Note 410, Superintendent of Documents, U.S. Government Printing Office (1966). For continuum states, see M. A. Melkanoff, J. S. Nodvik, D. Cantor and D. S. Saxon, *A Fortran Program for Elastic Scattering Analyses with the Nuclear Optical Model*, University of California Press (1961), especially pp. 24–29, and M. A. Melkanoff, T. Sawada and J. Raynal, "Nuclear Optical Model Calculations," in *Methods of Computational Physics*, Vol. 6, Academic Press (1966). This latter gives a particularly complete and up-to-date account.

[20] Only $\psi$ and $\dot{\psi}$ need be considered because all higher derivatives are at once expressible in terms of these two using equation (102).

[21] If the normalization of the state function is taken to be arbitrary, only the *ratio* of $\psi(0)$ to $\dot{\psi}(0)$ need be specified in actuality.

$$\psi(x_{n+1}) = \psi(x_n + \epsilon)$$

and expand the right side in a Taylor series. We obtain

$$\psi(x_{n+1}) = \psi(x_n) + \epsilon\dot{\psi}(x_n) + \frac{\epsilon^2}{2}\ddot{\psi}(x_n) + 0(\epsilon^3),$$

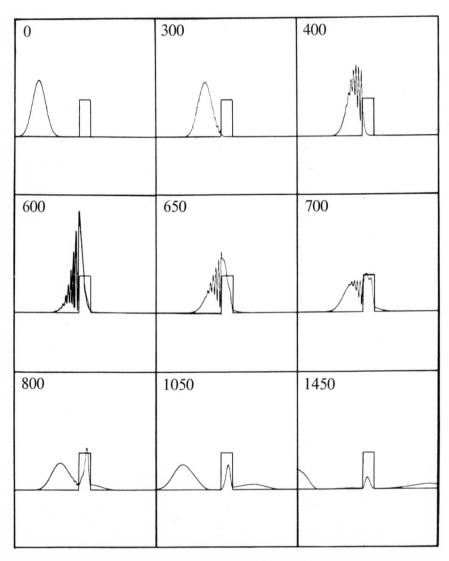

FIGURE 12(a).  Gaussian wave packet scattering from a square barrier. The average energy is equal to the barrier height. Numbers denote the time of each configuration in arbitrary units. Note the resonance effect in which a part of the probability distribution remains for a long time in the region of the potential.

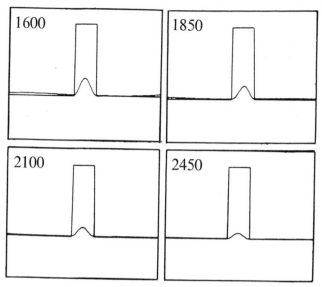

FIGURE 12(b).   Details of the decay of the resonant state seen in Figure 12(a).

whence, using equation (102),

$$\psi(x_{n+1}) = \psi(x_n) + \epsilon\dot\psi(x_n) - \frac{\epsilon^2}{2} f(x_n)\psi(x_n) + 0(\epsilon^3). \qquad (105)$$

This equation permits us to calculate $\psi(x_{n+1})$, provided $\psi(x_n)$ and $\dot\psi(x_n)$ are known. We now need a similar equation for $\dot\psi(x_{n+1})$. This is easily obtained from the identity

$$\dot\psi(x_{n+1}) = \dot\psi(x_n) + \int_{x_n}^{x_n+\epsilon} \ddot\psi(x) \, dx$$

$$= \dot\psi(x_n) - \int_{x_n}^{x_n+\epsilon} f(x)\psi(x) \, dx,$$

whence

$$\dot\psi(x_{n+1}) = \dot\psi(x_n) - \epsilon f(x_n)\psi(x_n) + 0(\epsilon^2). \qquad (106)$$

This completes the formalism because, given $\psi(0)$ and $\dot\psi(0)$, we use (105) and (106) to compute $\psi(\epsilon)$ and $\dot\psi(\epsilon)$. The result permits us to calculate $\psi(2\epsilon)$ and $\dot\psi(2\epsilon)$, and so we march through the set of mesh points. Evidently the minimum error in $\psi$ is of order $\epsilon^3$ and in $\dot\psi$ of order $\epsilon^2$.

This scheme is simple enough to make the basic idea clear, but much too crude to be useful. The scheme commonly used, called the Runge–Kutta method, effectively uses Simpson's rule for the numerical integrations leading to equations (105) and (106).[22] The resulting expressions

---

[22] Equation (105) is equivalent to a trapezoidal rule of integration.

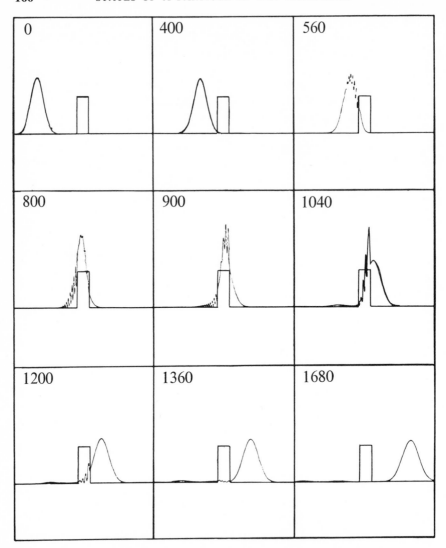

FIGURE 13. Gaussian wave-packet scattering from a square barrier. The average energy is twice the barrier height. Numbers denote the time of each configuration in arbitrary units.

are too complicated to write out, but they are very well-suited to machine calculation and they permit use of much larger values of the mesh size $\epsilon$ than do the simple expressions given above.

The actual construction of a numerical solution proceeds quite differently for the bound state and continuum state problems, and we now take up these two cases in order.

**Bound States.**[23] Here the unknown discrete energy eigenvalues must be determined, and this is usually done in the following way. Suppose the true energy eigenvalues in increasing order are $E_n$, $n = 0, 1, 2, \ldots$. The *mathematical* solution to Schrödinger's equation then has the property that if $E < E_0$, $\psi$ has no nodes, if $E_0 < E < E_1$, $\psi$ has one node, and so on. The eigenvalues are thus marked by the appearance (or disappearance) of a node at infinite values of $x$, and this in turn is signaled by a change in sign of the sense in which $\psi$ diverges as $E$ passes across an eigenvalue. Thus one simply guesses a value of $E$ and numerically integrates out from the origin until this solution begins to diverge. An increase in $E$ increases the curvature of $\psi$ and tends to increase the number of nodes, and conversely for a decrease in $E$. By appropriate adjustments in the assumed values of the energy, the eigenvalues can thus be readily bracketed. This behavior is illustrated in Figure 14 for the ground state of a symmetrical potential. In the figure, the numerical solutions are schematically shown for several assumed values of $E$. It is clear from the figure that the true ground state energy lies between $E_b$ and $E_c$, and is already quite closely bracketed by these two values.

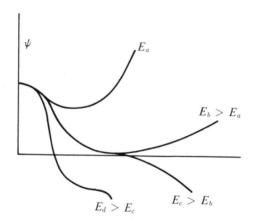

FIGURE 14. Schematic behavior of the solution of Schrödinger's equation for several trial values of $E$ in the neighborhood of the ground state energy $E_0$. The behavior of these solutions shows that $E_b < E_0 < E_c$.

**Continuum States.**[24] Here solutions are desired for some *given* value of $E$ in the continuum. These are obtained very simply by integrating outward from the origin and joining the numerical solution smoothly to the known pure de Broglie wave asymptotic state functions of equations

---

[23] Caswell, op. cit.

[24] Melkanoff, *et al.*, op. cit.

(91) and (92). The analysis is complicated considerably by the fact that the state function is complex and not real.

Finally, we remark on the numerical solution of the time-dependent equation.[25] Here one must introduce a mesh in the time coordinate as well as the space coordinate. The numerical problem is then the following. Given $\psi$ at every spatial mesh point at some time $t_n$, calculate $\psi$ at every spatial mesh point $t_{n+1}$, starting, of course, from the initial value of $\psi$ at $t = 0$, say. Note that here we must specify a complete *function* to define a solution in contrast to the time-independent case where only the two numbers $\psi(0)$ and $\dot\psi(0)$ were required. The problem is further complicated by the fact that special care must be taken to avoid numerical instabilities[26] and to ensure probability conservation. The resulting equations, which turn out to be second-order difference equations in the spatial coordinate and first-order in the time, are lengthy and complex, and we shall not bother to write them out.

---

**Problem 1.** Obtain the energy spectrum and eigenstates of a particle in a box from those for a particle in a square well potential of depth $V_0$ by passing to the limit in which $V_0 \to \infty$. Verify that your results are the same as those obtained directly in Chapter IV. Hint: Measure all energies from the bottom of the well, *before* going to the limit.

**Problem 2.** If $\psi_n$ is the $n$th harmonic oscillator eigenstate, evaluate:

    (a)  $\langle \psi_n | a^{\dagger s} | \psi_n \rangle$, $\langle \psi_n | a^s | \psi_n \rangle$

    (b)  $\langle \psi_n | x | \psi_n \rangle$, $\langle \psi_n | x^2 | \psi_n \rangle$, $\langle \psi_n | x^4 | \psi_n \rangle$

    (c)  $\langle \psi_n | p | \psi_n \rangle$, $\langle \psi_n | p^2 | \psi_n \rangle$, $\langle \psi_n | p^4 | \psi_n \rangle$

    (d)  $\langle \psi_m | a^{\dagger s} | \psi_n \rangle$, $\langle \psi_m | a^s | \psi_n \rangle$

    (e)  $\langle \psi_m | x | \psi_n \rangle$, $\langle \psi_m | x^2 | \psi_n \rangle$

    (f)  $\langle \psi_m | p | \psi_n \rangle$, $\langle \psi_m | p^2 | \psi_n \rangle$.

Hints: (1) Work in the creation space representation and use the known ortho-normality of the harmonic oscillator states. (2) Express $x$ in terms of $a$ and $a\dagger$, and similarly for $p$. Remarks: This problem is not hard if you know and understand what you are doing. By brute force methods, it's a mess!

**Problem 3.** A simple harmonic oscillator of frequency $\omega_0$ is in its ground state. The force constant is suddenly diminished at $t = 0$ to such a value that the oscillator frequency becomes (1) $\omega = \omega_0/2$; (2) $\omega = \omega_0/10$. For both cases:

---

[25] Goldberg, Schey and Schwartz, op. cit.

[26] A numerical scheme is called *unstable* if truncation and round-off errors accumulate in such a way that the numerical solution diverges uncontrollably from the true solution.

(a)  What is $\psi(x, t = 0)$ ?

(b)  Calculate $\psi(x, t)$. Sketch the behavior of $|\psi|^2$ over one period.

(c)  If the momentum of the particle is measured at some time $t > 0$, what is the probability that it will lie in the interval between $p$ and $p + dp$ ?

(d)  If, instead, the energy of the particle is measured, what is the probability that it will have the value $E_n = (n + 1/2)\hbar\omega$ ? (Use the integral representation of equation (61) to work this out.) For what value of $n$ will this probability be a maximum? Discuss your results briefly.

**Problem 4.**  A particle moves in a square well potential of width $2a$ and depth $V_0$. Consider the limit in which $2a$ tends to zero and $V_0$ to infinity in such a way that their product approaches the finite value $g$. By considering the square well solutions discussed in the text, show that such a potential has exactly one bound state and that

(a)  its binding energy is $\epsilon = g^2 m/2\hbar^2$.

(b)  its normalized eigenstate is $(2m\epsilon/\hbar^2)^{1/4} \, e^{-\sqrt{2m\epsilon}|x|/\hbar}$.

Next consider continuum states. From the solutions given in the text,

(c)  find the amplitude reflection and transmission coefficients and verify that probability is conserved.

(d)  Finally, noting that such a potential can be expressed as a delta function, $V(x) = -g\delta(x)$ [why?], see if you can deduce that the slope of $\psi$ must be discontinuous across such a potential. Specifically, show that $\psi$ is such that

$$\left.\frac{d\psi}{dx}\right|_{0_+} - \left.\frac{d\psi}{dx}\right|_{0_-} = -\frac{2mg}{\hbar^2}\,\psi(0).$$

Use this result to obtain the answers to parts (a), (b) and (c) directly.

**Problem 5.**  Consider a bouncing ball of mass $m$. Assume (1) that the motion is exactly vertical, (2) that collisions with the floor are perfectly elastic and (3) that the force of gravity is uniform.

(a)  Sketch the potential in which the ball moves. Write Schrödinger's equation and give the boundary conditions satisfied by $\psi$.

(b)  Show that the stationary states of the system can be expressed in terms of Bessel functions of order one-third.

(c)  Find, if you can, the transcendental equation which determines the allowed energies.

**Problem 6.**  A particle of mass $m$ moves in an attractive potential $V(x) = -V_0 \, e^{-|x|/L}$.

(a)  By making the substitution $z = e^{-x/2L}$ show that the stationary states can be expressed in terms of Bessel functions.

(b)  Noting that the stationary states can be classified according to their parity, find the transcendental equation which determines the energy of the bound states of the system.

(c)   Consider next continuum states, $E > 0$. Find expressions for the amplitude reflection and transmission coefficients. Verify that probability is conserved.

(d)   Find an expression for the time increment $\Delta t$ associated with passage of a wave packet through the potential. Compare your result with that to be expected classically.

**Problem 7.**  A particle moves in a square well potential of depth $V_0$ and width $2a$. Considering only continuum states in the limit $E \gg V_0$, and using the solutions obtained in the text:

(a)   Show that the amplitude transmission coefficient $\tau$ is given by

$$\tau(E) \simeq \exp[iV_0 \sqrt{m/2E}\, 2a/\hbar]$$

and calculate the time increment associated with the passage of a wave packet through the potential. Explain your results.

(b)   Show that for the reflected waves

$$\rho(E) \simeq \frac{V_0}{4E} \left[ \exp\left[2i(\sqrt{2mE} + 2V_0\sqrt{m/2E}\sqrt{a}/\hbar\right] - \exp[2i\sqrt{2mE}\,a/\hbar] \right].$$

(c)   Using the same rough method used in obtaining equations (100) and (101), show that there are two contributions to the reflected wave packet and calculate the time of arrival of each at $x = -x_1$, say. Explain your results.

**Problem 8.**  A particle moves in the uniform field of force $F$.

(a)   Write Schrödinger's equation in *momentum space*.

(b)   Find the stationary states $\phi_E(p)$.

(c)   Given an initial wave packet $\phi(p, t = 0)$, find $\phi(p, t)$. Do this by constructing the momentum space propagator $K(p', p; t)$.

(d)   Use your results to obtain an integral representation for $\psi(x, t)$, given $\psi(x, 0)$. See if you can verify that the wave packet accelerates in the expected way.

**Problem 9.**  A harmonic oscillator is in the state

$$\psi(x, t) = \frac{1}{\sqrt{2}} [\psi_0(x, t) + \psi_1(x, t)],$$

where $\psi_0$ and $\psi_1$ are the normalized ground and first excited harmonic oscillator states. Calculate $\langle E \rangle$, $\langle x \rangle$ and $\langle p \rangle$ and discuss the time dependence of each.

**Problem 10.**  Calculate the probability flux $j(x, t)$ for the state given in problem 9.

**Problem 11.**

(a)   Show that bound stationary state wave functions can always be chosen to be real functions with no loss of generality.

(b)   Show that the probability current $j(x, t)$ is zero for any bound stationary state.

**Problem 12.**  A 10 gm particle undergoes simple harmonic motion with frequency 2 cycles per second. If it is in its lowest state, what is the uncertainty in its position? in its momentum? Suppose it is set in motion with an amplitude of 10 cm. What is its energy? What is the order of magnitude of the quantum numbers appropriate to states of this energy?

**Problem 13.**  Show that a state function of definite parity in configuration space has the same parity in momentum space. What are the explicit properties of the parity operator in momentum space?

**Problem 14.**  A harmonic oscillator of mass $m$, charge $e$ and classical frequency $\omega$ is in its ground state in a uniform electric field. At time $t = 0$, the electric field is suddenly turned off.

(a)   Using the known properties of the propagator, find an exact, closed form expression for the state of the system at any time $t > 0$. Compare your result with that for a classical oscillator.

(b)   If a measurement is made at any time $t > 0$, what is the probability that the oscillator will be found in its $n$th state?

Hint: Introduce a shift of origin to find the exact initial state.

**Problem 15.**

(a)   A particle moves in a potential $V(x)$. The stationary states $\psi_E$ of the system have the following properties:

(i) The spectrum is discrete for $E < 0$, continuous for $E \geqslant 0$.
(ii) There are a denumberably infinite number of bound states.
(iii) For each bound state and for every integer $q$,

$$\langle \psi_E | x^{2q+1} | \psi_E \rangle = 0.$$

Sketch a simple smooth potential $V(x)$ consistent with all of these properties. *Briefly* explain your reasoning.

(b)   The same except that, instead of (ii), only a (small) finite number of bound states exist.

(c)   The same as (b), except that property (iii) is replaced by:

(iiia) The expectation value of $x$ is zero for the ground state but increases monotonically with excitation energy (bound states only).

(d)   Suppose that in case (b), the ground state energy of the system is $E_0 = -2\ eV$. Suppose also that $\langle \psi_{E_0} | x^2 | \psi_{E_0} \rangle = 4 \times 10^{-18}\ cm^2$. If the mass of the particle is $10^{-27}$ gm, estimate the strength of $V(x)$ in electron volts and its spatial extent in angstroms, that is, affix a rough numerical scale to your sketch for part (b). Hint: Use the uncertainty principle to make your estimate.

**Problem 16.** The Hamiltonian of a particle can be expressed in the form

$$H = \epsilon_1 \, a^+a + \epsilon_2 \, (a + a^+), \qquad (a, a^+) = 1,$$

where $\epsilon_1$ and $\epsilon_2$ are constants.

(a)  Find the energies of the stationary states. (You are not required to find the corresponding state functions.)

(b)  The same except that the commutator of $a$ and $a^+$ is

$$(a, a^+) = q^2,$$

where $q$ is a pure number.

Hint: Keeping the harmonic oscillator in mind, introduce new annihilation and creation operators $b$ and $b^+$ by writing

$$b = \alpha a + \beta, \qquad b^+ = \alpha a^+ + \beta$$

and choose the constants $\alpha$ and $\beta$ wisely.

**Problem 17.**  Let $\psi(x, t)$ denote an arbitrary, time dependent harmonic oscillator state function. Prove that

$$\langle x \rangle_t = \langle x \rangle_0 \cos\omega t + \frac{\langle p \rangle_0}{m\omega} \sin\omega t$$

$$\langle p \rangle_t = \langle p \rangle_0 \cos\omega t - m\omega \langle x \rangle_0 \sin\omega t,$$

in complete correspondence with the classical equations.

Do this in the following two ways:

(a)  By finding expressions for $d\langle p \rangle/dt$ and $d\langle x \rangle/dt$ and integrating the resulting coupled equations.

(b)  By direct evaluation of $\langle x \rangle_t$ and $\langle p \rangle_t$, using the propagator for the harmonic oscillator to express $\psi(x, t)$ in terms of $\psi(x, 0)$.

**Problem 18.**

(a)  Let $\psi(x, t)$ denote an arbitrary, time dependent harmonic oscillator state function. Prove that

$$\langle a \rangle_t = \langle a \rangle_0 \, e^{-i\omega t}, \qquad \langle a^\dagger \rangle_t = \langle a^\dagger \rangle_0 \, e^{i\omega t}.$$

Do this in the following three ways.

   i) By finding expressions for $d\langle a \rangle/dt$ and $d\langle a^\dagger \rangle/dt$ and integrating the resulting equations.

   ii) By direct evaluation of $\langle a \rangle_t$ and $\langle a^\dagger \rangle_t$ using the propagator, as given in closed form in equation (68), to express $\psi(x, t)$ in terms of $\psi(x, 0)$.

iii) By direct evaluation using equation (65) for the propagator.

(b)   Solve the classical oscillator problem using as classical dynamical variables $a$ and $a\dagger$ instead of $x$ and $p$.

**Problem 19.**   Using the numerical integration prescription of equations (105) and (106), integrate the differential equation

$$\frac{d^2\psi}{dx^2} + \psi = 0$$

from $x =$ zero to one for the following two cases:

(a)   $\psi(0) = 0$,     $\dot{\psi}(0) = 1$

(b)   $\psi(0) = 1$,     $\dot{\psi}(0) = 0$.

In both cases use a mesh size $\epsilon = 0.2$. Compare your results with the exact answers.

# VII

# *Approximation methods*

## 1. THE WKB APPROXIMATION

In the last chapter we studied the solutions of Schrödinger's equation for the harmonic oscillator and for a square well potential. These are two important examples of potentials for which exact solutions of the quantum mechanical equations can be found. While there are many such one-dimensional examples, more generally we must deal with potentials for which exact solutions cannot be found.[1] To enable us to handle this larger class of problems, we now take up a variety of approximation methods, each with its own domain of validity.

We first discuss a method for treating potentials which change very slowly in a de Broglie wavelength. For classical systems the wavelength approaches zero, so that this restriction is always satisfied for physically realizable potentials. This method thus gives essentially the classical limit and is therefore frequently called the *semiclassical approximation*. More often, however, it is called the *WKB approximation*, after Wentzel, Kramers and Brillouin, who first applied the method to quantum mechanical problems and are commonly credited by physicists with inventing it. Actually, the subject is an old one to mathematicians, under the heading of asymptotic solutions to differential equations, and dates back to Stokes, who originated the techniques in the middle of the 19th century. These were rediscovered by Wentzel, Kramers and Brillouin, and also, independently, by Jeffries. Therefore, the method is sometimes called the WKBJ approximation.

To proceed, consider a particle moving in a potential $V(x)$ so that Schrödinger's equation for the stationary states has the usual form,

---

[1] This is even more the case in three dimensions and in the study of systems of interacting particles. In both cases the class of exactly soluble problems is very limited indeed, as we shall see.

$$-\frac{\hbar^2}{2m}\frac{d^2\psi_E}{dx^2} + V(x)\psi_E(x) = E\psi_E(x). \qquad (1)$$

We now want to study the limit in which $\hbar$ can be regarded as tending to zero. However, we cannot do so in equation (1) as it stands, since $\psi_E$ oscillates so rapidly in this limit that the first term gives a finite contribution and not zero, as might at first be thought. To make this behavior explicit, we write $\psi_E(x)$ in the form

$$\psi_E(x) = A(x)\, e^{iS(x)/\hbar}, \qquad (2)$$

which is seen to be a generalized kind of de Broglie wave with amplitude $A(x)$ and phase $S(x)$, both of which are, as yet, unknown functions of position.[2] Upon differentiation twice, equation (2) becomes

$$\frac{d^2\psi_E}{dx^2} = \left[\frac{d^2A}{dx^2} + 2i\frac{dA}{dx}\frac{dS}{dx} + \frac{i}{\hbar}A\frac{d^2S}{dx^2} - \frac{A}{\hbar^2}\left(\frac{dS}{dx}\right)^2\right]e^{iS/\hbar},$$

whence we obtain, upon substituting into Schrödinger's equation, cancelling the common factor $e^{iS/\hbar}$ and regrouping the terms, the *exact* non-linear equation

$$A\left[\frac{1}{2m}\left(\frac{dS}{dx}\right)^2 + V(x) - E\right] - \frac{i\hbar}{2m}\left[2\frac{dA}{dx}\frac{dS}{dx} + A\frac{d^2S}{dx^2}\right] - \frac{\hbar^2}{2m}\frac{d^2A}{dx^2} = 0. \qquad (3)$$

Regarding $\hbar$ as a *parameter of smallness,* it is noted that the first term is of order unity, the second of order $\hbar$ and the last of order $\hbar^2$. We provisionally neglect the term in $\hbar^2$ and set the other two separately equal to zero, thus determining $S$ and $A$. From the first we have at once

$$\frac{dS}{dx} = \pm\sqrt{2m(E-V)} = \pm p(x)$$

or

$$S(x) = \pm \int p(x)\, dx. \qquad (4)$$

Using this result,[3] the second becomes

$$2\frac{dA}{dx}p(x) + A\frac{dp}{dx} = 0$$

or, multiplying through by $A$ and combining terms,

---

[2] This form may appear quite special, but no loss of generality is incurred by writing it. On the contrary, we have introduced *two* unknown functions to replace a single such function. We shall shortly take advantage of this redundancy by choosing $S$ and $A$ in a particularly convenient way.

[3] Equation (4) will be recognized by those familiar with the Hamilton–Jacobi formulation of classical mechanics (Reference [14]).

$$\frac{d}{dx}(A^2 p) = 0. \tag{5}$$

Note that, from equation (2), $\psi_E^* \psi_E = A^2$, for real $A$ and $S$. We thus see that this last equation is merely a statement of *conservation of probability*, since it is equivalent to $dj/dx = 0$, where the probability flux $j$ is given, in the approximation to which we are working, by

$$j = \rho v = \psi_E^* \psi_E \, p/m = A^2 p/m.$$

In any case, we have from equation (5)

$$A = c/\sqrt{p},$$

where $c$ is an arbitrary constant. Our approximate solution thus has the form

$$\psi_E^\pm(x) = \frac{c_\pm}{\sqrt{p}} \, e^{\pm i \int p \, dx/\hbar},$$

where

$$p = \sqrt{2m[E - V(x)]}.$$

Note that the answer is expressed in terms of an indefinite integral. However, it can easily be re-expressed in terms of a definite integral by referring $\psi_E(x)$ to its value at some given point $x_0$. Doing so, we obtain at once

$$\psi_E^\pm(x) = \psi_E^\pm(x_0) \sqrt{\frac{p(x_0)}{p(x)}} \exp\left[\pm i \int_{x_0}^x p \, dx/\hbar\right]. \tag{6}$$

The structure of the WKB result is easy to understand. We have already seen that the amplitude function is determined by the requirement that probability be conserved. The phase factor is interpreted in the following way. Consider the change in phase $\delta\varphi$ of a de Broglie wave of wavelength $\lambda$ when it advances a distance $\delta x$. Evidently

$$\delta\varphi = \frac{2\pi}{\lambda} \delta x.$$

If the wavelength is not constant, but changes with position, then the accumulated phase shift $\varphi$ for a finite advance is

$$\varphi = \int \frac{2\pi}{\lambda} dx = \int p \, dx/\hbar,$$

in agreement with our result. Now this argument is based on the notion of a position dependent wavelength. But the very concept of wavelength loses its meaning if the wavelength is not practically constant over a distance of the order of the reduced wavelength in size. Hence, roughly speaking, we expect our approximation to be valid only if the wavelength changes by a sufficiently small fraction of itself over such a distance. That is, we expect that $\delta\lambda/\lambda \ll 1$, where $\delta\lambda = (d\lambda/dx)\lambda$, or that, putting all this together,

$$\left|\frac{d\lambda}{dx}\right| \ll 1 \tag{7}$$

for the WKB approximation to hold. Expressing $\lambda$ in terms of $p$, this is equivalent to

$$\left|\frac{\hbar}{p^2}\frac{dp}{dx}\right| = \left|\frac{1}{p}\frac{dp}{dx}\lambda\right| \ll 1 \tag{8}$$

or, also, expressing $p$ in terms of $E$ and $V$,

$$\frac{\hbar}{2\sqrt{2m}}\left|\frac{1}{(E-V)^{3/2}}\frac{dV}{dx}\right| \ll 1. \tag{9}$$

Equation (8) states that the fractional change in the momentum in a wavelength must be small, while equation (9) states that, in a wavelength, the change in the potential energy, relative to the kinetic energy, must be small. An alternative way to put these conditions is that the wavelength must be small compared to the distance over which the momentum changes appreciably or over which the potential changes appreciably. Of course, equations (7), (8) and (9) are completely equivalent to each other and are merely different ways of stating the same condition.

Even though we have been guided by classical arguments in obtaining the WKB approximation, we note from equation (9) that, as a kind of bonus, the approximation is also valid in *classically forbidden* regions of *negative* kinetic energy, provided only that this negative kinetic energy is sufficiently large and that it changes slowly enough. On the other hand, we observe that in the neighborhood of a *classical turning point* where $E = V$, $p = 0$ and $\lambda = \infty$, the approximation breaks down completely. [We shall have more to say on both these points later.]

It still remains for us to give a more precise statement of the condition for the validity of the WKB approximation than that deduced by the rough arguments used above. To do so, we return to the exact equation (3), which was our starting point, and take into account the neglected term[4] in $\hbar^2$. With this term included, the phase function $S$ is determined

---

[4] We shall treat the $\hbar^2$ term as a correction term in the expression for $S$. However, it could equally well have been treated as a correction to $A$. We have chosen the former because it simplifies the subsequent analysis, but the final result is independent of which choice is made.

by the exact equation

$$A\left[\frac{1}{2m}\left(\frac{dS}{dx}\right)^2 + V(x) - E\right] - \frac{\hbar^2}{2m}\frac{d^2A}{dx^2} = 0$$

or

$$\frac{dS}{dx} = \pm\sqrt{p^2(x) + \frac{\hbar^2}{A}\frac{d^2A}{dx^2}}.$$

To first approximation, the effect of the small term in $\hbar^2$ is obtained by expanding the square root to give

$$\frac{dS}{dx} = \pm\left[p(x) + \frac{\hbar^2}{2pA}\frac{d^2A}{dx^2} + \cdots\right],$$

while, to this order, $A$ may still be taken as proportional to $p^{-1/2}$. Thus we have

$$\frac{dS}{dx} \simeq \pm\left[p + \tfrac{1}{2}\hbar^2 p^{-1/2}d^2(p^{-1/2})/dx^2\right]$$

or, integrating,

$$S = \pm\left(\int_{x^0}^{x} p\,dx + \Delta S\right),$$

where

$$\Delta S \simeq \frac{\hbar^2}{2}\int_{x_0}^{x} dx\, p^{-1/2}\,(d^2p^{-1/2}/dx^2).$$

With this choice for the limits of integration, we then have in place of equation (6),

$$\psi_E^{\pm}(x) = \psi_E^{\pm}(x_0)\sqrt{\frac{p(x_0)}{p(x)}}\exp\left[\pm i\left[\int_{x_0}^{x} p\,dx/\hbar + \Delta S/\hbar\right]\right].$$

Consequently, we conclude that the WKB approximation will be valid provided that

$$|\Delta S/\hbar| = \left|\frac{\hbar}{2}\int_{x_0}^{x} dx\, p^{-1/2}\,(d^2p^{-1/2}/dx^2)\right| \ll 1, \tag{10}$$

which is the precise condition we seek. Note that this condition involves more than the local behavior of $p(x)$; it takes into account the *accumulated* error in the phase function $S$ over the entire interval from the reference point $x_0$ to the point $x$ at which the state function is being calculated.

Unfortunately, equation (10) is somewhat complicated and difficult to interpret. We shall therefore give a rough, but adequate, upper bound to $\Delta S/\hbar$ which is much easier to understand. We have by inspection

$$\left|\frac{\Delta S}{\hbar}\right| \leq \frac{\hbar}{2\sqrt{p_{\min}}} \int_{x_0}^{x} dx \left|d^2 p^{-1/2}/dx^2\right|,$$

where $p_{\min}$ is the minimum value of $p$ over the domain of integration. For simplicity, suppose first that $d^2 p^{-1/2}/dx^2$ is monotonic over the interval from $x_0$ to $x$. The integration can then be performed at once, and we obtain

$$\left|\frac{\Delta S}{\hbar}\right| \leq \frac{\hbar}{2\sqrt{p_{\min}}} \left|\left(\frac{dp^{-1/2}}{dx}\right)_x - \left(\frac{dp^{-1/2}}{dx}\right)_{x_0}\right|.$$

Since, evidently,

$$\left|\left(\frac{dp^{-1/2}}{dx}\right)_x - \left(\frac{dp^{-1/2}}{dx}\right)_{x_0}\right| \leq 2\left|\frac{dp^{-1/2}}{dx}\right|_{\max},$$

where $\left|dp^{-1/2}/dx\right|_{\max}$ is the largest value attained by $\left|dp^{-1/2}/dx\right|$ over the interval from $x_0$ to $x$, we thus find

$$\left|\frac{\Delta S}{\hbar}\right| \leq \frac{\hbar}{\sqrt{p_{\min}}} \left|\frac{dp^{-1/2}}{dx}\right|_{\max}.$$

Suppose now that $(d^2 p^{-1/2}/dx^2)$ is not actually monotonic but that it changes sign $q$ times over the integration interval. The contribution to $|\Delta S/\hbar|$ over each subinterval in which $d^2 p^{-1/2}/dx^2$ is monotonic is bounded by twice the maximum value attained by $\left|dp^{-1/2}/dx\right|$ and hence, since there are $q + 1$ such subintervals,

$$\left|\frac{\Delta S}{\hbar}\right| \leq \frac{(q+1)\hbar}{\sqrt{p_{\min}}} \left|\frac{dp^{-1/2}}{dx}\right|_{\max},$$

where $\left|dp^{-1/2}/dx\right|_{\max}$ still denotes the largest value of $\left|dp^{-1/2}/dx\right|$ over the entire domain of integration. Since

$$\left|\frac{dp^{-1/2}}{dx}\right|_{\max} = \frac{1}{2}\left|\frac{1}{p^{3/2}}\frac{dp}{dx}\right|_{\max} \leq \frac{1}{2}\frac{1}{p_{\min}^{3/2}}\left|\frac{dp}{dx}\right|_{\max},$$

we have, finally,

$$\left|\frac{\Delta S}{\hbar}\right| \leq \frac{(q+1)\hbar}{2}\frac{1}{p_{\min}^2}\left|\frac{dp}{dx}\right|_{\max}$$

and, therefore, the WKB approximation is guaranteed to be valid, provided that

$$\frac{(q+1)\hbar}{2}\frac{1}{p_{\min}^2}\left|\frac{dp}{dx}\right|_{\max} \ll 1. \tag{8a}$$

Up to numerical factors of order unity, this is the same as the rough condition, equation (8), except that the minimum value of $p$ and the maximum value of $(dp/dx)$ over the relevant interval must be used. Note that if the interval is small enough, this distinction becomes unimportant and equation (8) is thus essentially the correct *local* condition as it stands.

We now give some examples of the application of the WKB approximation. We first discuss continuum states and then the more interesting case of discrete states. In the latter, we obtain essentially the Bohr quantization rules, as we shall see.

In our treatment of continuum states we shall restrict our attention to the case in which the potential vanishes at infinity at least as rapidly as $x^{-1}$. We shall also assume that the particle has energy $E$ which everywhere exceeds the potential energy so that there are no classical turning points. In that case, the two independent solutions of Schrödinger's equation are given by equation (6), and correspond to a particle traveling either toward the right or toward the left, *without reflection*. This latter feature is a consequence of the assumption that the potential changes very slowly. In general, the more slowly the potential changes, the smaller is the reflection coefficient, and in the WKB limit the reflection becomes negligibly small, in agreement with the classical behavior. The amplitude transmission coefficient $\tau$ then has magnitude unity, of course, but it contains a phase factor $\delta$ which is related to the time increment associated with the passage of a particle through the potential. We now obtain an explicit expression for $\tau$ and for this time increment. Considering a particle moving to the right, we have, from equation (6),

$$\psi_E(x) = \psi_E(x_0) \sqrt{\frac{p(x_0)}{p(x)}} \exp\left[i \int_{x_0}^{x} p\, dx/\hbar\right].$$

We now fix the reference point $x_0$ by taking it to be infinitely far to the left; this choice permits us to prescribe the form of the incident wave at once. In view of the assumed behavior of the potential at infinity, $V(x_0)$ is negligible for $x_0$ sufficiently remote and the momentum assumes the constant value

$$p(x_0) = \sqrt{2mE}.$$

The state function in this region is thus a pure de Broglie wave and we write

$$\psi_E(x_0) = f(E)\, e^{i\sqrt{2mE}\, x_0/\hbar},$$

where $f(E)$ is the (arbitrary) amplitude of the incident wave. Our expression for $\psi_E(x)$ can now be written in the form

$$\psi_E(x) = \lim_{x_0 \to -\infty} \left[ f(E)\left(\frac{\sqrt{2mE}}{p}\right)^{1/2} \exp\left[i\left(\int_{x_0}^{x} p\, dx + \sqrt{2mE}\, x_0\right)/\hbar\right]\right].$$

In order to pass to the limit, observe that

$$\sqrt{2mE} \ x_0 + \int_{x_0}^{x} p \ dx = \sqrt{2mE} \ x + \int_{x_0}^{x} (p - \sqrt{2mE}) \ dx.$$

Because $(p - \sqrt{2mE})$ vanishes more rapidly than $x_0^{-1}$, the limit can now be taken, and we thus obtain as our WKB state function

$$\psi_E(x) = f(E) \left(\frac{\sqrt{2mE}}{p}\right)^{1/2} \exp\left[i\left(\int_{-\infty}^{x} (p - \sqrt{2mE}) \ dx + \sqrt{2mE} \ x\right)/\hbar\right].$$

Finally, letting $x$ approach infinity on the right, we have

$$\psi_E(x) = f(E) \exp\left[i \int_{-\infty}^{\infty} (p - \sqrt{2mE}) \ dx/\hbar\right] e^{i\sqrt{2mE} \ x/\hbar}$$

$$= \tau f(E) \ e^{i\sqrt{2mE} \ x/\hbar},$$

where, as anticipated, the transmission coefficient $\tau$ has the form

$$\tau = e^{i\delta(E)}$$

with

$$\delta(E) = \int_{-\infty}^{\infty} (p - \sqrt{2mE}) \ dx/\hbar = \frac{\sqrt{2m}}{\hbar} \int_{-\infty}^{\infty} (\sqrt{E - V} - \sqrt{E}) \ dx.$$

Recall now that, according to equation (VI-101), the time increment $\Delta t$ associated with the passage of a particle across a potential is given by

$$\Delta t = \hbar \ d\delta/dE$$

whence, in WKB approximation, we obtain, upon performing the indicated differentiation,

$$\Delta t = \frac{1}{2} \sqrt{2m} \int_{-\infty}^{\infty} \left(\frac{1}{\sqrt{E - V}} - \frac{1}{\sqrt{E}}\right) dx$$

$$= \int_{-\infty}^{\infty} \left(\frac{1}{v(x)} - \frac{1}{v_0}\right) dx,$$

where $v(x) = \sqrt{2(E - V)/m}$ is the classical velocity of a particle of total energy $E$ in a potential $V(x)$ and where $v_0 = \sqrt{2E/m}$ is the velocity of a free particle of the same energy. We thus see that we obtain exactly the classical result for the difference in arrival time of a free particle of energy $E$ and a particle of the same energy which has traversed a potential $V$.

We next consider the bound states of a particle in a potential $V(x)$. Suppose that for given energy $E$, the classical turning points occur at $x_1$ and $x_2$, where for definiteness we take $x_1 < x_2$. In the interior of the potential, but not too close to the turning points, the WKB approximation to the solutions can be written as a linear combination of waves traveling to left and right. It is convenient to choose this linear combination to have the form

$$\psi_E(x) = \frac{c}{\sqrt{p}} \sin\left[\left(\int p(x)\,dx/\hbar\right) + \delta\right].\tag{11}$$

Since $\psi_E$ contains two arbitrary constants $c$ and $\delta$, this is still a perfectly general expression. Now, to the right of $x_2$, and not too close to it, the solution can also be treated in WKB approximation and is an exponentially damped function; similarly, to the left of $x_1$. None of these WKB solutions, however, are valid near the turning points, and the basic mathematical problem is to join together, or to connect, solutions on one side of a turning point with those on the other. This problem can be solved in a variety of ways and a general answer can be given. That answer is easy to understand, but the details of the derivation are complicated and we omit them.[5] To understand the answer, we first point out that, if the classically forbidden region were also forbidden quantum mechanically, the wave function would have to vanish at each turning point, and this could be accomplished by exactly fitting a half-integral number of de Broglie waves between the turning points, that is, by making the sinusoidal function of equation (11) vanish at each turning point. However, since the true wave function actually extends into the forbidden region a little, one finds instead that slightly less than an integral number of half-wavelengths must be fitted between the turning points. Specifically, at each extreme, it turns out that the wave function behaves *as if* it vanished exactly 1/8 of a wavelength into the forbidden region. Thus, the condition for a bound state is that there be an integral number of half-wavelengths diminished by 1/4 of a wavelength, or by 1/2 of a half-wavelength. Since the number of half-wavelengths in the interval between $x_1$ and $x_2$ is

$$\int_{x_1}^{x_2} \frac{dx}{\lambda/2} = \frac{2}{h}\int_{x_1}^{x_2} p\,dx,$$

we thus have, according to our rule,

$$m - \frac{1}{2} = \frac{2}{h}\int_{x_1}^{x_2} p\,dx, \qquad m = 1, 2, 3, \ldots$$

or equally well, replacing $m$ by $n + 1$,

[5] The standard procedure is the following: Suppose the WKB solution to be valid except in some interval of length $L$ about the turning point. If the potential changes slowly enough, it can be treated as linear over this interval and Schrödinger's equation can then be solved exactly in terms of Bessel functions of order one-third. This solution can then be joined smoothly to the WKB form at the extremities of the interval. For a detailed derivation see Reference [19].

$$\oint p \, dx = 2 \int_{x_1}^{x_2} p \, dx = \left(n + \frac{1}{2}\right)h, \qquad n = 0, 1, 2, \ldots, \qquad (12)$$

where the integral on the left is the conventional integral over one period of the classical motion and is thus twice the integral between turning points, as indicated.[6]

This result is recognized as a modified version of the Bohr quantization condition, modified by the inclusion of the zero point energy, which arises from the extra $h/2$ on the right. It is an obvious improvement on the Bohr rule, and in addition some estimate of its domain of applicability can be given. Specifically, since the WKB approximation requires that the wavelength be short compared to the distance over which the potential changes appreciably, equation (12) is seen to be increasingly applicable as the quantum number $n$ increases and the wavelength correspondingly decreases.

In applying equation (12), we have, of course, that

$$p = \sqrt{2m(E - V)} = p(E, x),$$

while the turning points also depend on the unknown energy $E$. The left-hand side is thus a function of $E$ and the solutions of equation (12) then give the allowed energies in WKB approximation.

As an example, we briefly work out the energy eigenvalues for the harmonic oscillator, where equation (12) *happens* to give the exact answers for *all* states. We have

$$p = \sqrt{2m(E - m\omega^2 x^2/2)},$$

and the turning points are seen to occur at

$$x = \pm\sqrt{2E/m\omega^2}.$$

Thus equation (12) becomes

$$2 \int_{-\sqrt{2E/m\omega^2}}^{\sqrt{2E/m\omega^2}} \sqrt{2m(E - m\omega^2 x^2/2)} \, dx = \left(n + \frac{1}{2}\right)h.$$

Introducing a new variable $\theta$ by writing

$$x = \sqrt{2E/m\omega^2} \cos\theta$$

we obtain

$$\left(n + \frac{1}{2}\right)h = \frac{4E}{\omega} \int_0^\pi \sin^2\theta \, d\theta = 2\pi E/\omega,$$

whence

[6] Note that equation (12) is expressed in terms of $h$ and not $\hbar$.

$$E = \left( n + \frac{1}{2} \right) \hbar \omega,$$

in fortunate agreement with the correct result for all $n$. Although the WKB approximation gives the correct *energy*, it does not, of course, yield the correct wave function. For a comparison between exact and WKB harmonic oscillator state functions, see Problem VII-17.

## 2. THE RAYLEIGH–RITZ APPROXIMATION

There is a distinct and significant difference in the problem of obtaining approximate solutions to Schrödinger's equation for bound states as compared to continuum states. In the latter case, the problem is finding the stationary state functions for some preassigned energy in the continuum. In the former, the discrete, allowed energies must also be determined. We now present a variational method which optimizes the approximate determination of the allowed energies of discrete states, at least for the lowest few of these states. The technique involved was originated by Rayleigh and by Ritz toward the end of the 19th century in the analysis of classical boundary value problems.

Consider the time independent Schrödinger's equation

$$H\psi_E = E\psi_E \tag{13}$$

for a system described by some Hamiltonian $H$. Now multiply through by $\psi_E{}^*$ and integrate over all space. The result can be written in the form

$$E = \frac{\int \psi_E{}^* H\psi_E \, dx}{\int \psi_E{}^* \psi_E \, dx}, \tag{14}$$

where it is convenient to assume that $\psi_E$ is not necessarily normalized. For the correct eigenfunctions of $H$, this equation is just an identity, of course. Our problem, for the bound states under consideration, is that neither $E$ nor $\psi_E$ is known. Suppose, however, we consider some approximation to $\psi_E$, say $\psi$. Replacement of $\psi_E$ by $\psi$ in equation (14) then gives some approximation to $E$, say $E'$. In other words, $E'$ is *defined*, for any physically admissible $\psi$ and $\psi^*$, by

$$E' \equiv \frac{\int \psi^* H\psi \, dx}{\int \psi^* \psi \, dx}. \tag{15}$$

Evidently, $E'$ can be interpreted as the expectation value of the Hamiltonian for the approximate state $\psi$.

We now show that the error in $E'$ is of second order in the errors in the state functions or, equivalently, that *the energy is stationary with respect to arbitrary, independent variations of $\psi$ and $\psi^*$ about their* *true values.* To carry this out we need to define a class of arbitrary functions $\psi$ which includes the true functions. A general way to do this is simply to introduce the one-parameter family

$$\psi = \psi_E + \alpha\phi,$$

which merely expresses the deviation of $\psi$ from the true value $\psi_E$ as $\alpha\phi$. The parameter $\alpha$ then serves as a convenient measure of this deviation. Similarly, we write

$$\psi^* = \psi_E^* + \beta\chi,$$

where normally, of course, $\beta = \alpha^*$ and $\chi = \phi^*$. However, it turns out, and we might as well make it explicit, that the stationary properties hold even if $\psi^*$ is *not* the complex conjugate of $\psi$, but is an *entirely* independently chosen function. In any case, equation (15) can now be rewritten, after clearing of fractions, in the form

$$E'(\alpha, \beta) \int (\psi_E^* + \beta\chi)(\psi_E + \alpha\phi) \ dx$$

$$= \int (\psi_E^* + \beta\chi) H(\psi_E + \alpha\phi) \ dx, \quad (16)$$

where, of course, $E'(\alpha = 0, \beta = 0) = E$. Differentiating with respect to $\beta$, and then setting $\alpha = \beta = 0$, we obtain

$$\frac{\partial E'}{\partial \beta}\bigg|_{\alpha,\beta=0} \int \psi_E^*\psi_E \ dx + E \int \chi\psi_E \ dx = \int \chi H\psi_E \ dx$$

or

$$\frac{\partial E'}{\partial \beta}\bigg|_{\alpha,\beta=0} = \frac{\int \chi(H\psi_E - E\psi_E) \ dx}{\int \psi_E^*\psi_E \ dx}.$$

Hence, since $\psi_E$ satisfies equation (13) by definition, we have finally

$$\frac{\partial E'}{\partial \beta}\bigg|_{\alpha,\beta=0} = 0. \quad (17)$$

Similarly, using the Hermitian character of $H$, we find that

$$\frac{\partial E'}{\partial \alpha}\bigg|_{\alpha,\beta=0} = 0. \quad (18)$$

Equations (17) and (18) together mean that $E'$ has the form $E'(\alpha, \beta) = E +$ terms quadratic and higher in $\alpha$ and $\beta$ and thus that, as asserted, the errors in the approximate energy are of second order in the errors in the approximate wave function when the energy is computed from equation (15).

---

**Exercise 1.**   Derive equation (18).

---

Conversely, if we start from equation (15) and demand that $\psi$ and $\psi^*$ be such that $E'$ is stationary with respect to independent variations in these functions, it can be shown that $\psi$ and $\psi^*$ must be eigenfunctions of $H$. In other words, equation (15), regarded as a stationary expression, is entirely equivalent to Schrödinger's equation or, in the language of the calculus of variations, Schrödinger's equation is the Euler's equation of the variational problem.

How can this be used as an approximation method? The simplest procedure is simply to introduce some physically reasonable $\psi$ as a trial function. The calculation of $E'$ then gives an optimum estimate of the correct energy $E$. To proceed more systematically, introduce a trial function which explicitly depends upon some set of free parameters. Then calculate $E'$ as a function of these parameters using equation (15). Finally, make $E'$ stationary by differentiating with respect to each parameter in turn and setting the result of each differentiation equal to zero. The solution of the resulting set of simultaneous equations in the unknown parameters then yields an approximate value of the energy. This value is the best that can be attained with a trial function of the initially chosen form. In a moment we shall give an illustrative example.

The actual utility of this scheme in practice is greatly enhanced by the fact that the *ground state energy* calculated in this way provides an *upper bound* to the true ground state energy. Thus, while equation (15) provides a variational principle for *any* eigenstate of the Hamiltonian, it provides a *minimum principle* for the ground state. The proof is simple. Any trial function $\psi$ can be expressed as a superposition of the complete set of eigenstates of $H$. Thus we can write

$$\psi = \sum_E c_E \psi_E,$$

and hence, upon substitution into equation (15),

$$E' = \frac{\sum E |c_E|^2}{\sum |c_E|^2}.$$

If $E_0$ denotes the ground state energy, then for each term in the summation $E \geq E_0$, and hence

$$E' \geq E_0.$$

Furthermore, the equality holds only for $\psi = \psi_{E_0}$, assuming the ground state to be nondegenerate.

Obviously, the minimum principle does not hold for higher states unless the trial function is *orthogonal to all the exact lower-lying states*. To see this, consider the $n$th state with exact energy $E_n$. Suppose $\psi$ is orthogonal to all $\psi_E$ for $E < E_n$. Then, if $\psi$ is expressed as a superposition of the exact states, no terms for $E < E_n$ enter by hypothesis. Hence, we have

$$\psi = \sum_{E \geq E_n} c_E \psi_E$$

and, as claimed,

$$E_n{}' = \frac{\sum\limits_{E \geq E_n} E|c_E|^2}{\sum\limits_{E \geq E_n} |c_E{}^2|} \geq E_n .$$

This condition of orthogonality is obviously difficult to achieve, in general, because it requires knowledge of all the exact state functions for $E < E_n$. Fortunately, symmetry properties can often be used in practice to ensure such exact orthogonality for some low-lying states. For example, in a symmetrical potential even states and odd states are automatically orthogonal to each other. Hence the lowest even state and lowest odd state each satisfy a minimum principle. More generally, the best one can do is to make the trial function orthogonal to the approximately known lower states, but the method quickly loses reliability as the quantum number of the state increases. Since the WKB method works best for large quantum numbers, the variational and WKB methods are thus seen to complement each other to some degree.

The variational method requires that some specific function or other be introduced as a trial function, and we have suggested that this function should be physically reasonable. How should one actually construct or choose such a function, and what do we mean by "physically reasonable"? It is difficult to give a precise answer, but perhaps the following remarks will help. Suppose, first, that we are considering the ground state of a particle in a symmetrical potential. Ground state wave functions have no nodes and, like all bound state functions, they vanish rapidly at large distances. The simplest trial function is thus a smooth function centered at the origin, because it must be symmetrical, which falls off over some characteristic distance. This latter can be treated as a parameter to be determined variationally. A specific and frequently used example is a Gaussian,

$$\psi = e^{-x^2/a^2} .$$

Other examples, which have similar properties, are csch $x/a$, $(x^2 + a^2)^{-m}$, and so on. More general functions can be constructed by multiplying any of the above by polynomials, with coefficients which are to be determined variationally. Unless one has access to a computing machine, one usually seeks functions for which the requisite integrals can be readily performed, provided that these trial functions have the general features indicated by the above examples.

Considering excited states, one again builds into the trial function the correct (and known) number of nodes, the correct symmetry and the expected overall behavior. For the first excited state in a symmetrical potential, which is odd and thus has a node at the origin, any of the above mentioned functions multiplied by $x$ would be a suitable trial function. For the second excited state, which is even and has two nodes, any of the above functions could be multiplied by the polynomial $(b^2 - x^2)$. The location of the nodes at $x = \pm b$ could then be determined variationally by treating $b$ as a variational parameter.

In summary, then, one constructs trial functions which are reasonably simple to work with and which contain all of the known specific and general features of the exact functions, or at least as many as possible.

As an example, we now use the variational method to estimate the ground state energy of the harmonic oscillator. Use of a Gaussian yields the exact result, of course, because the exact ground state is Gaussian. We shall thus use a polynomial over the region $|x| \leq a$ which has the right general shape and which smoothly vanishes at the extremities. For $|x| > a$, we shall take the trial function to be zero and, of course, we shall determine $a$ variationally. Specifically, we choose the function

$$
\begin{aligned}
\psi &= (a^2 - x^2)^2, &\quad |x| &\leq a, \\
&= 0, &\quad |x| &\geq a.
\end{aligned}
\tag{19}
$$

Recalling that the harmonic oscillator Hamiltonian is

$$
H = \frac{p^2}{2m} + \frac{1}{2} m\omega^2 x^2,
$$

equation (15) becomes

$$
E'(a^2) = \frac{-\dfrac{\hbar^2}{2m} \displaystyle\int \psi^* \frac{d^2\psi}{dx^2} \, dx + \frac{1}{2} m\omega^2 \displaystyle\int x^2 \psi^* \psi \, dx}{\displaystyle\int \psi^* \psi \, dx}.
$$

We now evaluate the various integrals involved. Considering first the denominator, we have

$$
\int_{-a}^{a} \psi^* \psi \, dx = 2 \int_{0}^{a} (x^2 - a^2)^4 \, dx = 2a^9 \left( \frac{1}{9} - \frac{4}{7} + \frac{6}{5} - \frac{4}{3} + 1 \right) = \frac{256 a^9}{9 \cdot 7 \cdot 5}.
$$

In the same way, the potential energy term gives

$$\int_{-a}^{a} x^2 \psi^* \psi \, dx = \frac{256a^{11}}{11 \cdot 9 \cdot 7 \cdot 5}.$$

The kinetic energy term is most easily evaluated as follows:

$$\int_{-a}^{a} \psi^* \frac{d^2\psi}{dx^2} = \int_{-a}^{a} \frac{d}{dx}\left(\psi^* \frac{d\psi}{dx}\right) dx - \int_{-a}^{a} \frac{d\psi^*}{dx} \frac{d\psi}{dx} \, dx$$

or, since the first term vanishes,

$$\int_{-a}^{a} \psi^* \frac{d^2\psi}{dx^2} = -2 \int_{0}^{a} \frac{d\psi^*}{dx} \frac{d\psi}{dx} dx = -2 \int_{0}^{a} [4x(x^2 - a^2)]^2 \, dx = -\frac{256a^7}{7 \cdot 5 \cdot 3}.$$

Putting these results together, we find

$$E'(a^2) = \frac{\dfrac{\hbar^2}{2m} \dfrac{256a^7}{7 \cdot 5 \cdot 3} + \dfrac{1}{2} m\omega^2 \dfrac{256a^{11}}{11 \cdot 9 \cdot 7 \cdot 5}}{256a^9/9 \cdot 7 \cdot 5} = \frac{3}{2} \frac{\hbar^2}{ma^2} + \frac{1}{22} m\omega^2 a^2.$$

Note the standard competition between the kinetic and potential energy terms which we referred to in our earlier qualitative discussion of the characteristics of bound states, using the uncertainty principle as a guide. The smaller $a^2$ is, the lower the potential energy, but the larger the kinetic energy because of the increased localization of the wave function. The variational procedure makes this competition quite explicit in the present instance. In any case, we obtain, upon differentiating $E'$ with respect to the variational parameter $a^2$ and setting the result equal to zero,

$$\frac{dE'}{da^2} = 0 = -\frac{3}{2} \frac{\hbar^2}{ma^4} + \frac{1}{22} m\omega^2,$$

whence

$$a^2 = \sqrt{33} \, \hbar/m\omega$$

and

$$E' = \frac{1}{2} \hbar\omega \sqrt{12/11},$$

in error by only 5 percent and larger than the true ground state energy, as it must be.

We leave further examples to the problems.

## 3. STATIONARY STATE PERTURBATION THEORY

We now come to the most important method for obtaining approximate solutions to Schrödinger's equation, a method based on the theory of perturbations. This method is applicable to any situation in which the Hamiltonian $H$ describing the system under consideration is not too different from the Hamiltonian $H_0$, say, describing some similar but simpler system, simple enough that the full set of eigenfunctions and eigenvalues of $H_0$ are known exactly. Such circumstances may seem too special to be of consequence, but in fact they occur very commonly, as later applications make clear, and it is this fact which gives the method its practical importance. In the present section, we discuss the simplest case, namely that in which the states are *nondegenerate, stationary* and *discrete*.

We start by expressing the perturbed Hamiltonian $H$ in terms of the unperturbed Hamiltonian $H_0$ by writing, in complete generality,

$$H = H_0 + \lambda H'. \tag{20}$$

The quantity $H'$ is called the *perturbation term* in the Hamiltonian. The dimensionless parameter $\lambda$, which is redundant, may be taken to have the value unity for the actual physical problem. It is introduced for convenience as a visible parameter of smallness and also to permit us to gradually pass from the physical problem to the unperturbed problem, in a well-defined way, simply by letting $\lambda$ tend to zero.

We seek the eigenfunctions and eigenvalues of $H$, which we define by

$$E_n\psi_n = H\psi_n = (H_0 + \lambda H')\,\psi_n. \tag{21}$$

The known, unperturbed eigenfunctions and eigenvalues are similarly defined by

$$H_0\phi_n = \mathscr{E}_n\phi_n. \tag{22}$$

We assume that the $\phi_n$ and $\psi_n$ are both nondegenerate. We now make the fundamental assumption that, as $\lambda$ approaches zero, the set of energies $E_n$ approaches the set $\mathscr{E}_n$. In view of the nondegeneracy of the states, each $\psi_n$ thus approaches some particular $\phi_n$ and we now make this one-to-one correspondence explicit by labeling the states in such a way that, for each $n$,

$$\lim_{\lambda \to 0} E_n = \mathscr{E}_n, \qquad \lim_{\lambda \to 0} \psi_n = \phi_n. \tag{23}$$

We next express the unknown perturbed states $\psi_n$ as a superposition of the complete set of unperturbed states. Thus we write

$$\psi_n = \sum_i c_{in}\,\phi_i, \tag{24}$$

and we note for future reference that, according to equation (23),

$$\lim_{\lambda \to 0} c_{in} = \delta_{in}. \tag{25}$$

Now substitute the superposition equation (24) into equation (21), giving

$$E_n \sum_i c_{in}\phi_i = (H_0 + \lambda H') \sum_i c_{in}\phi_i = \sum_i c_{in}\mathscr{E}_i\phi_i + \lambda \sum_i c_{in} H'\phi_i,$$

and then multiply through by $\phi_j{}^*$ and integrate. Using the orthonormality of the $\phi_n$, we obtain, *for each value of* $j$,

$$c_{jn}(E_n - \mathscr{E}_j) = \lambda \sum_i H'_{ji} c_{in}, \tag{26}$$

where the numbers $H'_{ji}$, called the *matrix elements* of $H'$, are defined by

$$H'_{ji} = \int \phi_j{}^* H'\phi_i \, dx = \langle \phi_j | H' | \phi_i \rangle. \tag{27}$$

Note that $H'_{ji} = H'_{ij}{}^*$, since $H'$ is Hermitian.

For a given $n$, we must now solve the exact set of equations (26), one for each $j$, where $\mathscr{E}_j$, $H'_{ji}$ and $\lambda$ are to be regarded as known. If this infinite set of homogeneous equations is to have a solution, its determinant must vanish, and this yields, in principle, the exact eigenvalues and eigenfunctions of the problem. In practice, it is not possible to carry through this procedure, and we thus seek an approximate solution for small enough $\lambda$. Now equations (23) and (25) tell us that, as $\lambda$ approaches zero, $E_n \to \mathscr{E}_n$, while $c_{nn} \to 1$ and all other $c_{jn}$ approach zero. Thus we separate out the dominant terms by rewriting equation (26) in the form

$$j = n: c_{nn}(E_n - \mathscr{E}_n) = \lambda c_{nn}H_{nn}' + \lambda \sum_i{}' H'_{ni} c_{in}$$

$$j \neq n: c_{jn}(E_n - \mathscr{E}_j) = \lambda c_{nn}H_{jn}' + \lambda \sum_i{}' H'_{ji} c_{in},$$

where the prime on the summation symbol means that the term $i = n$ is omitted, since that term has already been explicitly separated out. Dividing through by $c_{nn}$ and rearranging, we finally obtain

$$j = n: E_n = \mathscr{E}_n + \lambda H'_{nn} + \lambda \sum_i{}' H'_{ni} \frac{c_{in}}{c_{nn}} \tag{28}$$

and

$$j \neq n: \frac{c_{jn}}{c_{nn}} = \frac{\lambda H'_{jn}}{E_n - \mathscr{E}_j} + \frac{\lambda}{E_n - \mathscr{E}_j} \sum_i{}' H'_{ji} \frac{c_{in}}{c_{nn}}. \tag{29}$$

If the perturbed state function is to be normalized, then these equations must be supplemented by the condition, $\sum_i |c_{in}|^2 = 1$ or, equivalently,

$$|c_{nn}|^2 = \frac{1}{1 + \sum_i{}' |c_{in}/c_{nn}|^2}. \tag{30}$$

Now equation (29) tells us that $c_{jn}/c_{nn}$ is of order $\lambda$, and hence that the summation terms contribute, in all three equations, (28), (29) and (30), terms of order $\lambda^2$ and higher. We can thus develop a power series in $\lambda$

for the eigenvalues $E_n$ and for the state function $\psi_n$ by a scheme of successive approximations, as follows. In zeroth order, obtained by setting $\lambda = 0$, we obtain just the unperturbed solution, $E_n \simeq \mathscr{E}_n$, $c_{jn} \simeq \delta_{jn}$ and, of course, $\psi_n \simeq \phi_n$. Substitution of this zero-order result into the right sides of equations (28), (29) and (30) then gives the solution to first order in $\lambda$, or in brief, the first-order solution. Thus,

$$E_n \simeq \mathscr{E}_n + \lambda H'_{nn} \tag{31}$$

and

$$c_{nn} \simeq 1$$

$$c_{jn} \simeq \frac{\lambda H'_{jn}}{\mathscr{E}_n - \mathscr{E}_j}, \qquad j \neq n, \tag{32}$$

whence, using these expressions for the expansion coefficients, we obtain the first-order state function from equation (24),

$$\psi_n \simeq \phi_n + \lambda \sum_j{}' \frac{H'_{jn}}{\mathscr{E}_n - \mathscr{E}_j} \phi_j. \tag{33}$$

Note that the first-order expression for $E_n$ actually involves only the zeroth-order wave function since $H'_{nn} = \langle \phi_n | H' | \phi_n \rangle$. This is easily understood from the variational principle, because equation (31) is simply the Rayleigh–Ritz approximation to $E_n$ for the trial function $\psi_n = \phi_n$. From equation (32), we see that a necessary condition for the applicability of the perturbation method for the $n$th state is that

$$\lambda H'_{jn} \ll \mathscr{E}_n - \mathscr{E}_j$$

for all values of $j \neq n$ or, in other words, that the energy differences between the unperturbed energies be much larger than the corresponding matrix elements of the perturbation term in the Hamiltonian.

To carry the process one step further, the second-order approximation can now be obtained by substituting the first-order results, equations (31) and (32), into the right sides of equations (27), (28) and (29). We shall write explicitly only the second-order approximation to the energy, which is seen to be

$$E_n = \mathscr{E}_n + \lambda H'_{nn} + \lambda^2 \sum_i{}' \frac{H'_{ni} H'_{in}}{\mathscr{E}_n - \mathscr{E}_i}. \tag{34}$$

Because $H'$ is Hermitian, $H'_{ni} = H'_{in}{}^*$, and the result can be rewritten in the form

$$E_n = \mathscr{E}_n + \lambda H'_{nn} + \lambda^2 \sum_i{}' \frac{|H'_{in}|^2}{\mathscr{E}_n - \mathscr{E}_i}, \tag{35}$$

whence the energy is manifestly real, as it must be. Note that only the *first-order* approximation to the $c_{in}$, and hence to $\psi_n$, is required for the computation of this *second-order* expression for the energy. Clearly this is a general feature; knowledge of the wave function to a given order always permits calculation of the energy to one higher order.[7]

Still higher-order corrections can be obtained by systematically continuing this procedure, but the algebra becomes increasingly tedious and the results increasingly difficult to actually evaluate. As a result, the perturbation method is not very useful unless it converges rapidly enough that terms of higher order than the second are negligible. Indeed, even the second-order correction is usually omitted in practice unless the first-order result happens to vanish identically.[8]

A *physical* description of the perturbation corrections is the following: The perturbation term in the Hamiltonian is generated by forces which act on the unperturbed system and which tend to distort it from its unperturbed configuration. Because it is calculated from the *unperturbed* state function, the first-order energy is obtained by ignoring this distortion and thus treating the system as if it were rigid. The computation is entirely analogous to the classical calculation of the interaction energy of a *prescribed* charge distribution in an external electric field or of a *prescribed* mass distribution in an external gravitational field. Because it utilizes perturbed state functions, the second-order calculation takes into account, at least to first approximation, the *distortion* of the system by the perturbing forces. The matrix elements $H'_{jn}$ are a measure of the strength of the distorting forces, and the energy demonimator, $\mathcal{E}_n - \mathcal{E}_j$, measures the resistance to distortion, or the rigidity, of the system. These distortions are analogous to the classical *polarization* of a charge distribution by an external electric field or to the strain *deformation* produced in a material object by an external gravitational field.

The algebraic signs of the terms in the perturbation theory expression for the energy determine whether the energy of the system is increased or decreased as a result of the perturbation. The first-order correction can have either sign, depending simply upon whether the perturbation term is attractive or repulsive on the average; it is negative in the former case, positive in the latter. What about the second-order terms? We have just argued that these terms arise because of distortion effects, that is to say, because of the adjustment of the system to the influence of the perturbing forces. Now the response of any system to a perturbing force is to deform at the expense of that force and hence to *decrease* the potential energy

[7] This is most directly seen from equation (28). Because $\lambda$ multiplies the summation term, we observe at once that if $c_{in}/c_{nn}$ is given to order $\lambda^q$, the energy is given to order $\lambda^{q+1}$.

[8] To put it more generally, perturbation theory corrections are normally calculated only to the *lowest non-vanishing order*. It is rather uncommon that both the first- and second-order terms vanish and that higher-order terms are thus required.

of interaction. Consequently, we expect the second-order correction to lower the energy compared to its first-order value. Unfortunately, we are unable to specify the general conditions under which these expectations are realized, except for the special case of the *ground· state*. For that case, the second-order correction indeed always decreases the energy, no matter the character of the perturbation. This is readily seen from equation (35). The ground state energy $\mathcal{E}_0$, say, is less than $\mathcal{E}_j$ for every $j \neq 0$, hence every term in the summation, and therefore the entire $\lambda^2$ term in that equation, is negative. To summarize, on physical grounds we *expect* the second-order correction to lower the energy in general but have *proved* this to be the case only for the ground state.

The behavior of the matrix elements appearing in equation (35) is too dependent upon details to permit any very broad generalizations. However, the symmetry properties of the system do have a profound influence which is relatively easy to discern. Suppose, in particular, that the unperturbed Hamiltonian, $H_0$, is symmetrical, as is generally the case, and that the perturbation term $H'$ has a definite symmetry, either even or odd. It is then easy to see that half of the matrix elements of $H'$ are identically zero. The argument is based on the fact that the unperturbed states necessarily have a definite symmetry when the unperturbed Hamiltonian is symmetrical. If $H'$ is even, all matrix elements between states of opposite parity then automatically vanish,

$$\langle \phi_{\text{even}} | H'_{\text{even}} | \phi_{\text{odd}} \rangle = \langle \phi_{\text{odd}} | H'_{\text{even}} | \phi_{\text{even}} \rangle = 0, \qquad (36)$$

and, conversely, if $H'$ is odd, all matrix elements between states of the same parity automatically vanish,

$$\langle \phi_{\text{even}} | H'_{\text{odd}} | \phi_{\text{even}} \rangle = \langle \phi_{\text{odd}} | H'_{\text{odd}} | \phi_{\text{odd}} \rangle = 0. \qquad (37)$$

Such general and pervasive rules are special examples of what are called *selection rules* in spectroscopy. Observe that the selection rules for *odd $H'$* have the immediate consequence that the *first-order correction to the energy vanishes identically for every state*.[9]

We conclude this general discussion with some remarks on the *convergence* of the perturbation results. We have already remarked that the method breaks down for a given state, $n$, unless $\lambda H'_{jn} \ll \mathcal{E}_n - \mathcal{E}_j$ for all values of $j \neq n$. This is clearly a *necessary* condition for convergence, but hardly a *sufficient* one in view of the infinite summations appearing in

[9] The importance of this result may be judged from the example of an atomic or nuclear system perturbed by a uniform external electric field. In that example, $H_0$ is even and $H'$ is odd, so that equation (37) applies and the first-order correction to the energy of the ground state vanishes. There is thus no term in the energy *linear* in the applied electric field, which means that such systems exhibit no permanent electric dipole moment. This at once accounts for the observed fact that atoms and nuclei indeed do not possess such moments. For a somewhat more detailed discussion, see Section 3 of Chapter VIII.

equations (33) and (34). We now provide a rather crude, but more or less sufficient, condition for convergence by finding an *upper bound* to the magnitude of the second-order correction. We start by writing the inequality, obtained by replacing each term in the summation of equation (35) by its absolute value,

$$\left| \sum_i{}' \frac{|H'_{ni}|^2}{\mathscr{E}_n - \mathscr{E}_i} \right| \leq \sum_i{}' \frac{|H'_{ni}|^2}{|\mathscr{E}_n - \mathscr{E}_i|} . \qquad (38)$$

Next, introduce $\Delta \mathscr{E}_n$, the difference in energy between $\mathscr{E}_n$ and its *nearest* neighbor. Replacing each denominator on the right in equation (38) by $|\Delta \mathscr{E}_n|$, which only *increases* the inequality, we then obtain

$$\left| \sum_i{}' \frac{|H'_{ni}|^2}{\mathscr{E}_n - \mathscr{E}_i} \right| \leq \frac{1}{|\Delta \mathscr{E}_n|} \sum_{i \neq n} |H'_{ni}|^2 ,$$

where we have made explicit our convention that a prime on the summation symbol means that the term with $i = n$ is omitted. Adding and subtracting this omitted term, we then obtain

$$\left| \sum_i{}' \frac{|H'_{ni}|^2}{\mathscr{E}_n - \mathscr{E}_i} \right| \leq \frac{1}{|\Delta \mathscr{E}_n|} \left\{ \sum_i |H'_{ni}|^2 - |H'_{nn}|^2 \right\} .$$

Next, we observe that the summation on the right can be evaluated in the following way. From the definition of the matrix elements we have

$$\sum_i |H'_{ni}|^2 = \sum_i H'_{ni} H'_{in}$$

$$= \sum_i \int \phi_n^*(x') H'(x') \phi_i(x') \, dx' \int \phi_i^*(x) H'(x) \phi_n(x) \, dx.$$

Performing the summation over $i$ and using the closure property of the $\phi_i$, we obtain

$$\sum_i |H'_{ni}|^2 = \int \int \phi_n^*(x') H'(x') \, \delta(x - x') \, H'(x) \, \phi_n^*(x) \, dx \, dx',$$

whence, performing the integration over $x'$, we find[10]

$$\sum_i |H'_{ni}|^2 = \langle \phi_n | H'^2 | \phi_n \rangle .$$

Finally then, using $|H_{nn}'|^2 \equiv \langle \phi_n | H' | \phi_n \rangle^2$, we obtain

$$\left| \sum_i{}' \frac{|H'_{ni}|^2}{\mathscr{E}_n - \mathscr{E}_i} \right| \leq \frac{\langle \phi_n | H'^2 | \phi_n \rangle - \langle \phi_n | H' | \phi_n \rangle^2}{|\Delta \mathscr{E}_n|} , \qquad (39)$$

which states that the second-order correction to the energy of a given state is less in magnitude than the ratio of the *mean square deviation* of the perturbation Hamiltonian, about its average for the state in ques-

---

[10] See the next section, equation (47) with $A = B$, for a derivation of this result using the methods of matrix algebra.

tion, to the energy separation of that state from its nearest neighbor. Admittedly, this is not a very refined estimate, but it is a simple and useful one, as we shall shortly demonstrate.

As a first example of the use of perturbation theory, we consider an harmonic oscillator with a flattened bottom, produced by a Gaussian perturbation. Specifically, we write

$$H = \frac{p^2}{2m} + \frac{1}{2} m\omega^2 x^2 + V e^{-\alpha x^2}$$

so that, with $\lambda = 1$,

$$H' = V e^{-\alpha x^2}.$$

The first-order correction to the ground state energy, which is just the ground state expectation value of $H'$, is easily evaluated and we obtain almost at once

$$E_0 = \frac{\hbar\omega}{2} + V \left( \frac{m\omega/\hbar}{\alpha + m\omega/\hbar} \right)^{1/2}.$$

Observe that for $\alpha \ll m\omega/\hbar$, when the perturbation term is practically constant over the domain of the ground state, the energy is shifted upward by $V$, which is just what *should* happen when a constant potential of that magnitude is added to the Hamiltonian. Observe also that when $\alpha \gg m\omega/\hbar$, so that the Gaussian is narrow, the correction is small even when $V$ is comparable to $\hbar\omega$ in magnitude. Hence the first-order result appears to be satisfactory in the limits of small and large $\alpha$, for values of $V$ which are not necessarily small, and it thus appears to be similarly satisfactory for all values of $\alpha$. One way to test these conjectures is to calculate the second-order corrections, but this turns out to be very difficult to carry through. Instead, therefore, we shall use the simple upper bound to these corrections provided by equation (39). Because we are here dealing with the ground state, the second-order term must be negative (why?) and hence, after an elementary calculation, we can write our result in the form

$$E_0 = \frac{\hbar\omega}{2} + V \left( \frac{m\omega/\hbar}{\alpha + m\omega/\hbar} \right)^{1/2} (1 - \epsilon)$$

where $\epsilon$, the ratio of the second- to first-order correction, is such that

$$\epsilon \leq \frac{V}{\hbar\omega} \left\{ \left( \frac{\alpha + m\omega/\hbar}{2\alpha + m\omega/\hbar} \right)^{1/2} - \left( \frac{m\omega/\hbar}{\alpha + m\omega/\hbar} \right)^{1/2} \right\}.$$

When $\alpha$ is small, the case of a nearly constant perturbing potential, the right side is of the order of $\alpha^2 V/\hbar\omega$ and hence is indeed negligible, as we expected, even for appreciable values of $V/\hbar\omega$. On the other hand, for

$\alpha \gg m\omega/\hbar$, the right side is of the order of $V/\hbar\omega$ and hence convergence of the perturbation series is *not* assured unless $V/\hbar\omega \ll 1$, contrary to our earlier guess. The true situation for large $\alpha$ is then the following: The correction to the energy is small, as we indeed found it to be, but the first-order estimate for this small correction is unreliable unless $V/\hbar\omega$ is also small.

As our second, and final, example, we consider the physically much more interesting case of an anharmonic oscillator described by the Hamiltonian

$$H = \frac{p^2}{2m} + \frac{1}{2} m\omega^2 x^2 \left( 1 + \frac{x^2}{b^2} \right).$$     (40)

For the unperturbed Hamiltonian we write

$$H_0 = \frac{p^2}{2m} + \frac{1}{2} m\omega^2 x^2$$

so that, with $\lambda = 1$,

$$H' = \frac{1}{2} \frac{m\omega^2}{b^2} x^4.$$

To obtain the energy to first order we must calculate $\langle \phi_n | x^4 | \phi_n \rangle$, where $\phi_n$ is the $n$th harmonic oscillator state function, which, from equation VI–47, is given by

$$\phi_n = \frac{c_n}{c_0} a^{\dagger n} \phi_0.$$

Now from equations (VI–34) and (VI–35), we can write

$$x = \sqrt{\hbar/2m\omega} \, (a + a\dagger),$$

so that $x^4$ is a quartic in the annihilation and creation operators. Recalling that $a\dagger$ operating on a harmonic oscillator state produces the next higher state and $a$ the next lower state, we note that operating $q$ times on $\phi_n$ with $a$ and $4 - q$ times with $a\dagger$, in any sequence, gives the state $\phi_{n+4-2q}$. In view of the orthogonality of the $\phi_n$, we thus see that a non-vanishing contribution to the matrix element is obtained only for $q = 2$. In other words, only those terms in the quartic contribute which contain $a$ twice and $a\dagger$ twice. Explicitly,

$$x^4 = \left( \frac{\hbar}{2m\omega} \right)^2 [a^2 a\dagger^2 + aa\dagger aa\dagger + aa\dagger^2 a + a\dagger a^2 a\dagger + a\dagger aa\dagger a + a\dagger^2 a^2$$

$$+ \text{noncontributing terms}].$$

Consider a typical term, say the first. We have

$$\langle \phi_n | a^2 a^{\dagger 2} | \phi_n \rangle = \langle \phi_n | \frac{c_n}{c_0} | a^2 a^{\dagger n+2} \phi_0 \rangle = (n+2)(n+1) \langle \phi_n | \phi_n \rangle$$

$$= (n+2)(n+1),$$

where we have used the fact that, in the creation operator representation, $a = d/da\dagger$, which is to say that the effect of operating with $a$ is equivalent to differentiation with respect to $a\dagger$. Thus we have at once, using the same technique on the remaining terms,

$$\langle \phi_n | x^4 | \phi_n \rangle = \left( \frac{\hbar}{2m_\omega} \right)^2 [(n+2)(n+1) + (n+1)^2 + n(n+1) +$$

$$+ n(n+1) + n^2 + n(n-1)]$$

$$= 3 \left( \frac{\hbar}{2m_\omega} \right)^2 [1 + 2n(n+1)].$$

Finally, since

$$E_n \simeq \mathscr{E}_n + \langle \phi_n | H' | \phi_n \rangle = \left( n + \frac{1}{2} \right) \hbar\omega + \frac{m\omega^2}{2b^2} \langle \phi_n | x^4 | \phi_n \rangle,$$

we obtain

$$E_n \simeq \hbar\omega \left( n + \frac{1}{2} + \frac{3\hbar}{8m\omega b^2} [1 + 2n(n+1)] \right). \tag{41}$$

The validity of this approximation requires that the correction term be small compared to the spacing between states or, explicitly, that

$$\frac{3\hbar}{8m\omega b^2} [1 + 2n(n+1)] \ll 1.$$

The result is thus valid for sufficiently large $b$, but no matter how large $b$ is, the approximation is seen to become steadily worse as $n$ increases. This is not unexpected, because the higher states extend to larger and larger values of $x$ and, for sufficiently large $x$, the perturbation clearly becomes enormous, however large $b$ may be.

Although equation (41) provides an entirely unambiguous relation between the perturbation term in the Hamiltonian and the corrections to the energy states, the *experimental* detection of the presence of such anharmonic terms is possible only because of the term quadratic in $n^2$. This can be made clearer by rewriting the result in the completely equivalent form,

$$E_n = \hbar(\omega + \delta\omega)(n + \tfrac{1}{2}) + n^2\hbar \, \delta\omega, \tag{42}$$

where

$$\delta\omega = 3\hbar/4\pi m b^2.$$

Hence, the anharmonic perturbation is seen to have two distinct effects: first, it increases the frequency by $\delta\omega$,[11] and second, it makes the spacing between levels increase with increasing energy. The spacing between levels is what the spectroscopist observes, of course, and it is just this systematic increase which tells him that he is dealing with an anharmonic oscillator. The experimentalist's problem of unraveling his data is actually more complicated than we have indicated, because it so happens that a *cubic* perturbation produces *exactly* the same qualitative behavior as the quartic. It should be remarked that a cubic perturbation contributes only in second order; the first-order contribution is zero (why?).[12] It turns out, incidentally, that this correction is always negative, for all states, as we argued on physical grounds that such second-order terms should be.

We next turn our attention to the remarkably interesting case in which the quadratic perturbation has a *negative* sign. In a certain sense, to be defined shortly, all of our results can be taken over by merely replacing $b^2$ by its negative everywhere, and thus, for example, equation (42) applies with

$$\delta\omega = -3\hbar/4\pi mb^2.$$

We must now explain the sense in which our perturbation results are applicable, and it is these considerations which will lead us to the remarkable features of the problem. In Figure 1, we plot the potential energy

$$V = \frac{1}{2}\, m\omega^2 x^2 \left(1 - \frac{x^2}{b^2}\right), \tag{43}$$

and it is seen at once that the exact energy spectrum is *continuous*, that this spectrum is *unbounded* in either direction, and that therefore the system *does not even have a ground state*. These remarks apply, no matter how small the perturbation in the neighborhood of the origin. Among other things, this means that we have violated our fundamental assumption that the exact and unperturbed states can be placed in one-to-one correspondence and that these states approach each other as the perturbation tends to zero. It is clear that this is simply not true for $E < 0$ or for $E > V_m$, where $V_m$ is the maximum value attained by the potential, as shown in Figure 1.

What about the situation when $0 \leqslant E < V_m$? There are two sharply different kinds of states:

(1)  *Continuum* states in which the particle is almost exclusively out-

---

[11] This frequency shift has *nothing* to do with the amplitude dependent alteration of the classical period of the oscillator, as the presence of $\hbar$ in the expression for $\delta\omega$ makes abundantly clear. The classical period can be shown to be related to the spacing between levels, but we shall not pursue the subject further.

[12] For a discussion of cubic and quartic anharmonicities, see page 136 of Reference [24].

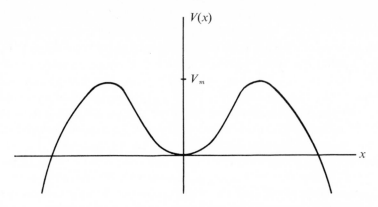

FIGURE 1. The potential energy function $V(x)$ for the anharmonic oscillator of equation (43). The maximum value of the potential is denoted by $V_m$.

side of the potential well and which therefore cannot be placed in one-to-one correspondence with the unperturbed states.

(2) States in which the particle is almost exclusively inside of the potential well.
It is these states which correspond to the discrete unperturbed states, and it is for these states alone that our perturbation approximation holds. We shall call these states *quasi-bound* or *metastable*.

Classically, these two kinds of states are absolutely distinct, but quantum mechanically they are not. Because of the tunnel effect, a particle outside of the well has a very small but non-zero part of its wave function extending into the well. Conversely, the stationary state function of a particle inside the well has a very small but non-zero amplitude in the external region. This means that the probability of finding the particle in this region is finite. Hence, given a sufficiently large ensemble of such systems, or a sufficiently long time, particles will occasionally be observed emerging from the interior region. We have here, in short, just an analog of the process of $\alpha$-radioactivity.

Although the stationary state description of the above system reveals its main features, it is instructive to re-examine it in terms of time dependent states. Let us thus consider a very special wave packet, one which is made up of a superposition of states in the immediate neighborhood of one of the quasi-bound states and which is so constructed that the particle is initially inside the well with absolute certainty. Such a superposition of continuum states is required in order to cancel the tail of the quasi-bound state, which otherwise would extend into the external region. This superposition has some characteristic width $\Delta E$ and the rate at which the wave packet, because of tunneling, gradually emerges into the exterior domain is entirely governed by this width. More precisely, one sees that in a time $\tau$ such that $\tau\Delta E/\hbar$ is of the order of unity, the very special phase relations which were built into the initial packet will have been significantly altered and the packet will thus have begun to emerge from the central region. The time $\tau$ is just the mean life of the state and it is given by

$$\tau = \hbar/\Delta E.$$

This relation between mean life time and the energy width of a wave packet is very general; it is, as we have stated before, another manifestation of the uncertainty principle. The feature which makes the present case so special is that, unlike the situation in free space, the presence of the potential well makes it possible to construct a wave packet which is confined to a very small domain, and is thus well localized in space, and which *simultaneously* has an extremely small energy spread. From this point of view, a true *bound* stationary state is the limiting case in which the spatial domain is small and the energy spread is exactly zero. Any such state thus has an infinite mean life. Note that the *momentum* spread in these wave packets is always very large because of the presence of the potential, so that no violation of the uncertainty relation between position and momentum is encountered.

Coming back, finally, to the applicability of perturbation theory, we see that it is only these quasi-bound or quasi-stationary states for which the theory can be used. If the correction to some particular unperturbed state is small, then the width of that state is guaranteed to be narrow. It is in this sense that our perturbation theory analysis goes through without modification and that equation (41) applies with $b^2$ negative.

## 4. MATRICES

In our development of perturbation theory, we saw that the matrix elements of the perturbation term in the Hamiltonian played an essential role. We now briefly discuss the matrix representation of operators generally and give some elements of matrix algebra.

Consider some operator $A$ and let it operate upon one of a complete orthonormal set of functions $\phi_n$. The result is that some new function is produced which can be expressed as a superposition of the complete set. That is, we can write

$$A\phi_n = \sum_m A_{mn}\phi_m, \tag{44}$$

where the notation is intended to make explicit the fact that the coefficient of $\phi_m$ in the superposition depends on the operator $A$ and on the state $\phi_n$ upon which it operates. Given the orthonormality of the $\phi_m$, we see that

$$A_{mn} = \int \phi_m^* A \phi_n \, dx = \langle \phi_m | A | \phi_n \rangle. \tag{45}$$

The array of numbers thus defined is called a matrix, and any particular number is called an element of the matrix, or a matrix element. *This array is entirely equivalent to the operator A in the sense that these numbers completely define the effect of operating with A upon any arbitrary function.* This follows, since such a function can always be expanded in the complete set $\phi_n$, and equation (44) describes the precise result of operating on each $\phi_n$. The matrix, with elements $A_{mn}$, thus is a specific *realization* or *representation* of the operator $A$. A convenient way of visualizing this array is to arrange the numbers in rows and columns with $A_{mn}$ the entry in the $m$th row and $n$th column. Thus we write

$$A = \begin{pmatrix} A_{11} & A_{12} & \cdots & A_{1n} & \cdots \\ A_{21} & A_{22} & \cdots & A_{2n} & \cdots \\ \cdot & \cdot & & \cdot & \\ \cdot & \cdot & & \cdot & \\ A_{m1} & A_{m2} & \cdots & A_{mn} & \cdots \\ \cdot & & & \cdot & \\ \cdot & & & \cdot & \end{pmatrix} \tag{46}$$

We emphasize that any particular matrix representation of a given operator depends upon the particular set of functions $\phi_n$, or *basis functions*, which are used to construct the $A_{mn}$. If one uses a different set, say $\psi_n$, a new set of matrix elements will be obtained, but the new matrix still represents the same operator. Transformations from one representation to another form an important subject of study in quantum mechanics, but we shall not pursue it here.[13]

Suppose, we consider the product $C$ of two operators $A$ and $B$,

$$C = AB.$$

We have

$$C_{mn} = \int \phi_m^* C \phi_n \, dx = \int \phi_m^* A B \phi_n \, dx.$$

But now,

$$B \phi_n = \sum_s B_{sn} \phi_s,$$

and

$$A B \phi_n = \sum_s B_{sn} A \phi_s = \sum_{s,k} B_{sn} A_{ks} \phi_k,$$

[13] See Reference [19] or any of References [22]–[28].

whence

$$C_{mn} = \int \phi_m{}^* \sum_{s,k} A_{ks} B_{sn} \phi_k$$

and, finally, since the $\phi_m$ are orthonormal,

$$C_{mn} \equiv (AB)_{mn} = \sum_s A_{ms} B_{sn}. \tag{47}$$

This rule for multiplying matrices is easy to remember. The matrix elements are written in the same order as the operators, the first index of the first factor and the second of the last factor label the matrix element of the product, and the summation is taken over the common interior index. Pictorially, the $m$th, $n$th matrix element of the product is obtained by multiplying the $m$th row of the first matrix by the $n$th column of the second, element by element. Note that multiplication is defined only if the number of elements in the rows of the first and the columns of the second are equal.

The generalization to a product of more than two matrices is trivial. By successive application of equation (47) we obtain at once

$$(ABC \cdots R)_{mn} = \sum_{i,k,l,\ldots,s} A_{mi} B_{ik} C_{k1} \cdots R_{sn}. \tag{48}$$

The noncommutativity of operator multiplication is quite explicit using these matrix representations. Indeed, from our general rule we have

$$(BA)_{mn} = \sum_s B_{ms} A_{sn},$$

which is clearly different from $(AB)_{mn}$, as comparison with equation (47) demonstrates.

From the definitions, it follows trivially that

$$(A + B)_{mn} = (B + A)_{mn} = A_{mn} + B_{mn}$$

and, further, that if two matrices are equal, then each element of the first equals the corresponding element of the second.

With this understanding of matrix algebra, we see that all relations which hold between operators also hold between their matrix representations. For example, if

$$(A, B) = C,$$

then

$$(A, B)_{mn} \equiv (AB)_{mn} - (BA)_{mn} = C_{mn},$$

whence matrices representing commuting operators commute.

The matrix representation of the adjoint of an operator is related very simply to the matrix representation of the operator itself. Recalling the definition of the adjoint, equations (V–11a, 11b), it follows almost at once that

$$(A\dagger)_{mn} = A_{nm}{}^*. \tag{49}$$

Introducing the symbol $\tilde{A}$ to denote the *transposed* matrix

$$(\tilde{A})_{mn} \equiv A_{nm},$$

we see that equation (49) can be written in the representation-independent form

$$A\dagger = \tilde{A}^*.$$

---

**Exercise 2.** Deduce equation (49).

---

Observe that for an Hermitian operator, $A = A\dagger$, equation (49) yields the result found earlier for the particular case of the Hamiltonian operator,

$$A_{mn} = A_{nm}{}^*.$$

Thus the diagonal elements, $A_{nn}$, of an Hermitian matrix are real. This must be so, of course, since $A_{nn}$ is simply the expectation value of $A$ in the state $\phi_n$.

A particularly simple representation of an operator is obtained when its own complete set of eigenfunctions is used as a basis. Thus, if $\phi_n$ are the eigenfunctions of $A$,

$$A\phi_n = a_n\phi_n,$$

we have at once

$$A_{mn} = a_m\delta_{mn}. \tag{50}$$

In words, *in the basis formed by its eigenfunctions, the matrix representation of an operator is diagonal and its matrix elements are just the eigenvalues of the operator.* From this point of view, the problem of finding the eigenvalues and eigenfunctions of an operator is equivalent to that of transforming to a basis in which its matrix representation is diagonal.

## 5. DEGENERATE OR CLOSE-LYING STATES

The perturbation theory approximation method developed in Section 3 required for its applicability that the matrix element $H'_{nj}$ connecting the $n$th and $j$th states be small compared to the energy separation, $\mathcal{E}_n - \mathcal{E}_j$, between these states, for all values of $j$. We now discuss the modification required when this condition is violated for some one state, say the $l$th. A particular example of great importance is that for which the $n$th and $l$th states are degenerate, $\mathcal{E}_n = \mathcal{E}_l$. In any case, if $\mathcal{E}_n - \mathcal{E}_l$ is not necessarily large compared to $H'_{nl}$, we can no longer safely assume that $c_{ln}$, the amplitude of the $l$th state, is small. In analyzing equation (26), which is exact, of course, we must therefore treat $c_{nn}$ and $c_{ln}$ on an equal footing. All the remaining $c_{jn}$ may still be regarded as small, and we thus rewrite equation (26) in the form

$$j = n: \ (E_n - \mathcal{E}_n - \lambda H'_{nn})\ c_{nn} - \lambda H'_{nl}\ c_{ln} = \lambda \sum_{i \neq n,\, l} H'_{ni}\ c_{in}$$

$$j = l: \ -\lambda H'_{ln}\ c_{nn} + (E_n - \mathcal{E}_l - \lambda H'_{ll})\ c_{ln} = \lambda \sum_{i \neq n,\, l} H'_{li} c_{in} \qquad (51)$$

$$j \neq l, n: \ (E_n - \mathcal{E}_j)\ c_{jn} = \lambda H'_{jn}\ c_{nn} + \lambda H'_{jl}\ c_{ln} + \lambda \sum_{i \neq n,\, l} H'_{ji}\ c_{in}.$$

From the last of these equations we see that the right side of each of the first two equations is of order $\lambda^2$, and hence can be neglected to first order. We thus obtain to this approximation,

$$(E_n - \mathcal{E}_n - \lambda H'_{nn})\ c_{nn} - \lambda H'_{nl} c_{ln} = 0$$

$$-\lambda H'_{ln}\ c_{nn} + (E_n - \mathcal{E}_l - \lambda H'_{ll})\ c_{ln} = 0,$$

whence the determinant of this pair of homogeneous equations in the unknown $c_{nn}$ and $c_{ln}$ must vanish,

$$(E_n - \mathcal{E}_n - \lambda H'_{nn})(E_n - \mathcal{E}_l - \lambda H'_{ll}) - \lambda^2 H'_{ln} H'_{nl} = 0.$$

After some elementary algebra, we thus find

$$E_n^{\pm} = \frac{\mathcal{E}_n + \mathcal{E}_l + \lambda (H'_{nn} + H'_{ll})}{2}$$

$$\pm \sqrt{\left[ \frac{\mathcal{E}_n - \mathcal{E}_l + \lambda (H'_{nn} - H'_{ll})}{2} \right]^2 + \lambda^2 H'_{ln} H'_{nl}}, \qquad (53)$$

while from equation (43),

$$\left( \frac{c_{ln}}{c_{nn}} \right)^{\pm} = \frac{\lambda H'_{ln}}{E_n^{\pm} - \mathcal{E}_l - \lambda H'_{ll}} = \frac{E_n^{\pm} - \mathcal{E}_n - \lambda H'_{nn}}{\lambda H'_{nl}}. \qquad (54)$$

Note first that we obtain *two* perturbed energies and states, corresponding to the plus and minus sign in equation (53). The reason is that we have treated two unperturbed states, the $l$th and the $n$th, on an equal basis, and have thus simultaneously determined the first-order energy of each. Supposing that the states are not actually degenerate, the correctness of this interpretation can be seen by going to the limit of vanishingly small $\lambda$. In that limit, the term in $\lambda^2$ under the square root sign in equation

(53) can be treated as small. We then obtain, using the plus sign in equation (53),

$$E_n^{(+)} = \mathcal{E}_n + \lambda H'_{nn}$$

$$\left(\frac{c_{ln}}{c_{nn}}\right)^{(+)} = \frac{\lambda H'_{ln}}{\mathcal{E}_n - \mathcal{E}_l},$$

while, using the minus sign,

$$E_n^{(-)} = \mathcal{E}_l + \lambda H'_{ll}$$

$$\left(\frac{c_{ln}}{c_{nn}}\right)^{(-)} = \frac{\mathcal{E}_l - \mathcal{E}_n}{\lambda H'_{nl}}.$$

Thus the former is the proper first-order perturbation theory result for the $n$th state, and the latter is the same for the $l$th.

The qualitative features of equation (53) can be made clearer by rewriting it in the following way. Introduce first the *average* of the first-order energies of the two levels,

$$\bar{E} \equiv \tfrac{1}{2}(\mathcal{E}_n + \lambda H'_{nn} + \mathcal{E}_l + \lambda H'_{ll}).$$

Next, introduce the relative first-order *separation* of the energies,

$$\Delta \equiv \frac{[(\mathcal{E}_n + \lambda H'_{nn}) - (\mathcal{E}_l + \lambda H'_{ll})]}{\bar{E}}.$$

Equation (53) can now be expressed in the form

$$E_n^{\pm}/\bar{E} = 1 \pm \sqrt{(\Delta^2/4) + \lambda^2 H'_{ln} H'_{nl}/\bar{E}^2},$$

which shows the levels to be equally spaced about their *first-order* average. Note that the second-order terms always *increase* the relative separation of the levels and this by an amount which is larger the more nearly degenerate the states are to first order, or otherwise stated, the smaller is $\Delta$. This general tendency of close-lying levels to *repel* each other is illustrated in Figure 2, where $E_n^{\pm}/\bar{E}$ is plotted as a function of $\Delta$. Observe, in particular, that the second-order contribution is always such as to prevent exact degeneracy and thus to make it impossible for a pair of close-lying levels to cross-over as the strength of the perturbation term is altered.

We turn next to the situation in which the *unperturbed* states are *exactly* degenerate, $\mathcal{E}_n = \mathcal{E}_l$. In that case, equations (53) and (54) become

$$E_n^{\pm} = \mathcal{E}_n + \frac{\lambda(H'_{nn} + H'_{ll})}{2} \pm \lambda \sqrt{\left(\frac{H'_{nn} - H'_{ll}}{2}\right)^2 + H'_{ln} H'_{nl}} \quad (55)$$

and

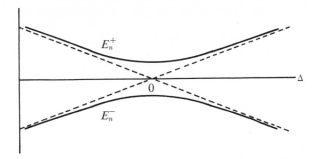

FIGURE 2.   Second-order energy of a pair of nearly degenerate states as a function of first-order relative separation $\Delta$. The magnitude of the second-order correction is the distance between the solid curves and the dotted lines. These latter are thus the energies obtained by neglecting the second-order corrections.

$$\left(\frac{c_{ln}}{c_{nn}}\right)^{\pm} = \frac{H'_{ln}}{\dfrac{H'_{nn} - H'_{ll}}{2} \pm \sqrt{\left(\dfrac{H'_{nn} + H'_{ll}}{2}\right)^2 + H'_{ln}H'_{nl}}} . \tag{56}$$

Note that the ratio of amplitudes $c_{ln}/c_{nn}$ *no longer depends on* $\lambda$. This fact is a consequence of the arbitrariness inherent in any specification of a set of degenerate states. Because of this arbitrariness, without further information we cannot establish *a priori* a unique general connection between perturbed and unpertubed states. Indeed, equation (56) tells us which particular linear combinations of the unperturbed degenerate states are in one-to-one correspondence with the perturbed states, and the answer is seen to depend on the character of the perturbation. As a special example, consider the case in which

$$\begin{aligned} H'_{nn} &= H'_{ll} = 0 \\ H'_{ln} &= H'_{nl}{}^* \neq 0. \end{aligned} \tag{57}$$

In that case we obtain the simple result

$$\begin{aligned} E_n^{\pm} &= \mathscr{E}_n \pm \lambda |H'_{ln}| \\ (c_{ln}/c_{nn})^{\pm} &= \pm H'_{ln}/|H'_{ln}|. \end{aligned} \tag{58}$$

Note that the original degenerate states are thoroughly mixed, since $|c_{ln}| = |c_{nn}|$. Such an example occurs for an antisymmetric perturbation acting on degenerate states of definite but opposite parity. On the other hand, if the $l$th and $n$th states have the same parity, then it is easy to see that $H'_{ln}$ is zero for an antisymmetric perturbation and the states remain degenerate to first order. In that case, a second-order calculation is required to find the splitting of the levels produced by the perturbation.

As a second special example, consider the case in which

$$H'_{ln} = H'_{nl}{}^* = 0,$$

so that the states are not connected to first order. In that case, the two states are described by

$$E_n = \mathcal{E}_n + \lambda H'_{nn}, \qquad c_{nn} \simeq 1$$

and

$$E_n = \mathcal{E}_n + \lambda H'_{ll}, \qquad c_{ln} \simeq 1,$$

and the degeneracy plays no role in the analysis. Examples of this sort occur whenever the perturbation commutes with the set of operators used to define the unperturbed states. Otherwise stated, the analysis reduces to that of conventional perturbation theory whenever the unperturbed states can be uniquely specified by a set of operators, each of which commutes with the perturbation term.

## 6. TIME DEPENDENT PERTURBATION THEORY

As a final approximation method, we now consider the case in which a system, in some definite initial state, is subjected to some external force which depends on the time. An example would be the oscillating force exerted on an atom by a light wave passing through it. We shall assume the external force to be weak enough that a perturbation method can be applied. We write the Hamiltonian in the form

$$H = H_0 + H'(t), \tag{59}$$

without bothering to introduce a parameter of smallness $\lambda$. We now seek approximate solutions of the time dependent Schrödinger's equation

$$H\psi \equiv [H_0 + H'(t)]\psi = -\frac{\hbar}{i} \frac{\partial \psi}{\partial t}.$$

Again denoting the eigenfunctions and eigenvalues of the unperturbed Hamiltonian by $\phi_n$ and $\mathcal{E}_n$, we express $\psi$ as the superposition,

$$\psi(x, t) = \sum_n a_n(t) \phi_n(x),$$

where the expansion coefficients must now be regarded as functions of time. If $H'$ were zero, the $a_n$ would be proportional to $e^{-i\mathcal{E}_n t/\hbar}$, and it is accordingly convenient to write, without loss of generality,

$$a_n(t) = c_n(t) e^{-i\mathcal{E}_n t/\hbar},$$

so that

$$\psi(x, t) = \Sigma c_n(t) \phi_n(x) e^{-i\mathcal{E}_n t/\hbar}. \tag{60}$$

We thus have

$$H\psi = (H_0 + H')\psi = \Sigma\, c_n(t)\, \mathcal{E}_n\phi_n e^{-i\mathcal{E}_n t/\hbar} + \Sigma\, c_n(t)\, e^{-i\mathcal{E}_n t/\hbar} H'\phi_n$$

and

$$-\frac{\hbar}{i}\frac{\partial\psi}{\partial t} = -\frac{\hbar}{i}\Sigma\,\frac{dc_n}{dt}\,\phi_n\, e^{-i\mathcal{E}_n t/\hbar} + \Sigma\, c_n(t)\,\mathcal{E}_n\phi_n\, e^{-i\mathcal{E}_n t/\hbar},$$

whence Schrödinger's equation becomes

$$-\frac{\hbar}{i}\Sigma\,\frac{dc_n}{dt}\,\phi_n\, e^{-i\mathcal{E}_n t/\hbar} = \Sigma\, c_n\, e^{-i\mathcal{E}_n t/\hbar} H'\phi_n.$$

Multiplying through by $\phi_m{}^*$ and integrating, we finally obtain, for each $m$,

$$\frac{dc_m}{dt} = -\frac{i}{\hbar}\Sigma\, H'_{mn}(t)\, e^{-i(\mathcal{E}_n - \mathcal{E}_m)t/\hbar}\, c_n(t). \tag{61}$$

This simultaneous set of coupled first-order differential equations is exact and is completely equivalent to the time dependent Schrödinger's equation. Of course, the initial conditions must still be specified. We shall consider only the case in which $\psi(t = 0) = \phi_k$, that is, in which the system is in the $k$th unperturbed state initially. Thus we have

$$c_n(t = 0) = \delta_{nk}.$$

We now assume that the perturbation is weak enough, or that we restrict our attention to short enough times, that all $c_n$ except $c_k$ remain small. To first order, we then have

$$m = k: \quad \frac{dc_k}{dt} = -\frac{i}{\hbar}H'_{kk}(t)\, c_k$$

$$m \neq k: \quad \frac{dc_m}{dt} = -\frac{i}{\hbar}H'_{mk}\, e^{-i(\mathcal{E}_k - \mathcal{E}_m)t/\hbar}\, c_k,$$

where we have retained only the $k$th term in the summation on the right side of equation (61). The first of these equations then gives at once

$$c_k = \exp\left[-i\int_0^t H'_{kk}(t)\, dt/\hbar\right] \simeq 1 - \frac{i}{\hbar}\int_0^t H'_{kk}(t)\, dt.$$

Neglecting the deviation of $c_k$ from unity, since this deviation contributes only terms of higher order than the first, the second equation gives

$$m \neq k: \quad c_m \simeq -\frac{i}{\hbar}\int_0^t H'_{mk}(t)\, e^{-i(\mathcal{E}_k - \mathcal{E}_m)t/\hbar}\, dt. \tag{62}$$

As a first example, suppose that $H'$ is independent of time, except for suddenly having been "switched on" at $t = 0$, and just as suddenly "switched off" at time $t$. We then obtain

$$c_k = e^{-iH'_{kk}t/\hbar},$$

which corresponds to an energy shift $H'_{kk}$, in agreement with our first-order stationary perturbation theory result. For the $c_m$ we have

$$c_m(t) = H'_{mk} \frac{e^{-i(\mathcal{E}_k - \mathcal{E}_m)t/\hbar} - 1}{\mathcal{E}_k - \mathcal{E}_m}$$

or

$$|c_m(t)|^2 = 4|H'_{mk}|^2 \frac{\sin^2(\mathcal{E}_k - \mathcal{E}_m)t/2\hbar}{(\mathcal{E}_k - \mathcal{E}_m)^2}. \tag{63}$$

For *non-degenerate* states, we thus see that if $H'_{mk}$ is small enough, then $|c_m|^2$ remains small for all $t$. For larger $H'_{mk}$, however, our approximation is valid only for small enough $t$. In either case, the interpretation of our result is the following. The probability that the system, initially in the $k$th unperturbed state, will be found in the $m$th unperturbed state after the interaction has occurred is $|c_m(t)|^2$. From this point of view, the role of the perturbation, acting over the time interval from 0 to $t$, is to produce *transitions* between the unperturbed states.

Consider next the case of degenerate, or nearly degenerate, states. For $\mathcal{E}_m \simeq \mathcal{E}_k$, we see that

$$|c_m|^2 \simeq |H'_{mk}|^2 t^2/\hbar^2, \tag{64}$$

and hence that the transition probability grows rapidly with time and eventually becomes of order unity. This follows no matter how small is $H'_{mk}$, just so long as it is not identically zero. The reason for this behavior is easy to understand from our previous stationary perturbation theory results. As we saw, when the $m$th and $k$th states are degenerate and $H'_{mk}$ differs from zero, the perturbed states are particular linear combinations of the degenerate unperturbed states with relative magnitudes of order unity. The growth of $|c_m|^2$ thus reflects the frantic (but unsuccessful) attempt of the system to achieve just the right linear combination in response to the perturbation. Since $c_m$ eventually becomes comparable to $c_k$, our approximation evidently holds only for short enough times. However, results valid for arbitrary times can be obtained by treating $c_m$ and $c_k$ simultaneously and on equal basis in equation (61). The resulting coupled equations are easy to solve, but the details are left to the problems.

One aspect of equation (63), which is rather surprising, is that $|c_m|^2$ initially grows as $t^2$. We would instead expect the probability of transitions to grow linearly with time, that is, the effect of the perturbation should be describable in terms of a *rate* at which it induces transitions. Clearly this is not the case for transitions to a single degenerate final state, but our expectation is realized for transitions to a dense group of final states, as in the continuum. To understand this important case, we now consider the transition probability $P$ to such a group of states,

$$P = \sum_{\mathscr{E}_m \simeq \mathscr{E}_k} |c_m(t)|^2 = \sum_{\mathscr{E}_m \simeq \mathscr{E}_k} 4|H'_{mk}|^2 \frac{\sin^2[(\mathscr{E}_k - \mathscr{E}_m)t/2\hbar]}{(\mathscr{E}_k - \mathscr{E}_m)^2}. \quad (65)$$

The structure of this summation is already very interesting. For $t$ large and fixed, the matrix element $H'_{mk}$ can be regarded as varying rather slowly from term to term compared to the remaining factor. Hence, regarded as a function of $\mathscr{E}_m$, the quantity $|c_m|^2$ is sharply peaked about $\mathscr{E}_m = \mathscr{E}_k$, as shown in Figure 3 below. From the figure it is clear that, for $t$ approaching infinity, transitions occur with *appreciable* probability only to a narrower and narrower band of states, with energy $E$ centered about the initial state energy $\mathscr{E}_k$. This expresses the fact of conservation of energy, since only those final states are appreciably populated whose

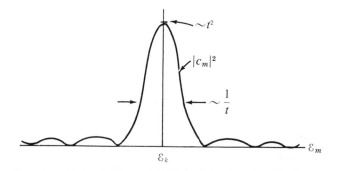

FIGURE 3.   Transition probability against energy.

energy differs infinitesimally from that of the initial state. The non-conservative effects of switching the interaction on and off thus become negligible if the time interval between these acts is long enough.

From the figure we note that the width of the peak in the transition probability curve is inversely proportional to $t$, while the height of the curve is proportional to $t^2$. Thus $P$, which is proportional to the area under the curve, increases linearly with $t$, as asserted. To make this explicit, recalling that the final states are assumed to be dense, we convert the summation in equation (65) to an integral over an interval $\Delta E$ surrounding $\mathscr{E}_k$, by expressing the number of states in the energy interval between $E$ and $E + dE$ as $\rho(E)\, dE$. Thus $\rho(E)$ is the *density of states* at $E$. In this way we obtain

$$P \simeq 4 \int_{\mathscr{E}_k - \Delta E/2}^{\mathscr{E}_k + \Delta E/2} \rho(E)\, dE\, |H'_{mk}|^2 \frac{\sin^2[(\mathscr{E}_k - E)t/2\hbar]}{(\mathscr{E}_k - E)^2},$$

where now $H'_{mk}$ means the matrix element between the initial state and a *typical* final state in the interval $\Delta E$. Neglecting the variation of $\rho(E)$ and of $|H'_{mk}|^2$ over this small interval, we then obtain

$$P \simeq 4\rho(E) \; |H'_{mk}|^2 \int_{\mathscr{E}_k - \Delta E/2}^{\mathscr{E}_k + \Delta E/2} \frac{\sin^2 [(\mathscr{E}_k - E)t/2\hbar]}{(\mathscr{E}_k - E)^2} \; dE.$$

Letting $x = (E - \mathscr{E}_k)t/2\hbar$, we next have

$$P \simeq \frac{2t}{\hbar} \, \rho(E) \; |H'_{mk}|^2 \int_{-t\Delta E/4\hbar}^{t\Delta E/4\hbar} \frac{\sin^2 x}{x^2} \; dx.$$

For $t \to \infty$, the integral can be shown to equal $\pi$, and hence we find

$$P \simeq \frac{2\pi}{\hbar} \, \rho(E) \; |H'_{mk}|^2 \, t, \tag{66}$$

and $P$ is proportional to $t$, as predicted. Introducing the transition rate $W = dP/dt$, we finally obtain

$$W = \frac{2\pi}{\hbar} \, \rho(E) \; |H'_{mk}|^2. \tag{67}$$

This is an extremely important and useful result. It is usually called, after Fermi, the *golden rule* of quantum mechanics. In words, it states that *the number of transitions per unit time from an initial state to a dense group of final states which conserve energy is $2\pi/\hbar$ times the density of final states times the absolute square of the matrix element connecting initial and final states.*

The domain of validity of this result is determined by the following considerations. In the first place, the total probability $P$ that the system has undergone *some* transition from the initial state must be small, $P \ll 1$. Hence, from equation (66), we must have

$$\frac{2\pi}{\hbar} \rho(E) \; |H'_{mk}|^2 t \ll 1, \tag{68}$$

which serves to restrict the time interval over which the results apply. On the other hand, the details of our derivation make clear that this interval cannot be too short. To see this, recall that the width of the peak in Figure 3 is inversely proportional to $t$. If we denote this width by $\epsilon$, we have specifically

$$\epsilon t/2\hbar \simeq 1. \tag{69}$$

Recall also that our evaluation of the sum over final states in equation (64) in effect assumes that $\epsilon$ is small enough that both the density of states and the matrix elements can be treated as constant. If $\epsilon_0$ is a measure of the maximum width over which this assumption is justified, then we must have

$$\epsilon_0 t/2\hbar \gtrsim 1. \tag{70}$$

The only way that equations (68) and (70) can be simultaneously satisfied, in general, is for $H'_{mk}$ to be small. In summary, then, we are not surprised to see that *the golden rule applies only for weak perturbations.*

One other aspect of equation (69) is worthy of comment. The quantity $\epsilon$ is a measure of the uncertainty in the energy of those final states which are reached by the system because of the perturbation. The product of this uncertainty and of the time for which the perturbation acts is thus of the order of $\hbar$, in agreement with the uncertainty principle.

So far, we have considered only perturbations which are independent of time, except for being switched on or off. We now generalize to the case of harmonic time dependence by considering a perturbation of the form

$$H'(x, t) = H'(x)e^{\pm i\omega t}, \tag{71}$$

which is of the type describing the effects of the electromagnetic radiation field.[14] It then follows at once from equation (53) that all of our previous results apply, except that $\mathscr{E}_k - \mathscr{E}_m$ is replaced by $\mathscr{E}_k - \mathscr{E}_m \mp \hbar\omega$. In particular, equation (63) becomes

$$|c_m(t)|^2 = 4|H'_{mk}|^2 \; \frac{\sin^2\left[(\mathscr{E}_k - \mathscr{E}_m \mp \hbar\omega)t/2\hbar\right]}{(\mathscr{E}_k - \mathscr{E}_m \mp \hbar\omega)^2}. \tag{72}$$

To make the notation clear, we remark that the upper sign in equation (72) goes with the upper sign in the exponent of equation (71), and similarly for the lower signs.

We now see that transitions from the $k$th to the $m$th state take place with appreciable probability only when

$$\mathscr{E}_m = \mathscr{E}_k \mp \hbar\omega, \tag{73}$$

which means that transitions are preferentially induced in which a *quantum* of energy $\hbar\omega$ is either absorbed or emitted by the system. The former is called *resonance absorption,* the latter is called *stimulated* or *induced emission.* If $\omega$ is such that equation (73) is satisfied, and if $k$ and $m$ refer to discrete, non-degenerate states, the transition probability is given by equation (64) as before, and hence is still seen to grow quadratically with the time. The result is thus valid only for short enough times, no matter how weak the perturbation. However, just as in the case of a time-independent perturbation, results valid for arbitrary times can be obtained by treating $c_m$ and $c_k$ on an equal footing in equation (61). The resulting equations are again easy to solve, but the details are left to the problems.

---

[14] Strictly speaking, the harmonic time dependence is trigonometric, say $\cos(\omega t + \delta)$. As a consequence, interference occurs between the positive and negative frequency components of the perturbation, but its contribution is negligible for those transitions which take place with appreciable probability, namely those near resonance.

Considering next transitions to (or from) a dense set of degenerate states in the continuum, the derivation of equation (67) is unaltered except that the initial and final state energies are related by equation (73). One can still speak of the process as conserving energy if it is understood that the energy referred to is that of the *total* system, the material part and the radiation field together. Thus in absorption, the material system absorbs a quantum $\hbar\omega$ at the expense of the radiation field. In emission, the radiation field gains a quantum at the expense of the material system.[15] With this understanding, the golden rule and our interpretation of it both stand without modification.

It must be emphasized that, in our development, the electromagnetic field has been treated as a strictly *classical* entity. The fact that absorption and emission are quantized is an automatic consequence of the quantum properties of the material system; it is not directly related to the quantum properties of the field, although it is clearly consistent with those properties. To take these latter into account, the field amplitudes must be regarded as operators acting upon the state function describing the field. These amplitudes can be expressed in terms of harmonic oscillator variables for each mode, and the state of the field is characterized by the number of quanta in each. The zero point energy of these harmonic oscillators plays a crucial role in the interaction of material systems with the electromagnetic field. These vacuum fluctuations, as they are called, are responsible for *spontaneous emission,* the emission of radiation by a system when, as is usually the case, no externally imposed radiation field is present. Our classical treatment, valid only for large quantum numbers, contains no reference to vacuum fluctuations and does not, therefore, account for spontaneous emission.[16]

We remark without proof that the golden rule also applies to the description of induced transitions to a definite, discrete final state from an initial state which is a dense mixture of degenerate states in the continuum. This process is thus the inverse of that described above.

As a specific example of the application of the golden rule, we now consider a particle of charge $e$ in its ground state $\psi_0$ in a potential $V(x)$, which is irradiated by light of frequency $\omega$ and thereby ejected into the continuum. The process is thus essentially that of *photo-ionization*. We shall make two simplifying assumptions in carrying out the calculation. The first of these is that the wavelength of the light is very great compared to the size of the system. This is a very common and generally valid assumption since, except in the x-ray region and below, the wavelength

---

[15] Multiple emission and absorption processes, where two or more photons are involved, are described by second- and higher-order perturbation theory calculations.

[16] For a semiclassical discussion of spontaneous emission, see Reference [23].

of light is certainly very much greater than atomic dimensions. With this assumption, the electric field $\mathcal{E}$ which acts on the particle can be taken to be uniform in space, but harmonic in time, and the perturbation term then has the form

$$H'(x, t) = H'(x)\ e^{-i\omega t} = -e\ \mathcal{E}x\ e^{-i\omega t}, \qquad (74)$$

which is just the instantaneous potential energy of the particle in such a field.[17] The second simplifying assumption we shall make is that the frequency $\omega$ is large enough that the energy of the final state $E$ is very large compared to $V(x)$. This assumption allows us to treat the final state of the particle as a free particle state, that is, as a state of definite momentum as well as energy.

The only points of difficulty in the calculation arise in evaluating the density of final states, since these lie in the continuum, and in the related problem of properly normalizing such states. Both can be solved by the trick of supposing the system to be enclosed in a large box of length $L$, and then eventually letting $L$ approach infinity. A discrete, normalizable set of states is easily obtained, by imposing appropriate boundary conditions, when the system is in such a box. For example, for a physical box, the wave function would have to vanish at the walls. However, the state functions so defined would not be states of definite momentum, even in the limit $L \to \infty$, because states of definite momentum never vanish. Thus we introduce, as a purely mathematical device, the nonphysical periodicity condition that $\psi(x_0 + L) = \psi(x_0)$, where the walls are assumed to lie at $x_0$ and $x_0 + L$. For obvious reasons, this requirement is called a *periodic boundary condition*.

Since the free-particle simultaneous eigenfunctions of momentum and energy are pure de Broglie waves, $e^{i\sqrt{2mE}\,x/\hbar}$, we thus require that

$$\exp\left[i\sqrt{2mE}\,(x_0 + L)/\hbar\right] = \exp\left[i\sqrt{2mE}\,x_0/\hbar\right]$$

and hence that

$$\sqrt{2mE}\ L/\hbar = 2n\pi, \quad n = 0, \pm 1, \pm 2, \ldots.$$

The normalized states are then seen to be

$$\psi_E = \frac{1}{\sqrt{L}}\ e^{i\sqrt{2mE}\,x/\hbar}, \qquad (75)$$

where

---

[17] We have neglected magnetic forces because these are of order $v/c$ smaller than the electric forces, where $v$ is the particle velocity. We have also omitted the positive frequency term because we are interested only in absorption.

$$\sqrt{2mE} = p = \frac{2n\pi\hbar}{L}, \qquad n = 0, \pm 1, \pm 2, \pm. \ldots \qquad (76)$$

As $L$ becomes larger and larger, we see that the spectrum becomes more and more closely spaced, and in the limit the states indeed become the usual free particle continuum states of definite momentum. It is just this fact which both motivates and justifies the introduction of periodic boundary conditions.

We now evaluate the density of states $\rho(E)$. Recall that this density is defined as the number of states with energy between $E$ and $E + \Delta E$. Now from equation (76), the total number of states $N(E)$ with energy less than or equal to $E$ is $2n + 1$, or equivalently, expressing $n$ in terms of $E$,

$$N(E) = \frac{L}{\pi\hbar} \sqrt{2mE} + 1.$$

Consequently,

$$N(E + \Delta E) = \frac{L}{\pi\hbar} \sqrt{2m(E + \Delta E)} + 1 \simeq N(E) + \frac{L\sqrt{2m}}{2\pi\hbar \sqrt{E}} \Delta E.$$

Hence, neglecting higher-order terms,

$$\Delta N = N(E + \Delta E) - N(E) = \frac{L}{2\pi\hbar} \sqrt{2m/E} \, \Delta E,$$

but by definition, $\Delta N = \rho(E)\Delta E$, and thus,

$$\rho(E) = \frac{L}{2\pi\hbar} \sqrt{2m/E}. \qquad (77)$$

Next we write the form of the matrix element of the perturbation $H'(x)$ defined in equation (74). To make explicit the fact that this matrix element is taken between the initial and final states, we denote it by $H'_{if}$. The initial state for our problem is the normalized ground state $\psi_0(x)$ with energy $E_0 = -\epsilon$, where $\epsilon$ is the binding energy. The final state is the normalized free-particle state defined by equation (75), with energy $E_f$ given by

$$E_f = E_0 + \hbar\omega = \hbar\omega - \epsilon. \qquad (78)$$

The matrix element is then

$$H'_{if} = \int \psi_0(x) H'(x) \psi_{E_f}(x) \, dx$$

or, explicitly, using (74) and (75),

$$H'_{if} = -\mathcal{E} \int \psi_0(x) ex \frac{1}{\sqrt{L}} e^{i\sqrt{2mE_f}\, x/\hbar} \, dx.$$

The factor multiplying the electric field strength $\mathscr{E}$ is seen to be the matrix element of the *dipole moment* operator $ex$, and $H'_{if}$ is a quantum mechanical average of the energy of a dipole in a uniform field $\mathscr{E}$.

In any case, the transition probability per unit time is now given by

$$W = \frac{2\pi}{\hbar} \rho(E_f)|H'_{if}|^2$$
$$= \frac{1}{\hbar^2} \sqrt{2m/E_f} \; e^2 \mathscr{E}^2 \left| \int \psi_0(x) \; x \; e^{i\sqrt{2mE_f}\,x/\hbar} \; dx \right|^2. \tag{79}$$

Note that the normalization length $L$, as it must not, does not enter into the final result. The density of states is proportional to $L$ but, through the normalization of the continuum states, the square of the matrix element is inversely proportional to $L$ and their product is thus independent of normalization length. Hence equation (61) as it stands is the limit attained as $L$ becomes infinite.

It is of some interest to actually evaluate the matrix element for some specific example. We shall do so for a particle in its ground state in a square well potential. To simplify the calculation, we shall assume the width of the potential to be narrow enough that the ground state is the only bound state, and further, that this state is so slightly bound that the wave function extends far beyond the walls of the potential. Indeed, we shall take the limit in which the wave function inside the square well contributes negligibly to the integral in equation (61) and can be ignored. Specifically, we thus write, extrapolating the exterior wave function to the origin,

$$\psi_0(x) \simeq (\sqrt{2m\epsilon/\hbar^2})^{1/2} \; e^{-\sqrt{2m\epsilon}\,|x|/\hbar}, \tag{80}$$

where $\epsilon$ is the binding energy. This function, which is normalized, is illustrated in Figure 4, and is schematically compared with the true ground state wavefunction for a narrow potential.[18]

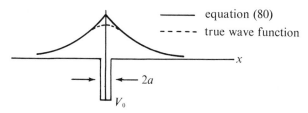

FIGURE 4. The wave function of equation (80) compared to the true wave function in the weakly bound ground state of a narrow square well potential.

---

[18] The wave function of equation (80) can be shown to be that obtained when $2a$ tends to zero and $V_0$ to infinity in such a way that $2aV_0$ approaches a constant, say $g$, that is, the limit in which the square well potential approaches $g\delta(x)$. The binding energy is given in terms of $g$ by $\epsilon = mg^2/2\hbar^2$, as may be verified by taking the limit of equation (VI–16). See also Problem 4 of Chapter VI.

With this final simplification, we then have

$$\int_{-\infty}^{\infty} \psi_0(x) x \, e^{i\sqrt{2mE_f}\,x/\hbar} \, dx$$

$$= \left(\frac{2m\epsilon}{\hbar^2}\right)^{1/4} \int_{-\infty}^{\infty} x \exp\left[-\sqrt{2m}(\sqrt{\epsilon}|x|-i\sqrt{E_f}x)/\hbar\right] dx$$

$$= \left(\frac{2m\epsilon}{\hbar^2}\right)^{1/4} \left\{ \int_{0}^{\infty} x \exp\left[-\sqrt{2m}\,(\sqrt{\epsilon}-i\sqrt{E_f})x/\hbar\right] dx \right.$$

$$\left. + \int_{-\infty}^{0} x \exp\left[\sqrt{2m}(\sqrt{\epsilon}+i\sqrt{E_f})x/\hbar\right] dx\right\}$$

$$= \left(\frac{2m\epsilon}{\hbar^2}\right)^{1/4} \frac{\hbar^2}{2m} \left\{\frac{1}{(\sqrt{\epsilon}-i\sqrt{E_f})^2} - \frac{1}{(\sqrt{\epsilon}+i\sqrt{E_f})^2}\right\}$$

$$= \left(\frac{2m\epsilon}{\hbar^2}\right)^{1/4} \frac{\hbar^2}{2m} \frac{4i\sqrt{E_f}\epsilon}{E_f^2 + \epsilon^2}.$$

Hence, equation (79) yields, for this example,

$$W = \frac{8\hbar e^2 \mathscr{E}^2}{m} \frac{\epsilon^{3/2} E_f^{1/2}}{(E_f^2 + \epsilon^2)^2},$$

where

$$E_f = \hbar\omega - \epsilon.$$

Since we have assumed from the outset that $\hbar\omega \gg \epsilon$, this can be reduced to

$$W = \frac{8e^2 \mathscr{E}^2}{m\hbar^{5/2}} \frac{\epsilon^{3/2}}{\omega^{7/2}}. \tag{81}$$

Subject to all the simplifying assumptions we have made, this result means that if an electromagnetic wave of frequency $\omega$ and electric vector $\mathscr{E}$ irradiates an ensemble of $N$ particles of charge $e$ and mass $m$ each in its ground state of binding energy $\epsilon$, in a square well potential, then the number of photoelectrons produced per second with energy $\hbar\omega - \epsilon$ is $NW$, where $W$ is given above.

The example we have worked out is entirely analogous to the calculation of the photo-ionization of atoms or of the photo (electric)-disintegration of nuclei. Of course, our calculation, nontrivial as it may be, is only one-dimensional, but the three-dimensional calculation is little more difficult. The differences between three dimensions and one are merely characteristic differences in the density of states. Other differences may also arise if the structure of the ground state wave function is different. The photo (electric)-disintegration of the deuteron isolates the effects of the change in density of states, since the deuteron ground state wave function is essentially the one we have used in our calculation. The result is that the transition probability per second turns out to be proportional to $\epsilon^{1/2}/\omega^{5/2}$ at high energies instead of to $\epsilon^{3/2}/\omega^{7/2}$. The photo-ionization of the hydrogen atom, on the other hand, also involves a change in the character of the ground state wave function, which is that appropriate to a Coulomb potential rather than to a square well, with the result that the transition probability falls off much more rapidly with frequency, being proportional to $\epsilon^{5/2}/\omega^{9/2}$.

As a final remark, we note that it is customary to describe these processes in terms of the probability of disintegration or ionization *per incident photon*, rather than in terms of the probability per unit time. The number of photons per cm² per second in a light beam of intensity $I$ is proportional to $I/\hbar\omega$, and $I$ in turn is proportional to the square of the electric field, $\mathscr{E}^2$. Hence the probability per incident photon per cm², or *cross-section* as it is called, is proportional to $\epsilon^{1/2}/\omega^{3/2}$ for photo (electric)-disintegration of the deuteron, and to $\epsilon^{5/2}/\omega^{7/2}$ for photo-ionization of the hydrogen atom.

---

**Problem 1.** If the classical turning point occurs at an infinite potential wall, the wave function actually vanishes at the turning point; it does *not* extend an effective one-eighth wavelength beyond the turning point. The quantum condition equation (12) must then be modified.

(a) What is the correct quantum condition for a particle in a box (infinite potential walls)?

(b) Use the WKB approximation to find the stationary states and compare with the exact results. Explain.

**Problem 2.** Consider the motion of a bouncing ball. Take the motion to be perfectly vertical and the collisions of the ball with the ground to be perfectly elastic. The potential is given in Figure 5, below.

(a) What are the correct quantum conditions?

(b) Use the WKB approximation to find the stationary state energies.

(c) What is the order of magnitude of the quantum number appropriate to the stationary state of a ball of mass 100 gm dropped from one meter?

FIGURE 5. Potential energy of a bouncing ball.

**Problem 3.** Apply the Rayleigh–Ritz variational method to the harmonic oscillator to find:

(a)   The ground state energy using a Gaussian of variationally determined width as a trial function.

(b)   The energy of the first excited state using the trial wave function, with $a^2$ as variational parameter,

$$\psi = \begin{cases} x(x^2 - a^2)^2, & |x| \leq a \\ 0, & |x| \geq a. \end{cases}$$

Does the minimum principle apply? Explain.

**Problem 4.**   Apply the Rayleigh–Ritz variational method to a particle in a box of width $L$ to find:

(a)   The ground state energy using a second-degree polynomial as trial function.

(b)   The ground state energy using a fourth-degree polynomial.

(c)   The energy of the first excited state using the simplest appropriate polynomial as a trial function.

Note: In each case choose your trial function to satisfy the correct boundary conditions at the walls.

**Problem 5.**   In nuclear physics, a frequently used potential is a harmonic oscillator cut off at $x = \pm b$, that is,

$$V(x) = \begin{cases} \frac{1}{2} m\omega^2 (x^2 - b^2), & |x| \leq b \\ 0, & |x| \geq b. \end{cases}$$

(a)   Use the Rayleigh–Ritz method to estimate the energies of the ground state and first excited state. As trial functions use the polynomial in the example in the text, equation (19), and that in Problem 3(b), in each case with $a^2$ as variational parameter. Choose $b$ such that $b > a$.

(b)   The same, for the case $b < a$.

(c)   Estimate the energies of the same two states using the WKB approximation (modified Bohr quantization condition).

(d)   From your variational results and your WKB results, estimate the maximum value of $b$ for which the potential contains only one bound state.

(e)   Compare your WKB and variational results in each case above and state which you think more reliable. Explain your reasoning.

**Problem 6.**   A particle is described by the Hamiltonian $H = H_0 - Fx$, where $H_0$ is the harmonic oscillator Hamiltonian. The system is thus that of a harmonic oscillator in a uniform gravitational field ($F = -mg$) or in a uniform electric field ($F = -e\mathcal{E}$).

(a)   Show that to first order the energy eigenvalues are unchanged.

(b)   Show that

$$x_{n, n+1} = x_{n+1, n} = \sqrt{(n + 1) \hbar/2m\omega}$$

$$x_{n, m} = 0, \; m \neq n \pm 1.$$

(c)  Calculate the energy eigenvalues to second order.

(d)  Find the *exact* solution to the problem. Hint: Introduce a shift of origin. Compare with the results of part (c).

**Problem 7.**

(a)  What is the matrix representing the operation of multiplication by unity? This matrix is called the unit matrix. What about multiplication by a constant $c$?

(b)  Verify by the rules of matrix multiplication that the unit matrix multiplying any matrix gives the expected result.

(c)  Given the two-by-two matrices

$$A = \begin{pmatrix} 1 & 2 \\ 2 & 3 \end{pmatrix}, \quad B = \begin{pmatrix} 2 & -1 \\ -1 & 1 \end{pmatrix},$$

find $A^2$, $B^2$, $A + B$, $AB$ and $BA$.

**Problem 8.**

(a)  The matrix elements $x_{nm}$ of $x$ for the harmonic oscillator are given in Problem 6(b). Using the rules of matrix multiplication find the matrix elements of $x^2$. Verify your results by direct computation of $x^2_{nm}$. (See Problem 2(b) of Chapter VI).

(b)  The same for $p$ and $p^2$. (See Problem 2(c) of Chapter VI.)

(c)  Use your results to verify that

$$(p, x) = \hbar/i.$$

**Problem 9.**  A particle moves with energy $E$ in a potential

$$V(x) = \frac{V_0}{\cosh^2 x/b}.$$

(a)  Assuming $E > |V_0|$, find as precisely as you can the conditions on $V_0$ and $b$ under which the WKB approximation is valid.

(b)  Under the conditions obtained in (a), calculate the WKB transmission coefficient $\tau(E)$ and the time increment associated with the passage of a particle of energy $E$ through the potential.

**Problem 10.**  Consider a perturbation $H'$ which is independent of time except for being switched on at $t = 0$ and switched off at time $t$. Suppose the $m$th and $n$th unperturbed states to be exactly degenerate, $\mathcal{E}_m = \mathcal{E}_n$.

(a)  Assuming that initially the system is in the $m$th unperturbed state, find its subsequent behavior. Start with equation (61) and neglect all states but the degenerate pair, but treat these without further approximation.

(b)  For what initial conditions will the system be in a stationary state, with no (first-order) transitions being induced by the perturbation?

(c)  Discuss the connection of your results with those obtained in stationary perturbation theory.

**Problem 11.** A particle of mass $m$ is placed in a one-dimensional box of width $2L$, centered at the origin, in the presence of a uniform gravitational field. Thus the Hamiltonian of the system is

$$H = \frac{p^2}{2m} + mgx, \qquad |x| \leq L.$$

Estimate the allowed energies of the system using:
(a)  WKB approximation.
(b)  Second-order perturbation theory (why not first-order?).
(c)  The Rayleigh–Ritz method (ground state only). Use the trial function $(1 - \alpha x) \cos \pi x/2L$. Explain why this is a reasonable trial function. Why is *any* function of *definite* symmetry inferior to the one suggested?

**Problem 12.** A particle of mass $m$ is confined inside the triangular potential well shown in Figure 6. Its Hamiltonian is thus $H = (p^2/2m) + V(x)$, where

$$V(x) = \infty, \qquad x < 0$$
$$= kx, \qquad x > 0.$$

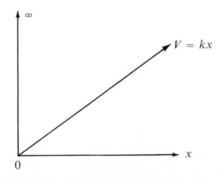

FIGURE 6.   Triangular potential well.

(a)  Estimate the ground state energy using the uncertainty principle.
(b)  Estimate the ground state energy using the Rayleigh–Ritz variational method. As trial function choose

$$\psi(x) = \begin{cases} 0 & , \qquad x \leq 0 \\ x^2 (x^2 - a^2), & \qquad 0 \leq x \leq a \\ 0 & , \qquad x \geq a, \end{cases}$$

and treat $a$ as a variational parameter.
(c)  Suggest a suitable trial function for the first excited states but do *not* carry out any calculations. Briefly justify your choice. Does the first excited state satisfy a minimum principle? Why?

**Problem 13.** The ground state energy of a system is estimated both by the Rayleigh–Ritz method and by a second-order perturbation theory calculation. The Rayleigh–Ritz result is found to be $-27.1$ eV and the perturbation theory result $-26.0$ eV. Which lies closer to the true ground state energy? Suppose the numbers had been reversed. Would it still be possible to decide which estimate is better? Explain your reasoning.

**Problem 14.** A particle of mass $m$ moves in a potential $V(x) = V_0(x/L)^{2s}$, where $s$ is a positive integer. (N.B. The problem is that of the harmonic oscillator for $s = 1$, of a square well for $s \to \infty$.)

(a) Estimate the magnitude of its ground state energy using the uncertainty principle.

(b) Estimate its ground state energy by the Rayleigh–Ritz method using as trial functions

$$\text{(i)} \quad \psi = \exp\left[-x^2/2a^2\right]$$

$$\text{(ii)} \quad \psi = \begin{cases} \sin \pi x/a, & x \le a \\ 0, & x > a. \end{cases}$$

Treat $a$ as a variational parameter in both cases. Which trial function gives better results for differing values of $s$? Discuss your result.

(c) Show that in WKB approximation the energy of the $n$th state can be expressed in the form

$$E_n = k(s) \left[\frac{(n + \frac{1}{2})^{2s} \hbar^{2s} V_0}{m^s L^{2s}}\right]^{\frac{1}{s+1}},$$

where $k(s)$ is a dimensionless quantity which is of order of magnitude unity for all $s$. Find an explicit expression for $k(s)$.

**Problem 15.** A harmonic oscillator of mass $m$, charge $e$ and classical frequency $\omega$ is in its ground state.

(a) A uniform electric field $\mathscr{E}$ is turned on at $t = 0$ and is then turned off at $t = \tau$. Use first-order time dependent perturbation theory to estimate the probability that the system is excited to its $n$th state.

(b) The same for a uniform but sinusoidally oscillating electric field, $\mathscr{E} = \mathscr{E}_0 \sin \omega_0 t$.

**Problem 16.** Let $\psi_1$ and $\psi_2$ be a pair of degenerate orthonormal states of some system. At time $t = 0$, a perturbation $H'$ is turned on and at time $t = \tau$ it is turned off. If the system is initially in the state $\psi_1$,

(a) Find $\psi(x, t)$ for $t > \tau$ and find the probability that a transition has occurred from $\psi_1$ to $\psi_2$. Neglect all other states except the degenerate pair. For simplicity, assume

$$\langle \psi_1 | H' | \psi_2 \rangle = \langle \psi_2 | H' | \psi_1 \rangle = \epsilon$$

$$\langle \psi_2 | H' | \psi_2 \rangle = \langle \psi_1 | H' | \psi_1 \rangle = 0.$$

(b)  Compare your result for the transition probability with that obtained using first-order time dependent theory and then find precise conditions for the validity of the latter in the present instance.

**Problem 17.** The harmonic oscillator reduces to a free particle system when its frequency tends to zero. It is easily verified that the harmonic oscillator propagator properly becomes that for a free particle in this limit. It is not so easy, however, to verify that the harmonic oscillator state functions reduce to free particle state functions.

(a)  Do so by letting $\omega \to 0$ and $n \to \infty$ in such a way that

$$E_n = (n + \tfrac{1}{2})\, \hbar\omega \to E$$

and examining the oscillator state function $\psi_n$ in this limit. [Hint: Use the integral representation of equation (VI–61) and evaluate the integral by a saddle point method.]

(b)  Consider the WKB approximation to the harmonic oscillator state functions (but not too close to the turning points). Show that, for large enough $n$, the state function is quite close to a free particle state over the central region.

(c)  Show that the WKB states reduce to the correct free particle states in the zero frequency limit described in part (a). Note that the turning points become infinitely remote in that limit.

**Problem 18.** A system with unperturbed eigenstates and energies $\phi_n$ and $E_n$, respectively, is subjected to a time dependent perturbation

$$H'(t) = \frac{A}{\sqrt{\pi}\,\tau}\, e^{-t^2/\tau^2},$$

where $A$ is a time *independent* operator.

(a)  If initially $(t = -\infty)$ the system is in its ground state $\phi_0$, show that, to first order, the probability amplitude that at $t = \infty$ the system will be in its $m$th state $(m \neq 0)$ is

$$c_m = -\frac{i\, A_{m0}}{\hbar}\, e^{-(E_0 - E_m)^2 \tau^2/4\hbar^2},$$

where

$$A_{m0} = \langle \phi_m | A | \phi_0 \rangle.$$

(b)  The limit $\tau^2 (E_1 - E_0)^2 / 4\hbar^2 \gg 1$ is called the *adiabatic limit*. Discuss the behavior of the system as $t$ progresses from minus to plus infinity in the adiabatic limit. Why do all transition probabilities tend to zero in this limit?

(c)  Next consider the limit of an impulsive perturbation, $\tau = 0$. Show that the probability $P$ that the system makes any transition whatsoever out of the ground state is

$$P = \frac{1}{\hbar^2} \left[ (A^2)_{00} - (A_{00})^2 \right]$$

$$= \frac{1}{\hbar^2} \left[ \langle \phi_0 | A^2 | \phi_0 \rangle - \langle \phi_0 | A | \phi_0 \rangle^2 \right].$$

Hint: Find the transition probability to the $m$th state and sum over all excited states using the methods of matrix algebra.

(d)  Show that the impulsive perturbation of part (c) is equivalent to

$$H'(t) = A \delta(t).$$

(e)  Integrate the time dependent Schrödinger's equation over an infinitesimal interval about $t = 0$ to show that $\psi(x, t)$ is discontinuous at $t = 0$. Show specifically that[19]

$$\psi(t = 0_+) = \left( 1 + \frac{iA}{2\hbar} \right)^{-1} \left( 1 - \frac{iA}{2\hbar} \right) \psi(t = 0_-),$$

and hence that the exact solution of the time dependent Schrödinger's equation is

$$t < 0: \qquad \psi = \phi_0 \, e^{-iE_0 t/\hbar}$$

$$t > 0: \qquad \psi = \Sigma \, c_m \, \phi_m \, e^{-iE_m t/\hbar}$$

$$c_m = \langle \phi_m | \left( 1 + \frac{iA}{2\hbar} \right)^{-1} \left( 1 - \frac{iA}{2\hbar} \right) | \phi_0 \rangle.$$

Show that this reduces to the first-order perturbation theory result if the perturbation is weak enough.

(f)  Verify that $\Sigma |c_m|^2 = 1$.

**Problem 19.**  Let $\psi_n$ denote the stationary bound states of a system corresponding to energy $E_n$. Choosing the $\psi_n$ to be real for convenience (see problem VI–11),

(a)  Show that

$$\langle \psi_n | px | \psi_n \rangle = \hbar/2i$$

$$\langle \psi_n | xp | \psi_n \rangle = ?.$$

(b)  Show that

$$\sum_q \left\{ \langle \psi_n | x H | \psi_q \rangle \langle \psi_q | x | \psi_n \rangle - \langle \psi_n | x | \psi_q \rangle \langle \psi_q | x H | \psi_n \rangle \right\}$$

$$= \sum_q (E_q - E_n) |x_{nq}|^2.$$

_____

[19] If B denotes some operator, its inverse $B^{-1}$ is *defined*, when it exists, by

$$B^{-1} B = B B^{-1} = 1.$$

(c)   Using the results of (a) and (b), and the methods of matrix algebra, show, finally, that

$$\sum_q (E_q - E_n)|x_{nq}|^2 = \hbar^2/2m.$$

[This is an extremely important example of what is called a *sum rule*. Its importance comes from the fact that the probability for a dipole radiative transition between the $q$th and $n$th states is proportional to $|x_{nq}|^2$. Thus these matrix elements are directly measurable quantities.]

(d)   Verify the sum rule for the harmonic oscillator by actual evaluation of the sum.

**Problem 20.** In Chapter VI we showed that an attractive square well has at least one bound state no matter how weak the potential. Use the Rayleigh–Ritz variational method to prove that this is a general property of *any* potential which is purely attractive. Do this by using the trial function

$$\psi = e^{-\alpha x^2}$$

and showing that $\alpha$ can always be so chosen that $E'(\alpha)$ is negative. (Why does this constitute a proof?)

# VIII

## *Systems of particles in one dimension*

### 1. FORMULATION

We now come to the first major generalization of our formulation of the laws of quantum mechanics. We have seen that the requirements of the correspondence principle led us to the Hamiltonian as the operator which determines the time development of a system consisting of a single particle moving in an external field of force. By similar arguments, it follows that the Hamiltonian operator for a system of particles plays exactly the same role, as we now show.

We consider a system of $A$ interacting particles, each of which may also be under the influence of some external force. The classical Hamiltonian for such a system is then

$$H(p_1, \ldots, p_A, x_1, \ldots, x_A) = \sum_{i=1}^{A} \frac{p_i^2}{2m_i} + \sum_{i=1}^{A} V_i(x_i) + \sum_{\text{pairs}} V_{ij}(x_i - x_j),$$

(1)

where $p_i^2/2m_i$ is the kinetic energy of the $i$th particle of mass $m_i$ and $V_i(x_i)$ is any *externally* imposed potential (such as gravity) in which the $i$th particle is moving. Finally,

$$V_{ij}(x_i - x_j) \equiv V_{ji}(x_j - x_i)$$

is the mutual potential energy of interaction between the $i$th particle at $x_i$ and the $j$th at $x_j$. As the notation shows, we take this interaction to depend only upon the coordinate difference $x_i - x_j$. The total energy of interaction is obtained by summing the two particle terms over all pairs of particles, as indicated. An explicit way of writing this term is

$$\sum_{\text{pairs}} V_{ij}(x_i - x_j) \equiv \sum_{\substack{i,j \\ i<j}} V_{ij}.$$

In summing over $i$ and $j$, the condition that $i$ is always less than $j$ insures that each pair is counted only once.

We now make the following assumptions about the quantum mechanical description of such a system, all of which are more or less obvious generalizations of the one particle description.

(a)   In configuration space, the state function of the system depends on the coordinate of each particle and on the time,

$$\psi = \psi(x_1, x_2, \ldots, x_A, t).$$

(b)   Assuming $\psi$ to be normalized,

$$\int dx\, dx_2 \cdots dx_A |\psi|^2 = 1, \tag{2}$$

the absolute probability density in configuration space is $|\psi|^2$. More explicitly, $|\psi|^2\, dx_1\, dx_2 \cdots dx_A$ is the probability that, at one and the same instant $t$, particle one is to be found in the interval between $x_1$ and $x_1 + dx_1$, particle two between $x_2$ and $x_2 + dx_2$, and so on for all $A$ particles. From this it follows that

$$\rho(x_1) = \int dx_2 \cdots dx_A |\psi(x_1, x_2, \ldots, x_A, t)|^2 \tag{3}$$

is the probability density for particle one alone, independent of the details of the behavior of the remaining particles. Thus $\rho(x_1)$ has the same meaning and the same properties as the familiar probability density for a single particle; in particular, the usual probability conservation laws must be satisfied.

(c)   The dynamical variables of each particle are operators satisfying the normal commutation law. However, since the dynamical variables of different particles represent *totally different* degrees of freedom, and thus noninterfering observables, the dynamical variables of different particles commute with each other. In brief, we thus write

$$[p_i, p_j] = [x_i, x_j] = 0$$

$$[p_i, x_j] = \frac{\hbar}{i}\,\delta_{ij}. \tag{4}$$

Specifically, in configuration space the $x_i$ are numbers and the $p_i$ are given by

$$p_i = \frac{\hbar}{i}\frac{\partial}{\partial x_i}. \tag{5}$$

In momentum space, conversely, the $p_i$ are numbers and the $x_i$ are given by

$$x_i = -\frac{\hbar}{i}\frac{\partial}{\partial p_i}. \tag{6}$$

From these assumptions it follows that the state function in momentum space $\phi(p_i, p_2, \ldots, p_A, t)$ is the $A$-fold Fourier transform[1] of $\psi$, that expectation values are computed in the usual way,

$$\langle f(x_i, \ldots, x_A; p_i, \ldots, p_A) \rangle = \langle \psi | f\left(x_i, \ldots, x_A; \frac{\hbar}{i} \frac{\partial}{\partial x_i}, \ldots, \frac{\hbar}{i} \frac{\partial}{\partial x_A}\right) | \psi \rangle$$

$$= \langle \phi | f\left(-\frac{\hbar}{i} \frac{\partial}{\partial p_i}, \ldots, -\frac{\hbar}{i} \frac{\partial}{\partial p_A}; p_1, \ldots, p_A\right) | \phi \rangle,$$

and that Schrödinger's equation must have the general form

$$H\psi = E\psi = -\frac{\hbar}{i} \frac{\partial \psi}{\partial t}, \tag{7}$$

where $H$ is some linear Hermitian operator. The first-order time derivative is required for probability conservation just as before. Using the fact that, from equations (4), (5) and (6),

$$[p_i, f(x_1, \ldots, x_A; p_1, \ldots, p_A)] = \frac{\hbar}{i} \frac{\partial f}{\partial x_i}$$

$$[f(x_1, \ldots, x_A, p_1, \ldots, p_A), x_i] = \frac{\hbar}{i} \frac{\partial f}{\partial p_i},$$

we then find that, by arguments exactly paralleling those used in the single-particle case (see Section 3 of Chapter V),

$$\frac{d}{dt} \langle x_i \rangle = \langle \frac{\partial H}{\partial p_i} \rangle$$

$$\frac{d}{dt} \langle p_i \rangle = -\langle \frac{\partial H}{\delta x_i} \rangle.$$

Associating expectation values with the classical variables in the sense of the correspondence principle, we see that these are the classical equations of motion in Hamiltonian form, provided that $H$ in equation (7) is the Hamiltonian operator, that is, the Hamiltonian of equation (1) regarded as a function of the quantum mechanical dynamical variables defined by (4), (5) and (6).

## 2. TWO PARTICLES: CENTER-OF-MASS COORDINATES

As a first application of our generalized formulation we consider two

---

[1] Specifically, this means that $\psi$ and $\phi$ are related by

$$\psi = (2\pi\hbar)^{-A/2} \int dp_1 \cdots dp_A \, \phi(p_1, \cdots, p_A) \exp [i(p_1 x_1 + p_2 x_2 + \cdots + p_A x_A)/\hbar]$$

and

$$\phi = (2\pi\hbar)^{-A/2} \int dx_1 \cdots dx_A \, \psi(x_1, \ldots, x_A) \exp [-i(p_1 x_1 + \cdots + p_A x_A)/\hbar].$$

particles, of mass $m_1$ and $m_2$, respectively, which interact through a potential $V(x_1 - x_2)$ in the absence of external forces. The Hamiltonian of the system is thus

$$H = \frac{p_1{}^2}{2m_1} + \frac{p_2{}^2}{2m_2} + V(x_1 - x_2), \tag{8}$$

and Schrödinger's equation in configuration space is then

$$\left[ -\frac{\hbar^2}{2m_1} \frac{\partial^2}{\partial x_1{}^2} - \frac{\hbar^2}{2m_2} \frac{\partial^2}{\partial x_2{}^2} + V(x_1 - x_2) \right] \Psi(x_1, x_2, t) = -\frac{\hbar}{i} \frac{\partial \Psi}{\partial t}. \tag{9}$$

The form of $V$ suggests the introduction of the distance between the particles as a new coordinate, and we therefore introduce the transformation

$$x = x_1 - x_2$$

$$X = \alpha x_1 + \beta x_2,$$

where $\alpha$ and $\beta$ are parameters of our choice. Of course $X$ will turn out to be the center-of-mass coordinate, but it is instructive to see how this works out. We now have

$$\frac{\partial}{\partial x_1} = \frac{\partial x}{\partial x_1} \frac{\partial}{\partial x} + \frac{\partial X}{\partial x_1} \frac{\partial}{\partial X} = \frac{\partial}{\partial x} + \alpha \frac{\partial}{\partial X}$$

$$\frac{\partial}{\partial x_2} = -\frac{\partial}{\partial x} + \beta \frac{\partial}{\partial X},$$

whence

$$\frac{1}{m_1} \frac{\partial^2}{\partial x_1{}^2} + \frac{1}{m_2} \frac{\partial^2}{\partial x_2{}^2} = \frac{1}{\mu} \frac{\partial^2}{\partial x^2} + \left( \frac{\alpha^2}{m_1} + \frac{\beta^2}{m_2} \right) \frac{\partial^2}{\partial X^2} + 2 \left( \frac{\alpha}{m_1} - \frac{\beta}{m_2} \right) \frac{\partial^2}{\partial x \partial X},$$

where $\mu$, the reduced mass of the system, is given by

$$\frac{1}{\mu} = \frac{1}{m_1} + \frac{1}{m_2}; \qquad \mu = \frac{m_1 m_2}{m_1 + m_2}. \tag{10}$$

To eliminate the cross-term, we choose $\beta$ such that

$$\beta = \frac{m_2}{m_1} \alpha,$$

and hence

$$\frac{1}{m_1} \frac{\partial^2}{\partial x_1{}^2} + \frac{1}{m_2} \frac{\partial^2}{\partial x_2{}^2} = \frac{1}{\mu} \frac{\partial^2}{\partial x^2} + \frac{\alpha^2}{m_1} (m_1 + m_2) \frac{\partial^2}{\partial X^2}.$$

The choice of $\alpha$ is still quite arbitrary; it merely serves to fix the scale of the $X$ coordinate. The most convenient choice is clearly that for which

$$\int \int dx_1 \, dx_2 \rightarrow \int \int dx \, dX,$$

and this requires that the Jacobian of the transformation be unity. Thus

$$1 = \left\| \begin{matrix} \dfrac{\partial x}{\partial x_1} & \dfrac{\partial x}{\partial x_2} \\[2mm] \dfrac{\partial X}{\partial x_1} & \dfrac{\partial X}{\partial x_2} \end{matrix} \right\| = \left\| \begin{matrix} 1 & -1 \\[2mm] \alpha & \dfrac{m_2}{m_1} \alpha \end{matrix} \right\| = \alpha \left( \dfrac{m_2}{m_1} + 1 \right),$$

and hence

$$\alpha = \frac{m_1}{m_1 + m_2} = \frac{m_1}{M},$$

where $M = m_1 + m_2$ is the total mass. Thus finally

$$x = x_1 - x_2, \qquad X = \frac{m_1 x_1 + m_2 x_2}{M} \tag{11}$$

and, conversely,

$$x_1 = X + \frac{m_2}{M} x, \qquad x_2 = X - \frac{m_1}{M} x, \tag{12}$$

whence Schrödinger's equation becomes

$$\left[ -\frac{\hbar^2}{2\mu} \frac{\partial^2}{\partial x^2} - \frac{\hbar^2}{2M} \frac{\partial^2}{\partial X^2} + V(x) \right] \Psi(x, X, t) = -\frac{\hbar}{i} \frac{\partial \Psi}{\partial t}. \tag{13}$$

We now give an alternative method for deriving Schrödinger's equation in the center-of-mass coordinate system, as the coordinates defined by equations (11) and (12) are called. The classical Hamiltonian, equation (8), can at once be written in center-of-mass coordinates in the familiar form[2]

$$H = \frac{P^2}{2M} + \frac{p^2}{2\mu} + V(x), \tag{14}$$

where

$$P = M \frac{dX}{dt} = m_1 \frac{dx_1}{dt} + m_2 \frac{dx_2}{dt} = p_1 + p_2$$

$$p = \mu \frac{dx}{dt} = \mu \left( \frac{p_1}{m_1} - \frac{p_2}{m_2} \right).$$

Consider now the quantum mechanical operators corresponding to these new coordinates and momenta. We have

---

[2] See References [14] through [17].

$$[p, x] = \left[ \mu \left( \frac{p_1}{m_1} - \frac{p_2}{m_2} \right), \ (x_1 - x_2) \right] = \frac{\mu}{m_1} [p_1, x_1] + \frac{\mu}{m_2} [p_2, x_2] = \frac{\hbar}{i}$$

$$[P, X] = \left[ (p_1 + p_2), \ \frac{m_1 x_1 + m_2 x_2}{M} \right] = \frac{m_1}{M} [p_1, x_1] + \frac{m_2}{M} [p_2, x_2] = \frac{\hbar}{i},$$

while, in the same way, it is easy to see that

$$[p, X] = [P, x] = 0.$$

The transformation from $p_1, x_1, p_2, x_2$ to $p, x, P, X$ thus leaves the commutation rules unchanged; such a transformation is called a *canonical transformation*.[3] From these commutation rules we can write, in configuration space,

$$p = \frac{\hbar}{i} \frac{\partial}{\partial x}, \qquad P = \frac{\hbar}{i} \frac{\partial}{\partial X}, \tag{15}$$

and hence Schrödinger's equation in the form of equation (13) follows at once from the Hamiltonian in center-of-mass coordinates, equation (14). Note that the center-of-mass momentum $P$ commutes with $H$ and hence is a constant of the motion. Just as in classical physics, *the total momentum of an isolated system is conserved*.

Schrödinger's equation in center-of-mass coordinates is clearly separable, which is the main point of the transformation. Thus, writing

$$\Psi(x, X, t) = \chi(X, t)\psi(x, t) \tag{16}$$

we have

$$\frac{1}{\psi} \left[ -\frac{\hbar^2}{2\mu} \frac{\partial^2 \psi}{\partial x^2} + V(x)\psi + \frac{\hbar}{i} \frac{\partial \psi}{\partial t} \right] = -\frac{1}{\chi} \left[ -\frac{\hbar^2}{2M} \frac{\partial^2 \chi}{\partial X^2} + \frac{\hbar}{i} \frac{\partial \chi}{\partial t} \right].$$

The separation constant is merely an additive constant in the total energy and hence can be set equal to zero without loss of generality.[4] With this choice, the state function of the center of mass of the system satisfies the free-particle Schrödinger's equation

$$-\frac{\hbar^2}{2M} \frac{\partial^2 \chi}{\partial X^2} = -\frac{\hbar}{i} \frac{\partial \chi}{\partial t}, \tag{17}$$

and can thus be taken to be a free-particle wave packet at rest or in uni-

[3] Such a transformation is the quantum analog of a classical canonical transformation, that is, a transformation which leaves the Hamiltonian equations unaltered in form. See, for example, Reference [14].

[4] More generally, the separation "constant" could be regarded as an arbitrary function of the time. This merely reflects the invariance of the total wave function $\Psi(x, X, T)$ to the transformation $\psi_f = \psi\, e^{if(t)}$, $\chi_f = \chi\, e^{-if(t)}$, by which we mean the simultaneous replacement of $\psi$ by $\psi\, e^{if(t)}$ and $\chi$ by $\chi\, e^{-if(t)}$. No physical result depends on $f$ and no loss of generality is incurred in choosing it to be zero.

form motion; it can also be taken to be an eigenstate of the total momentum, as circumstances warrant.

---

**Exercise 1.** From equations (14) and (16) verify that

$$\langle E_{\text{total}} \rangle = \langle E_{\text{cm}} \rangle + \langle E_{\text{rel}} \rangle,$$

where $E_{\text{cm}}$ and $E_{\text{rel}}$ respectively refer to the energy of the center of mass and of the relative motion.

---

The relative motion is governed by the equivalent single-particle Schrödinger's equation

$$-\frac{\hbar^2}{2\mu} \frac{\partial^2 \psi}{\partial x^2} + V(x)\psi = -\frac{\hbar}{i} \frac{\partial \psi}{\partial t}, \tag{18}$$

the solutions of which have already been thoroughly investigated. For example, for a quadratic interaction $V(x) = \mu\omega^2 x^2/2$, the states of relative motion are just those of the harmonic oscillator. Such a system is a kind of one-dimensional diatomic molecule, and the states in question are vibrational in character. As a second example, continuum states of the kind discussed earlier can be interpreted as describing the collision of two particles (in one dimension). The reflection and transmission coefficients give the probability that the particles bounce back from or pass through one another in the course of the collision. This transmission process may seem strange classically, since classical particles are normally regarded as impenetrable. However, even on the classical level, impenetrability implies an interaction potential which becomes infinitely repulsive at small distances of separation. For such a potential, the quantum mechanical transmission coefficient is easily seen to vanish, in agreement with the classical result.

Our discussion thus far has been concerned with only a pair of interacting particles. What about isolated systems of three or more particles? It is easy to demonstrate, just as in classical mechanics, that the center-of-mass motion can be separated out and that this motion is governed by the equations for a free particle. However, again just as in classical mechanics, the internal motion is extremely complicated. Even in the roughest approximation, the techniques involved in analyzing the states of such a system are too sophisticated and too special to make it profitable for us to pursue the subject further at this time.

## 3. INTERACTING PARTICLES IN THE PRESENCE OF UNIFORM EXTERNAL FORCES

We now generalize our considerations to the case in which external forces are present. In these circumstances, no particular simplification is guaranteed by a transformation to center-of-mass coordinates. Of course, the motion of the center of mass is governed by the net external force on the system, quantum mechanically as well as classically, but this net force depends in general upon the configuration of the system. Thus the center-of-mass motion and the internal motion are *coupled* or, to put it another way, Schrödinger's equation is not separable in the presence of arbitrary external forces. However, as the above argument implies, there is one exceptional situation in which separability is always possible, namely that in which the external forces are uniform, because the net force is then independent of configuration. We now take up this special case.

Specifically, we consider two particles, each of which is subject to a constant external force, say $F_1$ and $F_2$ respectively, and which interact with each other through a potential $V(x_1 - x_2)$. The Hamiltonian of the system is thus

$$H = \frac{p_1^2}{2m_1} + \frac{p_2^2}{2m_2} + V(x_1 - x_2) - F_1 x_1 - F_2 x_2.$$

Transforming to center-of-mass coordinates, using equation (12), we then have at once

$$H = \left[\frac{p^2}{2\mu} + V(x) - \left(\frac{F_1 m_2 - F_2 m_1}{M}\right) x\right] + \left[\frac{p^2}{2M} - (F_1 + F_2) X\right].$$

Schrödinger's equation is now easily seen to separate, and we obtain for the center-of-mass motion

$$\left[-\frac{\hbar^2}{2M} \frac{\partial^2}{\partial X^2} - (F_1 + F_2)X\right] \chi(X, t) = -\frac{\hbar}{i} \frac{\partial \chi}{\partial t} \qquad (19)$$

and, for the internal motion,

$$\left[-\frac{\hbar^2}{2\mu} \frac{\partial^2}{\partial x^2} + V(x) - \left(\frac{F_1 m_2 - F_2 m_1}{M}\right) x\right] \psi(x, t) = -\frac{\hbar}{i} \frac{\partial \psi}{\partial t}. \qquad (20)$$

As expected, equation (19) is the equation governing the motion of the center of mass under the influence of the net external force. Its solutions are, unfortunately, rather complicated, but they are merely the quantum mechanical transcription of motion under uniform acceleration.[5]

The additional term in equation (20) represents the differential effect of the external force on the motion of the particles relative to each other. If this term is small compared to the interaction potential, it can be treated as a perturbation in the usual way. Thus, for example, to first order the energy is shifted by the expectation value of the external force term. Of

[5] See Problem VI–8.

course, if $V(x)$ is symmetrical so that the unperturbed states have definite parity, this expectation value vanishes and the energy shift is of second order, that is, proportional to the *square* of the effective external forces.

The special case in which the interaction potential $V(x)$ is quadratic in $x$ is particularly simple, since the perturbing term can then be transformed away by a change of origin (see Chapter VII, Problem 6). The states are thus harmonic-oscillator states in which the minimum of the potential energy is shifted away from the origin and in which the energy of each state is decreased by a constant amount, proportional to the square of the effective external force. To give this example a physical cast, let us call the two-particle system a diatomic molecule. Further, let us particularize the external forces by considering first the case of such a molecule in a uniform gravitational field. In that case $F_1 = -m_1 g$, $F_2 = -m_2 g$ and the external force term in equation (20) is seen to vanish, while in equation (19) it is $(m_1 + m_2)g = MgX$. Thus, the internal motion is unaffected by the gravitational field and the center of mass accelerates in the expected way.

Consider next the case in which the external force is produced by a uniform electric field $\mathscr{E}$. If the particles have equal and opposite charge, of magnitude $e$, then $F_1 = -e\,\mathscr{E}$ and $F_2 = e\,\mathscr{E}$. The net external force thus vanishes, as it must since the molecule has no net charge, and the center-of-mass motion is that of a free particle. The additive term in equation (20) is $\mathscr{E}\,ex$, which is just the energy of an electric dipole of moment $ex$ in a uniform field $\mathscr{E}$. Recalling that the energy shift is quadratic in the external force, the ground state energy of the diatomic molecule has the form

$$E = E_0 - \tfrac{1}{2}\,\alpha\,\mathscr{E}^2, \tag{21}$$

where $\alpha$ is expressible in terms of the parameters of the harmonic oscillator potential. This energy shift is the energy of an *induced* dipole, of moment $\alpha\,\mathscr{E}$, in the field $\mathscr{E}$; the quantity $\alpha$ is called the *electric polarizability* of the molecule. Note, for comparison, that if the ground state of the system had not had a definite parity, then the energy shift would have been linear, rather than quadratic in $\mathscr{E}$. That is, the ground state energy would have had the form

$$E = E_0 - \mu_e\,\mathscr{E},$$

which is the energy of a *permanent* electric dipole, of moment $\mu_e$, in an external field. From this example, we see that if the ground state of a system has a definite parity, it has no permanent electric dipole moment. This is, overwhelmingly, the most common case in nature. For such systems, the action of an external electric field distorts the system, thereby producing a separation of charge proportional to the strength of the electric field which is expressed in terms of polarizability. We remark that the polarizability is a more or less directly measurable quantity, since the electric susceptibility of a gas can be simply expressed in terms of the electric polarizability of its constituent molecules.

## 4. COUPLED HARMONIC OSCILLATORS

Before going on to consider general external forces, we now briefly take up a second exactly soluble example, that of a pair of coupled harmonic oscillators. Specifically, we consider a two-particle Hamiltonian of the form[6]

$$H = \frac{p_1{}^2}{2m_1} + \frac{p_2{}^2}{2m_2} + \frac{1}{2} m_1\omega^2 x_1{}^2 + \frac{1}{2} m_2\omega^2 x_2{}^2 + \frac{1}{2} k(x_1 - x_2)^2. \tag{22}$$

It is readily verified that the *normal coordinates* for this problem are just the familiar center-of-mass and relative coordinates of equation (11).[7] Introduction of these coordinates transforms the Hamiltonian into the separable form

$$H = H_{\text{cm}} + H_{\text{rel}}, \tag{23}$$

where

$$H_{\text{cm}} = \frac{P^2}{2M} + \frac{1}{2} M\omega^2 X^2 \tag{24}$$

and

$$H_{\text{rel}} = \frac{p^2}{2\mu} + \frac{1}{2} \mu\bar{\omega}^2 X^2 \tag{25}$$

and where we have introduced the new frequency $\bar{\omega}$, defined by

$$\bar{\omega} = \sqrt{\omega^2 + \frac{k}{\mu}}. \tag{26}$$

Observe that if the interaction term is attractive, $k > 0$, we have $\bar{\omega} > \omega$, and conversely, for repulsive interactions, $k < 0$, we have $\bar{\omega} < \omega$.

Denoting the respective harmonic oscillator eigenfunctions of $H_{\text{cm}}$ and $H_{\text{rel}}$ by $\Phi_N(X)$ and $\phi_n(x)$, with energies $(N + \frac{1}{2})\hbar\omega$ and $(n + \frac{1}{2})\hbar\bar{\omega}$, we then have as the *exact* stationary states

$$\Psi_{Nn} = \Phi_N(X)\, \phi_n(x), \qquad N, n = 0, 1, 2, \ldots, \tag{27}$$

with energies

---

[6] For simplicity, we have chosen the somewhat special case in which the uncoupled oscillators each have the same frequency $\omega$.

[7] This and subsequent details are left to the problems.

$$E_{Nn} = (N + \tfrac{1}{2})\hbar\omega + (n + \tfrac{1}{2})\hbar\overline{\omega}. \tag{28}$$

The spectrum, which is seen to be non-degenerate if $\overline{\omega}$ and $\omega$ are noncommensurate, is a simple composition of two harmonic oscillator spectra. Its appearance, however, is strongly dependent on the relative magnitude of $\overline{\omega}$ and $\omega$. In Figure 1(a) the spectrum is shown for weakly attractive interactions, by which we mean $0 < k/\mu \ll \omega^2$. For this case, $\overline{\omega}$ only slightly exceeds $\omega$ and we obtain well-separated sets of states: an isolated ground state, a close-lying pair of first excited states, a close-lying triplet of second excited states, and so on. For weakly repulsive

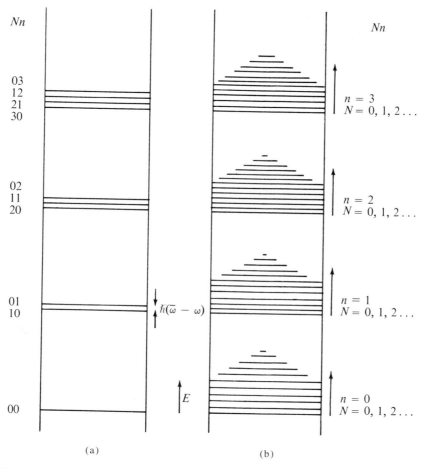

FIGURE 1. (a) Coupled oscillator spectrum for weak attractive forces. All close-lying states have the common spacing $\hbar(\overline{\omega} - \omega)$. (b) Coupled oscillator spectrum for dominantly strong attractive forces. Each infinite ladder of states starting on a given $n$ has equal spacing $\hbar\omega$.

states, the same structure appears, except that the order of the close-lying states is reversed. In Figure 1(b), the spectrum is shown for the opposite limit $k/\mu \gg \omega^2$, when the interaction term dominates. Here, the spectrum is seen to consist of a set of overlapping infinite ladders of closely-spaced states, the first ladder starting on the ground state, the next on the state $n = 1$, the next on $n = 2$, and so on.

The state function, although a reasonably simple function of the normal coordinates, is a rather complicated function of the particle coordinates $x_1$ and $x_2$, except for the ground state, which, properly normalized, can be expressed as

$$\psi_{00}(x_1, x_2) = \left(\frac{M\mu\omega\bar{\omega}}{\pi^2\hbar^2}\right)^{1/4}$$

$$\times \exp\left[-\frac{m_1\omega}{2\hbar}x_1{}^2 - \frac{m_2\omega}{2\hbar}x_2{}^2 - \frac{\mu(\bar{\omega} - \omega)}{2\hbar}(x_1 - x_2)^2\right]. \quad (29)$$

The excited states then consist of the same Gaussian factor multiplied by unwieldy polynomials of degree $(n + N)$ in $x_1$ and $x_2$.

The coupled oscillator problem has some intrinsic interest. For example, in the strong interaction limit it is a crude model of the states of a tightly bound diatomic molecule in a crystal. However, its main interest is that it serves as a guide to applying approximation methods to the study of properties of a system of particles (see Problem 9). In this connection, we remark that the utility of the coupled oscillator problem is not limited to the two-particle system we have been analyzing. Exact solutions can be found for any number of particles[8] and we shall, in fact, eventually discuss the three-particle case.

## 5. WEAKLY INTERACTING PARTICLES IN THE PRESENCE OF GENERAL EXTERNAL FORCES

We now return to the general case of nonuniform external forces where, as we emphasized at the outset, the center-of-mass motion cannot be separated out. Even for a two-particle system, this problem is a very difficult one, and we restrict our attention to the limit in which the external forces dominate over the internal forces, so that the latter can be treated as a perturbation. As a starting point we must thus discuss the unperturbed states, which are those of a set of noninteracting particles moving under the influence of external forces. We shall consider only two-particle systems in detail, but we will also briefly mention many-particle systems.

---

[8] For the classical case, see References [14] through [17]. For the quantum case, see I. Bloch and Y. Hsieh, *Physical Review* **96**, 382 (1954); **101**, 205 (1956).

We start with the two-particle Hamiltonian

$$H = \left[ \frac{p_1{}^2}{2m_1} + V_1(x_1) \right] + \left[ \frac{p_2{}^2}{2m_2} + V_2(x_2) \right] = H_1(p_1, x_1) + H_2(p_2, x_2). \quad (31)$$

Center-of-mass coordinates are clearly irrelevant for this system, which classically consists simply of two particles moving independently. Writing

$$\Psi(x_1, x_2, t) = \psi_1(x_1, t)\, \psi_2(x_2, t), \quad (32)$$

Schrödinger's equation is seen to separate, and we obtain the two single-particle equations

$$H_1\psi_1(x_1, t) = \left[ \frac{p_1{}^2}{2m_1} + V_1(x_1) \right]\psi_1 = -\frac{\hbar}{i}\frac{\partial\psi_1}{\partial t}$$

$$H_2\psi_2(x_2, t) = \left[ \frac{p_2{}^2}{2m_2} + V_2(x_2) \right]\psi_2 = -\frac{\hbar}{i}\frac{\partial\psi_2}{\partial t}.$$

The stationary states of the system are simply the product states

$$\psi_{nm} = \psi_{1n}(x_1)\psi_{2m}(x_2)$$

with energy

$$E_{nm} = E_{1n} + E_{2m},$$

where $\psi_{1n}$ and $\psi_{2m}$ are the orthonormal eigenfunctions of $H_1$ and $H_2$ with eigenvalues $E_{1n}$ and $E_{2m}$. An arbitrary general state can then be constructed as a superposition of such independent particle states. In particular, solutions of the product form of equation (32) are obtained as the product of an arbitrary superposition of the states of particle one and of a similar superposition of those of particle two. For such a product, we have from equation (32)

$$|\Psi|^2 = |\psi_1|^2|\psi_2|^2,$$

so that the particles are entirely uncorrelated and the time development of each proceeds just as independently as in the classical case.

We note, however, that it is possible to have states for which this complete independence does not hold. To see this, consider some arbitrary initial state $\psi(x_1, x_2, t = 0)$. The most general state function is expressible as

$$\Psi(x_1, x_2, t) = \Sigma\, c_{mn}\psi_{1n}(x_1)\psi_{2m}(x_2)\, \exp\left[-i(E_{1n} + E_{2n})t/\hbar\right],$$

and hence, in virtue of the orthonormality of the $\psi_{1n}$ and $\psi_{2m}$, we obtain, setting $t = 0$ and inverting,

$$c_{nm} = \int\!\int \Psi(x_1, x_2, t = 0)\, \psi_{1n}{}^*(x_1)\psi_{2m}{}^*(x_2)\, dx_1\, dx_2.$$

Now if the initial state is uncorrelated, that is, if it has the form $\psi_1(x_1)\psi_2(x_2)$, then $c_{nm}$ can be expressed as a product $c_{1n}c_{2m}$, and the state function remains uncorrelated for all time. However, if the initial state is correlated, for whatever reason, then these correlations, which have no classical counterpart, persist in the state function for all time. Such correlations play a fundamental role in determining the properties of systems of identical particles, as we later show.

We are now ready to consider the effects of interactions upon the stationary states, supposing these to be weak compared to the external forces. Denoting the interaction term by $\bar{V}(x_1 - x_2)$, the Hamiltonian has the form

$$H = \frac{p_1^2}{2m_1} + \frac{p_2^2}{2m_2} + V_1(x_1) + V_2(x_2) + \bar{V}(x_1 - x_2).$$

Treating $\bar{V}(x_1 - x_2)$ as a perturbation, the unperturbed states are just product states $\psi_{nm} = \psi_{1n}(x_1)\psi_{2m}(x_2)$ with energy $\mathcal{E}_{nm} = E_{1n} + E_{2m}$, as discussed previously. To first order, we then have at once

$$E_{nm} \simeq \mathcal{E}_{nm} + \langle \psi_{nm}|\bar{V}|\psi_{nm}\rangle, \tag{33}$$

while to second order,

$$E_{nm} \simeq \mathcal{E}_{nm} + \langle \psi_{nm}|\bar{V}|\psi_{nm}\rangle + \sum_{l,k}{}' \frac{\langle \psi_{nm}|\bar{V}|\psi_{lk}\rangle\langle \psi_{lk}|\bar{V}|\psi_{nm}\rangle}{\mathcal{E}_{nm} - \mathcal{E}_{lk}}, \tag{34}$$

where the matrix elements are explicitly given by

$$\langle \psi_{nm}|\bar{V}|\psi_{lk}\rangle = \int\int dx_1\, dx_2\, \psi_{1n}{}^*(x_1)\psi_{2m}{}^*(x_2)\bar{V}(x_1 - x_2)\psi_{1l}(x_1)\psi_{2k}(x_2).$$

The generalization to more than two particles is immediate. Consider first a three-particle system. The unperturbed states are products of three independent particle states, and in the same notation as before, we have for these states

$$\psi_{nml} = \psi_{1n}(x_1)\,\psi_{2m}(x_2)\,\psi_{3l}(x_3)$$

$$\mathcal{E}_{nml} = E_{1n} + E_{2m} + E_{3l}.$$

If the mutual interaction perturbation term is written

$$\bar{V} = V_{12}(x_1 - x_2) + V_{13}(x_1 - x_3) + V_{23}(x_2 - x_3),$$

then the first-order result is

$$E_{nml} = \mathcal{E}_{nml} + \langle \psi_{nml}|\bar{V}|\psi_{nml}\rangle.$$

The last term in this expression simplifies to a sum of two-body terms, exactly like those in equation (33). To see this, consider the contribution of, say, $V_{12}(x_1 - x_2)$. We have

$\langle \psi_{nml} | V_{12} | \psi_{nml} \rangle$

$$= \int dx_1\, dx_2\, dx_3 |\psi_{1n}(x_1)|^2\, |\psi_{2m}(x_2)|^2\, |\psi_{3l}(x_3)|^2\, V_{12}(x_1 - x_2)\,.$$

The $x_3$ integration can be performed at once, and because $\psi_{3l}$ is normalized, the result is, as we claimed,

$$\langle \psi_{nml} | V_{12} | \psi_{nml} \rangle = \langle \psi_{nm} | V_{12} | \psi_{nm} \rangle\,,$$

and it is independent of $l$. In the same way, the other two terms yield similar purely two-particle contributions.

For $A$ particles we have products of $A$ independent particle states and the perturbation corrections again take the form of a sum of two-particle contributions.

## 6. IDENTICAL PARTICLES AND EXCHANGE DEGENERACY

We now consider the extremely important case of a system of *identical particles*. By identical particles, we always mean particles which are *completely and absolutely* indistinguishable: No physical significance attaches to the labeling of such particles and no observable effects are produced when any two of them are interchanged. In particular, this means that the coordinates of all identical particles must appear in the Hamiltonian in *exactly* the same way. As a specific and simple example, consider two noninteracting identical particles under the influence of some external force. Each particle must feel exactly the same potential, and hence the Hamiltonian has the form

$$H = \frac{p_1^2}{2m} + \frac{p_2^2}{2m} + V(x_1) + V(x_2)\,. \tag{35}$$

The stationary state solutions are thus

$$\psi_{nq} = \psi_n(x_1)\psi_q(x_2) \tag{36}$$

with energy

$$E_{nq} = E_n + E_q\,,$$

where $\psi_s(x)$ is an eigenfunction, with eigenvalue $E_s$, of the *common single-particle Hamiltonian*

$$H_{\rm sp} = \frac{p^2}{2m} + V(x)\,.$$

Observe now that if $q$ and $n$ are different, this state is degenerate with respect to interchange of the particles, that is, the state

$$\psi_{qn} = \psi_q(x_1)\,\psi_n(x_2) \tag{37}$$

has the same energy as $\psi_{nq}$. This degeneracy is called *exchange degeneracy*, and it is a consequence of the evident invariance of the Hamiltonian, equation (35), to an exchange of the coordinates of the two particles.

We now generalize this very important result. Specifically, we show that exchange degeneracy is a property of the solutions of Schrödinger's equation for *any* system of identical particles, no matter what their mutual interactions, no matter what the external forces to which they are subjected, and no matter how many particles in the system. For this purpose we introduce the so-called *exchange operator* $P_{12} \equiv P_{21}$, which exchanges the coordinates of particles one and two when it acts upon any function of those coordinates.[9] Thus $P_{12}$ is defined by

$$P_{12} f(x_1, x_2, x_3, \ldots, x_A) = f(x_2, x_1, x_3, \ldots, x_A) \tag{38}$$

for arbitrary $f$. Similarly $P_{ij}$ exchanges the coordinates of the $i$th and $j$th particles. As further examples, we have

$$P_{13} f(x_1, x_2, x_3, \ldots, x_A) = f(x_3, x_2, x_1, \ldots, x_A)$$

and

$$P_{12} P_{13} f(x_1, x_2, x_3, \ldots, x_A) = P_{12} f(x_3, x_2, x_1, \ldots, x_A)$$
$$= f(x_3, x_1, x_2, \ldots, x_A),$$

and so on. Note carefully that, as this last example demonstrates, *the subscript on a coordinate labels the particle to which that coordinate refers*, and *not* the order in which a coordinate happens to appear in the state function.

Now $P_{ij}$ commutes with the Hamiltonian by the definition of indistinguishability. Hence, if $\psi_E(x_1, x_2, \ldots, x_N)$ is an eigenfunction of $H$ with eigenvalue $E$, so is $P_{ij}\psi_E$, inasmuch as

$$H P_{ij} \psi_E = P_{ij} H \psi_E = E P_{ij} \psi_E.$$

We thus see that these states are degenerate with respect to the interchange of the $i$th and $j$th particles, unless $P_{ij}\psi_E$ happens to be a multiple of $\psi_E$, as for example the independent particle state of equation (36) with $n = q$. Of course, the labels $i$ and $j$ can refer to any pair of particles and hence exchange degeneracy occurs when *any* two identical particles are exchanged. *For a system of $A$ particles the solutions of Schrödinger's equation may thus be as much as $A$!-fold degenerate, corresponding to the $A$! permutations of the order in which the $A$ particles can be labeled in writing the state function.* Any linear combination of the linearly independent states formed from the complete set of $A$! permutations is an eigenfunction of the Hamiltonian, and the matter of classifying these states is generally not an easy one, as we shall see. However, for two

---

[9] The algebraic properties of exchange operators are elaborated upon in Problem 1.

particles the problem is trivial and we now take up this simple, but important, special case.

## 7. SYSTEMS OF TWO IDENTICAL PARTICLES

For a two-particle state, we obviously need to consider only the single exchange operator $P_{12}$. Now $P_{12}^2$ is evidently unity and hence, just as for the parity operator, the eigenvalues of $P_{12}$ are $\pm 1$, corresponding to states which are *symmetrical* or *antisymmetrical under exchange*. Thus, two-particle states can *always* be classified according to their symmetry under exchange. As a particular example, for the independent particle states of equation (36), we obtain the (normalized) *correlated* states

$$n \neq q; \qquad \psi_{nq}^{\pm} = \frac{1}{\sqrt{2}} \left[ \psi_n(x_1)\psi_q(x_2) \pm \psi_n(x_2)\psi_q(x_1) \right], \quad (39)$$

and it is easily seen that, as it must,

$$P_{12}\,\psi_{nq}^{\pm} = \pm\,\psi_{nq}^{\pm}.$$

More generally, if $\psi_E(x_1, x_2)$ is an eigenfunction of some two-particle Hamiltonian with energy $E$, then the (unnormalized) *symmetrized states,* as they are called, are

$$\psi_E^{\pm}(x_1, x_2) = \left[ \psi_E(x_1, x_2) \pm \psi_E(x_2, x_1) \right], \qquad (40)$$

provided that $\psi_E(x_1, x_2)$ does not happen to be symmetrized or anti-symmetrized to start with.

The existence of such built-in symmetry might seem to be the exception, judging from the independent particle example where it occurs only in the special case $n = q$ of equation (36). We now show, however, that for real physical systems, that is, those containing interactions, we must expect the true two-particle bound states to automatically exhibit such definite symmetry properties. The argument is a simple one. The antisymmetric state function necessarily vanishes when $x_1 = x_2$, and hence the likelihood of finding the two particles in the immediate neighborhood of each other is reduced, compared to the similar likelihood for a symmetric state. The contribution of the interaction term in the Hamiltonian, whatever its sign, is thus numerically *larger* for symmetrical than for antisymmetrical states. We accordingly expect that symmetric and antisymmetric states do not form degenerate pairs, except perhaps by accident, and hence that the bound state spectrum consists of a set of discrete and nondegenerate states. Now we simply turn the argument around. If the spectrum is indeed nondegenerate, barring accidents, then we conclude, knowing that all states are classifiable by their symmetry under exchange, that the true states *must* have a definite symmetry, and hence will automatically turn out to be symmetrized.[10]

Our discussion of these properties of the states of two identical particles will be made clearer, perhaps, by consideration of the following simple and exactly soluble examples, first treated in Sections 3 and 4 for the case of distinguishable particles. We now take these up in order.

(a) *Uniform external forces.* Here the particles can interact arbitrarily with each other, but the external force is restricted to be uniform. The Hamiltonian has the form

$$H = \frac{p_1^2}{2m} + \frac{p_2^2}{2m} + V(x_1 - x_2) + F x_1 + F x_2.$$

Because the particles are identical, the external force $F$ acting on each particle is the same. Further, the interaction potential is necessarily a symmetrical function of its argument (why?). As a result, the states of relative motion are readily seen, from equation (20), with $m_1 = m_2 = m$ and $F_1 = F_2 = F$, to have definite symmetry with respect to the relative coordinate $x = x_1 - x_2$. Since the *center-of-mass* states are necessarily *symmetrical* under exchange (why?), the states of the system are seen to be automatically symmetrized, in agreement with our expectations.

---

**Exercise 2.** Work out the details of the arguments leading to the above conclusion.

---

(b) *Coupled Harmonic Oscillators.* This is a more interesting example, and we work it out in some detail. For identical particles, the Hamiltonian of equation (22) for coupled oscillators becomes

$$H = \frac{p_1^2}{2m} + \frac{p_2^2}{2m} + \frac{1}{2} m\omega^2 x_1^2 + \frac{1}{2} m\omega^2 x_2^2 + \frac{1}{2} k(x_1 - x_2)^2. \qquad (41)$$

The center-of-mass transformation now takes the simple form

$$X = \frac{x_1 + x_2}{2}, \qquad x = x_1 - x_2, \qquad (42)$$

and under this transformation $H$ becomes

---

[10] This argument is rather similar to our discussion in Chapter VI of the parity of states in a symmetrical potential. In that case, however, we were able to *prove* that the bound state spectrum in one dimension is *always* nondegenerate and hence that the bound states are *always* classifiable according to their symmetry with respect to parity. Here the argument is necessarily more tentative. Degeneracies *can* occur, after all, as they indeed do for almost all states in the independent particle case, or accidentally for the more realistic case of interacting particles. It should be noted that even for the realistic case we have given no *proofs*, but have merely presented some *plausibility* arguments.

$$H = \frac{P^2}{2M} + \frac{1}{2} M\omega^2 X^2 + \frac{p^2}{2\mu} + \frac{1}{2} \mu\bar{\omega}^2\chi^2 \qquad (43)$$

just as before, with $M = 2m$ and $\mu = m/2$. Also, as before,

$$\bar{\omega}^2 = \omega^2 + \frac{k}{\mu} = \omega^2 + \frac{2k}{m}.$$

The stationary states are given by

$$\psi_{Nn} = \Phi_N(X)\phi_n(x)$$

$$= \Phi_N\left(\frac{x_1 + x_2}{2}\right)\phi_n(x_1 - x_2), \qquad (44)$$

whence

$$P_{12}\psi_{Nn} = (-1)^n \psi_{Nn}, \qquad (45)$$

and hence these states are automatically symmetrized in agreement with expectations. The spectrum for weak interactions is that of Figure 1(a), and for dominantly strong interactions, that of Figure 1(b).

At this stage it might appear that, in spite of our earlier claims, no real complications result when identical particles are considered. On the contrary, the description is simplified, if anything, because of the more or less built-in symmetry properties. This conclusion is, in fact, a valid one for two-particle systems. Significant complications enter only for systems of three or more particles, as we show next.

## 8. MANY-PARTICLE SYSTEMS, SYMMETRIZATION AND THE PAULI EXCLUSION PRINCIPLE

We return now to the consideration of the properties of a system of $A$ identical particles. Such a system is described by the Hamiltonian

$$H = \sum_{i=1}^{A} \frac{p_i^2}{2m} + \sum_{i=1}^{A} V(x_i) + \sum_{\text{pairs}} \bar{V}(x_i - x_j), \qquad (46)$$

where $V(x)$ is the common external potential in which each and every particle moves and $\bar{V}(x_i - x_j) = \bar{V}(x_j - x_i)$ is the common mutual interaction potential of each and every pair of particles. To determine the symmetry characteristics of the states of such a system we must examine the properties of the exchange operators $P_{ij}$. Now as we have said before, $P_{ij}$ necessarily commutes with $H$ for every pair of particles, and hence the states of the system may be as much as $A$!-fold degenerate under exchange, corresponding to the $A!$ permutations of the order in which the $A$ particles can be labeled in writing the state function. That the classification of these states is a complicated matter follows at once from the observation that $P_{12}$ and $P_{13}$, say, or more generally $P_{ij}$ and $P_{ik}$, do *not* mutually commute. The symmetry of the state function with respect to exchange of the $i$th and $j$th particles and of the $i$th and $k$th particles cannot therefore be simultaneously and arbitrarily specified. We thus reach the important conclusion that, *in general, the solutions of Schrödinger's equation for a system of three or more identical particles do not have definite symmetry with respect to the exchange of each and every pair of particles.*

There are, however, two exceptional states, both of paramount importance in what follows, which do possess definite symmetry. One of these is symmetrical with respect to the exchange of each pair of particles and is called a *totally symmetric state*. The other is antisymmetric with respect to each particle interchange and is called *totally antisymmetric*. Both are very simply constructed from the unsymmetrized states $\psi_E(x_1, x_2, \ldots, x_A)$. Denoting the former by $\Psi_E^+$ and the latter by $\Psi_E^-$, these states can be symbolically exhibited as

$$\Psi_E^+ = \sum_{\text{permutations}} \psi_E(x_1, x_2, \ldots, x_A) \tag{47}$$

and

$$\Psi_E^- = \sum_{\text{permutations}} (-1)^r \psi_E(x_1, x_2, \ldots, x_A). \tag{48}$$

In each case the sum runs over the complete set of permutations. Any additional exchange merely shuffles these permutations and hence leaves the symmetric state unaltered, as it must. The index $r$ on the factor $(-1)^r$ in the antisymmetric state is the *number of two-particle exchanges required to achieve a given permutation.* The important point is that the sign of each term in the summation depends on whether this number is even or odd. Exchanging any pair of particles in the totally antisymmetric state function changes each even $r$ to an odd $r$ and vice versa, and hence changes the sign of $\Psi_E^-$, as it must. The fundamental symmetry property of these two states is now summarized by the statement that, for every particle pair $i$ and $j$,

$$P_{ij} \psi_E^\pm = \pm \psi_E^\pm. \tag{49}$$

Our discussion of the symmetry properties of many-particle systems has shown that the exchange degenerate states, which may be as much as $A!$-fold degenerate, possess no definite symmetry, with the exception of the very special totally symmetric and totally antisymmetric states.[11]

---

[11] In contrast to the two-particle case, even these highly symmetrized states do not appear automatically. They must be constructed according to the recipes of equations (47) and (48).

Fortunately, nature has enormously simplified the problem for us in that not all of these $A!$ solutions of Schrödinger's equation are, in fact, physically realizable. This is a feature of overriding importance for the structure of matter and of the very universe itself.

In order to make a precise statement of the restrictions imposed by nature, we need to mention that the elementary particles are not really structureless, but require certain internal coordinates for their complete specification. The relevant internal coordinate in the present discussion is the internal angular momentum, or *spin*, of a particle, which we shall denote by $s$. We shall take up this subject in some detail later; for now we merely remark that the spin angular momentum has a definite magnitude for a given kind of particle. These magnitudes are quantized and, in units of $\hbar$, can take on only half-integer or integer values. For example, electrons, protons and neutrons have spin one-half, the photon has spin one, the $\pi$-meson has spin zero. In any case, if we now agree that by exchange of coordinates we mean exchange of internal (spin) as well as external coordinates, then *all of our arguments about exchange degeneracy are unaffected by the presence of spin*, and the $A!$-fold degeneracy may still occur. Nature's simplification is then that *for a given type of particle* only *one* of these $A!$ states is ever observed. Specifically, *the state function for a system of integral spin particles must always be totally symmetric, while the state function for a system of half-odd-integral spin particles must always be totally antisymmetric*. The former is thus uniquely described by the special state of equation (47), the latter by that of equation (48).

The postulate that the state function for half-odd-integral particles is totally antisymmetric is nothing more than a precise and general version of the famous *Pauli exclusion principle*, as we shall make clear in a moment. Such particles satisfy the Fermi–Dirac statistics and are now commonly given the generic name *fermions*. On the other hand, particles of integral spin satisfy the Bose–Einstein statistics and are called *bosons*. These connections between spin and statistics, and the symmetry postulates themselves, were first discovered empirically, but they have since been deduced as a necessary consequence of the restrictions imposed upon the form of quantum mechanical laws by the requirements of relativistic invariance.

A better understanding of the structure of the state functions in equations (47) and (48) can be obtained by observing that, for a system without interactions, a stationary state solution of the $A$-particle system is simply a product of $A$ single-particle state functions. The totally antisymmetric state is then seen to be expressible as a determinant, called the *Slater determinant,* of these single-particle functions. Specifically, the normalized antisymmetric state is given by

$$\Psi_E^- = \frac{1}{\sqrt{A!}} \begin{Vmatrix} \psi_{m_1}(x_1, s_1) & \psi_{m_1}(x_2, s_2) & \cdots & \psi_{m_1}(x_A, s_A) \\ \psi_{m_2}(x_1 s_1) & \psi_{m_2}(x_2, s_2) & \cdots & \psi_{m_2}(x_A, s_A) \\ \vdots & & & \end{Vmatrix} , \qquad (50)$$

where the argument of the single-particle functions is written as $(x_i s_i)$ to indicate that both space and spin coordinates enter. The product of the diagonal elements gives the original unsymmetrized state and the other terms generate the complete set of permutations, each with its appropriate sign. In this independent-particle case we thus see that an antisymmetrized wave function vanishes if any two particles are in the same spatial and spin state, since two rows of the determinant are then identical. The familiar and elementary version of the Pauli principle, stating that no two particles can be in the same quantum state, is therefore seen to be a special case of the general antisymmetrization requirement.

The symmetric state cannot be written in such an elegant way, but it can be very simply pictured as a determinantal wave function in which the sign of every term is taken to be *positive*.

Note that symmetrized wave functions, such as that of equation (50), are *correlated states*, even though built out of uncorrelated products. Thus the probability is zero that any two particles will have identical coordinates, space and spin, in an antisymmetric state. Such correlations produce substantial effects, and they are responsible, for example, for the phenomenon of ferromagnetism.

The reduction from $A!$ to a single state is the end of the story for particles without spin. To give a very simple and specific example, consider the two-particle system described by the coupled oscillator Hamiltonian of equation (41). For spinless particles, the state function must be symmetric under exchange and hence, according to equation (45), only states with even $n$ are physically achievable. This means that only half of the *mathematical* (and perfectly well behaved) solutions of Schrödinger's equation are *physical* solutions and thus appear in the spectrum of Figure 1(a) (or of Figure 1(b)). For three or more particle systems, this reduction in the number of states is much greater. Because the states are not automatically symmetrized, the analysis of many-particle systems is considerably more difficult, as we illustrate shortly in the three-particle case.

For particles with spin one-half, such as electrons, the situation is more difficult still. The nature of these additional complications can be made quite explicit by noting that, except for relativistic effects, the Hamiltonian is typically independent of the spin coordinates. If so, the spin of each particle commutes with $H$ and hence each spin is separately a constant of the motion. This introduces into the state functions a new degeneracy, which may be expressed by the fact that any solution of the spin-independent Schrödinger's equation remains a solution when multiplied by an arbitrary function of spin coordinates. Now the symmetry requirements are conditions on the total wave function and not on the space or spin coordinates alone. Hence there are many ways in which symmetrization can be achieved. In the simplest case, that of a two-electron system, one can construct a totally antisymmetric state function in two ways, as a product of a symmetric space function and an antisymmetric spin function, or as a product of an antisymmetric space function and a symmetric spin function. As discussed in Chapter X, the former is a state of total spin zero, the latter of total spin one.[12] Again using the coupled oscillator as an example, this means that for spin one-half particles *all* of the states are physically realizable and the spectrum is that of Figure 1(a) (or Figure 1(b)) in its entirety. States with even $n$ are spin zero states (antisymmetric in the spins), those with odd $n$ are spin one states (symmetric in the spins).

Considering next a three-electron system, we find that there are *three* ways to construct a totally antisymmetric state function. The first is relatively simple. It is a state of total spin 3/2, which is symmetric in the spins and antisymmetric in the space coordinates. The other two are rather unpleasant. They are states of total spin 1/2, neither of which has definite symmetry with respect to space exchange or spin exchange separately.

To summarize, we have studied the states of a system of $A$ identical particles and have found that the most general of the stationary state solutions of Schrödinger's equation are $A!$-fold degenerate. However, depending on the spin of the particle, only certain combinations of these degenerate states are observed to occur. In particular, for particles without spin, the states must be totally symmetric with respect to exchange of any pair of particles. For particles with spin one-half, the state function must be totally antisymmetric with respect to interchange of the space *and* spin coordinates of any pair of particles.

## 9. SYSTEMS OF THREE IDENTICAL PARTICLES

In our discussion of the behavior of identical particles we have so far concentrated, at least for illustrative purposes, upon two-particle systems. The two-particle case is, however, deceptively simple because of the more or less automatic appearance of precisely the proper spatially symmetrized states, which, in fact, are the only states possible. To illustrate the new features which enter when more than two particles are in-

[12] The atomic states of helium are of this type. See Chapter XI.

volved, we now briefly discuss the properties of many-particle systems.

The essential question is the following: Given some particular many-particle system, how do we identify its *physically realizable* states? Quite generally, we do so, at least in principle, by first finding the solutions of Schrödinger's equation, *without regard to symmetrization*, and then symmetrizing the result according to the prescription of equation (47) for spin zero particles, or of equation (48) for spin one-half particles. We now demonstrate this procedure for a system of three identical particles, which we shall *first* take to be *spinless* for simplicity. We remark that the three-particle system is general enough to exhibit the relevant features of many-particle systems and yet simple enough to permit us to work out the details explicitly, as we shall see.

We begin by considering the simplest possible three-particle system, namely that in which the particles do not interact with each other.[13] The Hamiltonian is thus given by equation (46) with $\bar{V} = 0$. It can be expressed in the independent particle form,

$$H = \sum_{i=1}^{3} H_{\text{sp}}(x_i),$$  (51)

where the common single-particle Hamiltonian is

$$H_{\text{sp}} = \frac{p^2}{2m} + V(x).$$  (52)

Denoting the eigenfunctions and corresponding eigenvalues of $H_{\text{sp}}$ by $\phi_n$ and $E_n$, the stationary states can then be expressed as a product

$$\psi_{nqs}(x_1, x_2, x_3) = \phi_n(x_1)\, \phi_q(x_2)\, \phi_s(x_3)$$  (53)

with energy

$$\mathcal{E}_{nqs} = E_n + E_q + E_s.$$  (54)

Consider now an exchange degenerate state, say $P_{12}\, \psi_{nqs}$. We have

$$P_{12}\, \psi_{nqs}(x_1, x_2, x_3) = \psi_{nqs}(x_2, x_1, x_3) = \phi_q(x_1)\phi_n(x_2)\phi_s(x_3)$$
$$= \psi_{qns}(x_1, x_2, x_3),$$

and similarly for all the others. Because of the product form of the state function, we thus see that a permutation of the particle coordinates simply permutes the indices on the state function. Henceforth, for simplicity in writing, we shall omit the argument of the states whenever that argument is to be taken in the normal order $x_1, x_2, x_3$. The $3! = 6$ exchange degenerate states are thus simply expressed as $\psi_{nqs}, \psi_{nsq}, \psi_{qsn}, \psi_{qns}, \psi_{snq}, \psi_{sqn}$. Observe now that whether or not these states are linearly independent is

---

[13] This is an important case because it typically serves as the starting point for perturbation calculations. See Section 10.

determined by whether or not $n$, $q$ and $s$ refer to different single-particle states. To complete the description, we thus distinguish three cases, as follows:

(a) $n = q = s$. The state function is automatically symmetric and the physically realizable normalized states are

$$\psi_{nnn}{}^{+} = \phi_n(x_1)\phi_n(x_2)\phi_n(x_3) \tag{55}$$

with energy

$$\mathscr{E}_{nnn} = 3E_n.$$

The ground state of the system is necessarily included in this class.

(b) $n = q \neq s$. Only three of the six permuted states are distinct. The normalized totally symmetrical states, and thus the *only* physically realizable states of this type, are

$$\psi_{nns}{}^{+} = \frac{1}{\sqrt{3}}\,(\psi_{nns} + \psi_{nsn} + \psi_{snn}) \tag{56}$$

with energy

$$\mathscr{E}_{nns} = 2E_n + E_s.$$

The first excited state of the system is necessarily included in this class.

(c) $n \neq q \neq s$. Here the full set of six exchange degenerate states enters. The normalized totally symmetrical states, the *only* physically realizable states, are

$$\psi_{nqs}{}^{+} = \frac{1}{\sqrt{6}}\,(\psi_{nqs} + \psi_{nsq} + \psi_{qsn} + \psi_{qns} + \psi_{snq} + \psi_{sqn}) \tag{57}$$

with energy

$$\mathscr{E}_{nqs} = E_n + E_q + E_s.$$

Observe that in all three cases, the ordering of the indices for the symmetrized states carries no significance, that is,

$$\psi_{nqs}{}^{+} \equiv \psi_{nsq}{}^{+},$$

and similarly for every other permutation. Observe also that the degeneracy of the physical states now depends only upon the structure of the single-particle spectrum. If more than one distinct combination $(n, q, s)$ yields the same energy, the state are degenerate, otherwise not.

This completes the description for spinless particles. What modifications occur for spin one-half? Here an additional degeneracy is introduced because there are two possible orientations of the spin of the particle. Let $x_i$ denote the spin coordinate as well as the space coordinate and let $\phi_n{}^{\pm}$ denote the two possible single particle spin states.[14] If the Hamil-

---

[14] These may be thought of as corresponding to spin up or down with respect to some given reference axis, as discussed in Chapter X.

tonian is independent of spin, then the states $\phi_n{}^+$ and $\phi_n{}^-$ both have the same energy $E_n$. The solutions of Schrödinger's equation corresponding to energy $\mathscr{E}_{nqs}$ can now be expressed in the form

$$\psi_{nqs}{}^{abc} = \phi_n{}^a(x_1)\ \phi_q{}^b(x_2)\ \phi_s{}^c(x_3),\tag{58}$$

where $a$, $b$ and $c$ can each be either plus or minus symbols. The fact that $a$, $b$ and $c$ cannot possibly all be different has profound consequences for the system, as we now show. The totally antisymmetric state function can be expressed as the Slater determinant, equation (50),

$$(\psi_{nqs}{}^{abc})^- = \frac{1}{\sqrt{6}} \begin{Vmatrix} \phi_n{}^a(x_1) & \phi_n{}^a(x_2) & \phi_n{}^a(x_3) \\ \phi_q{}^b(x_1) & \phi_q{}^b(x_2) & \phi_q{}^b(x_3) \\ \phi_s{}^c(x_1) & \phi_s{}^c(x_2) & \phi_s{}^c(x_3) \end{Vmatrix}\tag{59}$$

Again, we distinguish the three cases discussed before.

(a) $n = q = s$. Because $a$, $b$ and $c$ cannot all be different, two rows of the determinant are necessarily identical and the determinant vanishes. There is thus no physically achievable state of this class in the spectrum, in agreement with the Pauli principle.

(b) $n = q \neq s$. If the determinant is not to vanish, we must have $b = -a$, whence a typical physically achievable state is, expanding out the determinant,

$$(\psi_{nns}{}^{+-+})^- = \frac{1}{\sqrt{6}}\ [\psi_{nns}{}^{+-+} - \psi_{nsn}{}^{++-}$$

$$+ \psi_{nsn}{}^{-++} - \psi_{nns}{}^{-++} + \psi_{snn}{}^{++-} - \psi_{snn}{}^{+-+}]\tag{60}$$

with energy

$$\mathscr{E}_{nns} = 2E_n + E_s.$$

The ground state of the system is included in this class of states, each of which is seen to be doubly degenerate. Note that these states have no definite symmetry with respect to space exchange alone or to spin exchange alone; they are antisymmetric only under the *simultaneous* exchange of both.

(c) $n \neq q \neq s$. Here, physically achievable states exist for any combination of $a$, $b$ and $c$. The states are thus *eight-fold degenerate*, corresponding to the eight possible assignments of $(a, b, c)$.[15] The simplest state of this type is totally symmetrical in the spins and antisymmetrical in the space coordinates. An example is

---

[15] This degeneracy is reduced if the Hamiltonian contains spin dependent terms. Specifically, the eight-fold state splits into a quartet and two doublets.

$$(\psi_{nqs}{}^{+++})^- = \frac{1}{\sqrt{6}} \, (\psi_{nqs}{}^{+++} - \psi_{nsq}{}^{+++} + \psi_{qsn}{}^{+++} - \psi_{qns}{}^{+++}$$

$$+ \psi_{snq}{}^{+++} - \psi_{sqn}{}^{+++}) . \quad (61)$$

**Exercise 3.** Find the degenerate companion state to that given in equation (60).

We now consider, as a second example, an exactly soluble three particle system *with* interactions. Specifically, we discuss a system of three coupled oscillators, described by the Hamiltonian

$$H = \sum_{i=1}^{3} \frac{p_i^2}{2m} + \sum_{i=1}^{3} \frac{1}{2} \, m\omega^2 x_i^2 + \frac{1}{2} \, k(x_1 - x_2)^2$$

$$+ \frac{1}{2} \, k(x_1 - x_3)^2 + \frac{1}{2} \, k(x_2 - x_3)^2 . \quad (62)$$

It is readily verified that the following[16] constitutes a set of normal coordinates for this problem:

$$X = \tfrac{1}{3} \, (x_1 + x_2 + x_3)$$

$$x = x_1 - x_2 \quad (63)$$

$$y = x_3 - \frac{x_1 + x_2}{2}$$

and, conversely,

$$x_1 = X + \frac{x}{2} - \frac{y}{3}$$

$$x_2 = X - \frac{x}{2} - \frac{y}{3} \quad (64)$$

$$x_3 = X + \tfrac{2}{3} \, y .$$

The physical significance of the normal coordinates is apparent; $X$ is the center-of-mass coordinate, $x$ is the relative coordinate of particles one and two, and $y$ is the coordinate of particle three relative to the center of mass of particles one and two. In any case, the introduction of these coordinates transforms the Hamiltonian into the separable form

$$H = \frac{P^2}{2M} + \frac{1}{2} \, M\omega^2 X^2 + \frac{p_x^2}{2\mu_x} + \frac{1}{2} \, \mu_x \bar{\omega}^2 x^2 + \frac{p_y^2}{2\mu_y} + \frac{1}{2} \, \mu_y \bar{\omega}^2 y^2 , \quad (65)$$

---

[16] Subsequent details are left to the problems.

where

$$M = 3m, \qquad \mu_x = \frac{m}{2}, \qquad \mu_y = \frac{2}{3}m \qquad (66)$$

and

$$\bar{\omega}^2 = \omega^2 + 3\frac{k}{m}, \qquad (67)$$

and where the new momenta are given by

$$P = p_1 + p_2 + p_3$$

$$p_x = \tfrac{1}{2}(p_1 - p_2) \qquad (68)$$

$$p_y = \tfrac{2}{3}\left(p_3 - \frac{p_1 + p_2}{2}\right).$$

It is not difficult to show that the canonical commutation relations are satisfied, whence the *exact* solutions of Schrödinger's equation are seen to be product states of the form

$$\psi_{N n_x n_y}(x_1, x_2, x_3) = \Phi_N\left(\frac{x_1 + x_2 + x_3}{3}\right) \phi_{n_x}(x_1 - x_2)\, \phi_{n_y}\left(x_3 - \frac{x_1 + x_2}{2}\right) \qquad (69)$$

with energies

$$E_{N n_x n_y} = \frac{\hbar}{2}(\omega + 2\bar{\omega}) + N\hbar\omega + (n_x + n_y)\hbar\bar{\omega}. \qquad (70)$$

Of course, $\Phi_N(X)$, $\phi_{n_x}(x)$ and $\phi_{n_y}(y)$ are the respective harmonic oscillator eigenfunctions for the $X$, $x$ and $y$ normal coordinates. We leave the identification of the physically realizable states and of the energy spectrum to the problems, but we make the following observations as a guide:

(1) The states associated with the center-of-mass motion are symmetrical under exchange and hence appear as a common factor in all symmetrized state functions. These states thus play no role in the symmetrization process.

(2) The maximum number of *exchange* degenerate states is three, and not six.[17]

(3) Totally symmetric states cannot be formed when $n_x$ is odd. Hence such states do not appear in the spectrum for spinless particles.

(4) In contrast to the independent particle example, the exchange degenerate solutions are not necessarily orthogonal.

[17] There are, however, additional degeneracies because the $x$ and $y$ normal modes have the same frequency.

**Exercise 4.** Verify the above assertions.

We bring this section to a close with a few remarks about the properties of three-particle systems in general. Let $\psi_E(x_1, x_2, x_3)$ denote an eigenfunction with energy $E$ of some three-particle Hamiltonian. The maximum possible number of linearly independent exchange degenerate states is $3! = 6$, but we have seen that this number is often reduced. This reduction is related to any symmetry properties which may be built into $\psi_E$ from the start. In particular, if an eigenstate is symmetric with respect to the exchange of one pair of particles, it exhibits only three-fold degeneracy under exchange. Further, if a state has definite symmetry with respect to two different pairs of particles, it is automatically a totally symmetrized state, and thus exhibits no exchange degeneracy.

**Exercise 5.** Prove the above statements.

## 10. WEAKLY INTERACTING IDENTICAL PARTICLES IN THE PRESENCE OF GENERAL EXTERNAL FORCES

In our discussion of identical particle systems we have concentrated upon the requirements of symmetrization and its consequences for the states of the system. In doing so, we have more or less taken the solution of the many-particle Schrödinger equation for granted. We now turn our attention to the matter of actually constructing such solutions, at least approximately, for interacting particles in the presence of some general external force. The problem is an extremely difficult one, of course, and we therefore restrict our attention to the simple situation in which the external forces dominate over the internal forces, so that the latter can be treated as a perturbation. We shall also restrict our considerations to two-particle systems.

Specifically, then, we consider the Hamiltonian

$$H = H_0 + \bar{V}(x_1 - x_2), \qquad (71)$$

where the unperturbed Hamiltonian has the independent particle form of equation (35),

$$H_0 = \frac{p_1^2}{2m} + \frac{p_2^2}{2m} + V(x_1) + V(x_2)$$

with the unperturbed solutions of equation (36),

$$\psi_{nm} = \psi_n(x_1)\psi_m(x_2),$$

corresponding to the unperturbed energies

$$\mathcal{E}_{nm} = E_n + E_m.$$

We must, of course, take into account the possible degeneracy of the unperturbed states. However, this is easy to work out in the two-particle case because, as we have already proved, the states can be classified by their symmetry under exchange. We now distinguish two cases:

(a)  *Perturbation of a Nondegenerate State.*  In this case we consider the effect of the interaction on a nondegenerate unperturbed state, that is, a state in which both particles are in the *same* single-particle state, such as the ground state. The unperturbed state function has the form

$$\psi_{nn}(x_1, x_2) = \psi_n(x_1)\psi_n(x_2) \tag{72}$$

and is automatically symmetric. To first order, we have at once

$$E_{nn} = \mathcal{E}_{nn} + \langle \psi_{nn}|\bar{V}|\psi_{nn}\rangle. \tag{73}$$

(b)  *Perturbation of a Degenerate State.*  Here we consider an unperturbed state involving two *distinct* single-particle states. We now choose the symmetrized combinations

$$\psi_{nm}^{(+)} = \frac{1}{\sqrt{2}} \left[ \psi_n(x_1)\psi_m(x_2) + \psi_n(x_2)\psi_m(x_1) \right] \tag{74}$$

and

$$\psi_{nm}^{(-)} = \frac{1}{\sqrt{2}} \left[ \psi_n(x_1)\psi_m(x_2) - \psi_n(x_2)\psi_m(x_1) \right], \tag{75}$$

the first of which is symmetric, the second antisymmetric. The matrix element of the perturbation connecting these states necessarily vanishes, since the perturbing interaction commutes with $P_{12}$. Hence we can use the methods of nondegenerate perturbation theory to obtain at once, for the symmetric state,

$$E_{nm}^{(+)} = \mathcal{E}_{nm} + \langle \psi_{nm}^{(+)}|\bar{V}|\psi_{nm}^{(+)}\rangle \tag{76}$$

and, for the antisymmetric state,

$$E_{nm}^{(-)} = \mathcal{E}_{nm} + \langle \psi_{nm}^{(-)}|\bar{V}|\psi_{nm}^{(-)}\rangle. \tag{77}$$

As we have argued earlier, the matrix elements in equations (76) and (77) are generally quite different, and the states are thus split by the interaction. Specifically, since $\psi_{nm}^{(-)}$ is zero when $x_1 = x_2$, and since we expect the interaction to be strongest when the particles are close together, the matrix element for the antisymmetric state is numerically smaller than for the symmetric state. Of course, the sign of the matrix element depends upon whether $\bar{V}$ is attractive or repulsive.

It is very instructive to examine the structure of the matrix elements in equations (76) and (77) in detail. Substituting the unperturbed functions of equations (74) and (75), we obtain, after rearranging and collecting like terms,

$$E_{nm}^{(+)} = \mathscr{E}_{nm} + J_{nm} + K_{nm} \tag{78}$$

and

$$E_{nm}^{(-)} = \mathscr{E}_{nm} + J_{nm} - K_{nm}, \tag{79}$$

where $J_{nm}$, the so-called *direct interaction energy*, is given by

$$J_{nm} = \int \int dx_1 \, dx_2 \; \psi_n^*(x_1)\psi_m^*(x_2)\bar{V}(x_1 - x_2)\psi_n(x_1)\psi_m(x_2) \tag{80}$$

and $K_{nm}$, the *exchange interaction energy*, is given by

$$K_{nm} = \int \int dx_1 \, dx_2 \; \psi_n^*(x_1)\psi_m^*(x_2)\bar{V}(x_1 - x_2)\psi_n(x_2)\psi_m(x_1), \tag{81}$$

where we have assumed the phases of $\psi_n$ and $\psi_m$ to be chosen in such a way that $K_{nm}$ is real, as we can always do without loss of generality.

The meaning of $J_{nm}$ is clear; the quantity $|\psi_n(x_1)|^2|\psi_m(x_2)|^2$ is just the joint probability that particle one is at $x_1$ and particle two is at $x_2$, and hence that their interaction energy is $\bar{V}(x_1 - x_2)$. The integral thus gives the mean interaction energy, as expected intuitively. The quantity $K_{nm}$, on the other hand, has no similar simple interpretation. It appears as a consequence of the correlations required by invariance under exchange, and it has no classical counterpart.

It is instructive to rederive these results using the methods of *degenerate state* perturbation theory, starting from the unsymmetrized unperturbed states $\psi_n(x_1)\psi_m(x_2)$ and $\psi_m(x_1)\psi_n(x_2)$. However, we leave this as an exercise.

---

**Exercise 6.** Taking the unperturbed states to be $\psi_n(x_1)\psi_m(x_2)$ and $\psi_m(x_1)\psi_n(x_2)$, derive equations (78) and (79). Why must degenerate state perturbation theory be used?

---

What now of the effect of the spin and statistics of the particles? As we have already established, for spin zero particles, only spatially symmetric states can occur, and hence only the state function of equation (74), with first-order energy given by equation (78), is permitted. The spectrum thus contains about half the states which are obtained as mathematical, well-behaved solutions of Schrödinger's equation. For spin one-half particles, on the other hand, both kinds of states are represented in the spectrum. The symmetric space state is associated with an antisymmetric spin state (spin zero, the spins of the particle are opposed) while the antisymmetric space state is associated with a symmetric spin state (spin one, the spins are aligned). The atomic states of helium furnish an excellent example of a system of this type.

---

**Problem 1.** Let $P_{ij}$ denote the exchange operator for the $i$th and $j$th particles. Verify each of the following statements:

    (a)  $P_{ij}$ is Hermitian.

    (b)  $P_{ij}^2 = 1$.

    (c)  The projection operators $P_{ij}^{\pm}$ for exchange are

$$P_{ij}^{\pm} = \frac{1 \pm P_{ij}}{2}$$

and have the properties

$$(P_{ij}^{\pm})^2 = P_{ij}^{\pm}$$

$$P_{ij}^{+} P_{ij}^{-} = P_{ij}^{-} P_{ij}^{+} = 0$$

$$P_{ij}^{+} + P_{ij}^{-} = 1.$$

    (d)  $P_{ij}$ and $P_{kl}$ commute if $(i,j)$ and $(k,l)$ refer to two different pairs of particles, but $P_{ij}$ and $P_{il}, j \neq l$, do *not* commute.

**Problem 2.**

    (a)  Consider a system of $A = 2N$ identical particles. Show that there are $N(2N-1)$ independent exchange operators but that only $N$ of these mutually commute. Exhibit one specific complete set of these mutually commuting operators.

    (b)  Suppose $A = 2N + 1$. How many independent exchange operators are there, and how many of these mutually commute?

**Problem 3.**

    (a)  Two noninteracting particles of mass $0.98m$ and $1.02m$, respectively, are placed in a box of width $L$. Draw an energy level diagram showing the first half-dozen or so states of the system.

    (b)  Suppose the particles have a weak, attractive interaction. Show qualitatively what happens to the states of the system.

**Problem 4.**

    (a)  Two identical noninteracting particles of mass $m$ are placed in a box of width $L$. Draw an energy level diagram (to the same scale as in Problem 3) showing the first half-dozen or so states of the system. Indicate the exchange degeneracy, if any, of each level.

    (b)  Same as part (b) of Problem 3.

(c)   What states would be observed for spin zero particles? For spin one-half particles?

**Problem 5.**   A system consists of a neutral particle and a particle of charge $e$. The interaction of the particles is described by the potential

$$V(x_1 - x_2) = \tfrac{1}{2}\,\mu\omega^2(x_1 - x_2)^2,$$

where $\mu$ is the reduced mass.

(a)   Find the ground state energy of the system when it is in a uniform electric field $\mathscr{E}$, directed along the $x$-axis.

(b)   What is the polarizability of the system?

(c)   Describe the motion of the center of mass of the system.

**Problem 6.**   A two-particle system is described by the Hamiltonian

$$H = \frac{p_1^2}{2m_1} + \frac{p_2^2}{2m_2} + \frac{1}{2}\,m_1\omega^2x_1^2 + \frac{1}{2}\,m_2\omega^2x_2^2 + V_0\,e^{-(x_1-x_2)2/a^2}.$$

(a)   Treating the Gaussian interaction term as a perturbation, obtain the energy of the ground state and first excited state to first order.

(b)   The same, but first transform to center-of-mass coordinates.

(c)   Use equation (VII–39) to find an upper bound to the second-order correction to the ground state energy. Use your result to determine the conditions under which the first-order result is valid.

(d)   Suppose the interaction term is strong enough to dominate over the oscillator terms, at least for low-lying states. Discuss qualitatively, but in as much detail as you can, the behavior of the system.

**Problem 7.**   The same as Problem 6, but for identical spinless particles.

**Problem 8.**   Consider the two identical particle, coupled harmonic oscillator Hamiltonian of equation (41).

(a)   Show that a transformation to center-of-mass coordinates leads to equation (43), and thus to the solutions given in equation (44).

(b)   Find the *classical* solutions to the problem and describe the normal modes in both the weak and strong interaction limits.

(c)   For spinless particles, sketch the spectrum corresponding first to Figure 1(a), then to Figure 1(b).

**Problem 9.**   A system of two spinless identical particles is described by the coupled oscillator Hamiltonian of equation (41).

(a)   Considering the interaction term to be small, use first-order perturbation theory to estimate the energy of the ground state, and of the first excited state.

(b)   The same, except treat the Hamiltonian as follows: Square out the interaction term and collect like terms. The Hamiltonian then becomes

$$H = \frac{p_1^2}{2m} + \frac{p_2^2}{2m} + \frac{1}{2}m\omega_1^2 x_1^2 + \frac{1}{2}m\omega_1^2 x_2^2 - kx_1 x_2,$$

where

$$\omega_1^2 = \omega^2 + \frac{k}{m}.$$

Now take the perturbation term to be $(-kx_1 x_2)$.

(c)   Use equation (VII–39) to find an upper bound on the second-order correction to your first-order results for the ground state. Do this for the perturbation method of *either* part (a) or part (b), whichever seems easier.

(d)   Use the Rayleigh–Ritz method to estimate the ground state energy, using a product of Gaussians, of variationally determined width, as a trial function. In all cases, compare your results with the exact results given in the text.

**Problem 10.** Consider three identical particles described by the coupled harmonic oscillator Hamiltonian of equation (62).

(a)   Show that the normal coordinate transformation of equation (63) yields the Hamiltonian of equation (65).

(b)   Find the *classical* solutions to the problem and describe the normal modes in both the weak and strong interaction limits.

(c)   Show that the momenta of equation (68) and the coordinates of equation (63) satisfy the canonical commutation relations.

(d)   Sketch the energy spectrum in both the weak and strong interaction limits, *without regard to symmetrization*. Give the degeneracies of the first few states.

(e)   Which of these states are physically achievable for spinless particles?

**Problem 11.** A system of three spinless identical particles is described by the coupled harmonic oscillator Hamiltonian of equation (62).

(a)   Considering the interaction terms to be small, estimate the energy of the ground state, and of the first excited state.

(b)   The same, except square out the interaction terms, collect like terms, and treat the quantity $(-k)(x_1 x_2 + x_2 x_3 + x_1 x_3)$ as the perturbation term. [See Problem 9(b).]

(c)   Use equation (VII–39) to find an upper bound on the second-order correction to your first-order results. Do this for the perturbation method of *either* part (a) or part (b), whichever seems easier.

(d)   Use the Rayleigh–Ritz method to estimate the ground state energy, using a product of Gaussians, of variationally determined width, as a trial function. In all cases, compare your results with the exact results given in the text.

**Problem 12.** Four identical particles are described by the Hamiltonian

$$H = \sum_{i=1}^{4} H_{\text{sp}}(x_i),$$

where

$$H_{\text{sp}} = \frac{p^2}{2m} + \frac{1}{2} m\omega^2 x^2.$$

(a) Find the eigenfunctions and energies of the system, *without regard to symmetrization*.

(b) Give the degeneracy of the lowest four states, including exchange degeneracy.

(c) The same, but for the physically realizable states of spinless particles.

(d) What is the ground state for spin one-half particles? What is its degeneracy?

# IX

## *Motion in three dimensions*

### 1. FORMULATION: MOTION OF A FREE PARTICLE

For simplicity, we have thus far considered only one-dimensional motion. We now generalize our treatment to three dimensions. No conceptual difficulties are encountered, but the mathematical aspects are considerably more complicated, as we shall see.

We consider the motion of a particle of mass $m$ in an external potential $V(\mathbf{r})$, where $\mathbf{r}$ is the position vector of the particle referred to some conveniently chosen origin. The Hamiltonian for this system is then

$$H = \frac{p^2}{2m} + V(\mathbf{r}),\qquad(1)$$

where $\mathbf{p}$ is the vector momentum of the particle. Introducing a set of unit vectors $\hat{e}_x$, $\hat{e}_y$ and $\hat{e}_z$, along the axes of rectangular coordinates, we have

$$\begin{aligned}\mathbf{r} &= x\hat{e}_x + y\hat{e}_y + z\hat{e}_z \\ \mathbf{p} &= p_x\hat{e}_x + p_y\hat{e}_y + p_z\hat{e}_z.\end{aligned}\qquad(2)$$

It is often convenient to use numerical indices to identify the components of vectors, and we shall therefore also write equation (2) in the equivalent form

$$\begin{aligned}\mathbf{r} &= x_1\hat{e}_1 + x_2\hat{e}_2 + x_3\hat{e}_3 \\ \mathbf{p} &= p_1\hat{e}_1 + p_2\hat{e}_2 + p_3\hat{e}_3,\end{aligned}\qquad(3)$$

where the subscripts 1, 2 and 3 are intended to represent the $x$, $y$ and $z$ components respectively.

Different spatial components of the motion represent different degrees of freedom and hence commute with each other. The general commutation relations can thus be compactly written

$$(p_i, p_j) = (x_i, x_j) = 0$$

$$(p_i, x_j) = \frac{\hbar}{i} \delta_{ij}. \tag{4}$$

In particular, in configuration space, the $x_i$ are numbers and

$$p_x = \frac{\hbar}{i} \frac{\partial}{\partial x}, \qquad p_y = \frac{\hbar}{i} \frac{\partial}{\partial y}, \qquad p_z = \frac{\hbar}{i} \frac{\partial}{\partial z},$$

whence, introducing

$$\nabla \equiv \hat{e}_x \frac{\partial}{\partial x} + \hat{e}_y \frac{\partial}{\partial y} + \hat{e}_z \frac{\partial}{\partial z}, \tag{5}$$

we have, in brief,

$$\mathbf{p} = \frac{\hbar}{i} \nabla. \tag{6}$$

The state function $\psi(\mathbf{r}, t) = \psi(x, y, z, t)$ satisfies the time dependent Schrödinger's equation

$$H\psi = E\psi, \tag{7}$$

where

$$E = -\frac{\hbar}{i} \frac{\partial}{\partial t}$$

and where the Hamiltonian operator, in configuration space, is given by

$$H = \frac{1}{2m} (p_x^2 + p_y^2 + p_z^2) + V(\mathbf{r})$$

$$= -\frac{\hbar^2}{2m} \left( \frac{\partial^2}{\partial x^2} + \frac{\partial^2}{\partial y^2} + \frac{\partial^2}{\partial z^2} \right) + V(\mathbf{r})$$

or, using (5),

$$H = -\frac{\hbar^2}{2m} \nabla^2 + V(\mathbf{r}). \tag{8}$$

The operator $\nabla^2$ is called the *Laplacian operator* or simply the Laplacian. In any case, Schrödinger's equation in configuration space is the partial differential equation

$$\left[ -\frac{\hbar^2}{2m} \nabla^2 + V(\mathbf{r}) \right] \psi(\mathbf{r}, t) = -\frac{\hbar}{i} \frac{\partial \psi}{\partial t}. \tag{9}$$

As usual, the stationary state solutions can be separated out by writing

$$\psi_E(\mathbf{r}, t) = \psi_E(\mathbf{r}) \, e^{-iEt/\hbar}, \tag{10}$$

and we obtain at once the time independent Schrödinger's equation

$$\left[ -\frac{\hbar^2}{2m} \nabla^2 + V(\mathbf{r}) \right] \psi_E(\mathbf{r}) = E\psi_E(\mathbf{r}). \tag{11}$$

The fact that this is a partial, rather than an ordinary, differential equation, as it was in one dimension, accounts for the greatly increased mathematical complexity of the three-dimensional problem.

Before going on to consider the solutions of equation (11), we remark that for normalized state functions, $\psi^*\psi$ is the probability density in ordinary three space and that expectation values are defined by

$$\langle \psi | A | \psi \rangle = \int \psi^*(\mathbf{r}, t) A\psi(\mathbf{r}, t) \, d^3\mathbf{r},$$

where the symbol $d^3\mathbf{r}$ means the three-dimensional volume element. We remark also that the probability flux or current density is given by

$$\mathbf{j} = \frac{\hbar}{2mi} (\psi^* \nabla \psi - \psi \nabla \psi^*), \tag{12}$$

which is an obvious generalization of equation (VI–95). It is not difficult to verify that probability is properly conserved.

Finally, we remark that the above equations apply not only to the motion of a particle in an external potential, but equally to the relative motion of a pair of isolated interacting particles (after the uninteresting center-of-mass motion has been separated out). In the latter case, the coordinate $\mathbf{r}$ is the distance between the particles, $V(\mathbf{r})$ is their interaction potential, $\mathbf{p}$ their relative momentum and $m$ their reduced mass.

The solutions of equation (11) are rather complicated in general and can only be obtained exactly and explicitly for a few simple cases. The simplest case of all, of course, is that of a free particle, when $V(\mathbf{r})$ vanishes. The free-particle solutions are just de Broglie waves traveling in some given direction, and are easily verified to be

$$\psi_\mathbf{p}(r) = e^{i\mathbf{p}\cdot\mathbf{r}/\hbar} = \exp\left[i(p_x x + p_y y + p_z z)/\hbar\right]$$

$$E = p^2/2m, \tag{13}$$

where, as our notation indicates, these are states of definite vector momentum $\mathbf{p}$. Note that such states are infinitely degenerate because there are infinitely many possible orientations of the momentum vector for a fixed energy $E$.

The momentum states form a complete set, and hence *any* state function can be expressed as a superposition of these functions. We thus have the three-fold Fourier integral representation

$$\psi(\mathbf{r}, t) = \left(\frac{1}{2\pi\hbar}\right)^{3/2} \int \int \int dp_x \, dp_y \, dp_z \, \phi(\mathbf{p}, t) \exp\left[i(p_x x + p_y y + p_z z)/\hbar\right]$$

or, in brief,

$$\psi(\mathbf{r}, t) = \left(\frac{1}{2\pi\hbar}\right)^{3/2} \int d^3\mathbf{p}\, \phi(\mathbf{p}, t)\, e^{i\mathbf{p}\cdot\mathbf{r}/\hbar}. \tag{14}$$

Here $\phi(\mathbf{p}, t)$, the probability amplitude in momentum space, can be expressed in terms of $\psi(\mathbf{r}, t)$ by the inversion

$$\phi(\mathbf{p}, t) = \left(\frac{1}{2\pi\hbar}\right)^{3/2} \int d^3\mathbf{r}\, \psi(\mathbf{r}, t)\, e^{-i\mathbf{p}\cdot\mathbf{r}/\hbar}. \tag{15}$$

Equations (14) and (15) are thus seen to be three-dimensional generalizations of our earlier results in one dimension.

Just as in the one-dimensional case discussed in Chapter IV, the free-particle Schrödinger's equation is trivially soluble in momentum space. The state function $\phi(\mathbf{p}, t)$ satisfies the *ordinary* differential equation

$$\frac{p^2}{2m}\, \phi(\mathbf{p}, t) = -\frac{\hbar}{i}\, \frac{\partial\phi}{\partial t}$$

with the general solution

$$\phi(\mathbf{p}, t) = \phi(\mathbf{p}, t_0)\, \exp\left[-ip^2(t - t_0)/2m\hbar\right].$$

This result constitutes the *complete* momentum space solution to the problem of the motion of an arbitrary free-particle wave packet. The corresponding solution in configuration space is expressible in terms of the three-dimensional free-particle propagator, which is easily seen to be a straightforward generalization of the one-dimensional expression, equation (IV–13). The details are left to the problems.

## 2. POTENTIALS SEPARABLE IN RECTANGULAR COORDINATES

The simplest three-dimensional problem is that in which the potential has the very special form

$$V(\mathbf{r}) = V_1(x) + V_2(y) + V_3(z), \tag{16}$$

since in that case Schrödinger's equation separates in rectangular coordinates, as we now show. The stationary state Schrödinger's equation for such a potential has the form

$$\left\{\left[-\frac{\hbar^2}{2m}\, \frac{\partial^2}{\partial x^2} + V_1(x)\right] + \left[-\frac{\hbar^2}{2m}\, \frac{\partial^2}{\partial y^2} + V_2(y)\right]\right.$$

$$\left. + \left[-\frac{\hbar^2}{2m}\, \frac{\partial^2}{\partial z^2} + V_3(z)\right]\right\} \psi_E = E\psi_E$$

and hence, writing

$$\psi_E(x, y, z) = \psi_{E_1}(x)\psi_{E_2}(y)\psi_{E_3}(z) = \psi_{E_1}(x_1)\psi_{E_2}(x_2)\psi_{E_3}(x_3),$$

we obtain, for each factor,

$$\left[-\frac{\hbar^2}{2m}\frac{\partial^2}{\partial x_i^2} + V_i(x_i)\right]\psi_{E_i}(x_i) = E_i\psi_{E_i}(x_i), \quad i = 1, 2, 3$$

$$E = E_1 + E_2 + E_3.$$

Thus, the states are expressible as a simple composition of familiar one-dimensional states.[1]

As a first example, consider the states in a rectangular box of sides $L_1$, $L_2$ and $L_3$ respectively. Taking the origin to be at one corner of the box we see that if $\psi_E$ is required to vanish at each of the walls, then the stationary state solutions are

$$\psi_E = \sqrt{8/V}\,\sin\frac{n_1\pi x}{L_1}\,\sin\frac{n_2\pi y}{L_2}\,\sin\frac{n_3\pi z}{L_3}, \qquad n_i = 1, 2, 3\ldots,$$

where $V = L_1L_2L_3$ is the volume of the box and where

$$E = \frac{\hbar^2}{2m}\left[\left(\frac{n_1\pi}{L_1}\right)^2 + \left(\frac{n_2\pi}{L_2}\right)^2 + \left(\frac{n_3\pi}{L_3}\right)^2\right] = E_{n_1n_2n_3}.$$

The spectrum is thus fairly complicated and has no particular regularities for general $L_1$, $L_2$ and $L_3$.

For a cubical box of side $L$ the situation is somewhat simpler, since in that case

$$E_{n_1n_2n_3} = \frac{\hbar^2\pi^2}{2mL^2}(n_1^2 + n_2^2 + n_3^2).$$

The lowest state is that for which $n_1 = n_2 = n_3 = 1$, and its energy is $E_0 = 3\hbar^2\pi^2/2mL^2$. The next state is that for which one of the $n_i$ is two and each of the remaining is unity. This state is thus three-fold degenerate, and its energy is $6\hbar^2\pi^2/2mL^2 = 2E_0$. The third state is that for which any two of the $n_i$ are two and the remaining one is unity. It is also three-fold degenerate, and its energy is $3E_0$. This regular pattern is broken with the fourth state, which occurs when one of the $n_i$ is equal to three and the other two are each equal to unity. This state has energy $11\hbar^2\pi^2/2mL^2 = 11E_0/3$, and it is three-fold degenerate. The fifth state, which is non-degenerate and has energy $4E_0$, occurs for $n_1 = n_2 = n_3 = 2$, while the sixth state has energy $14E_0/3$ and is six-fold degenerate corresponding to the permutations of its quantum numbers 1, 2 and 3. The spectrum obtained by continuing in this way is shown in Figure 1.

---

[1] The motion of a free particle can be regarded as a special case in which the potentials $V_i$ are identically zero. Observe that the free-particle stationary states of equation (13) are just such a composition of one-dimensional states.

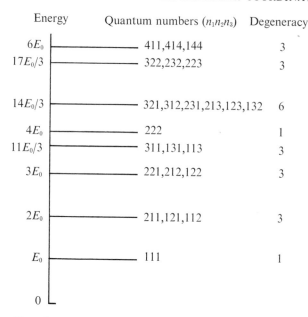

| Energy | Quantum numbers $(n_1 n_2 n_3)$ | Degeneracy |
|---|---|---|
| $6E_0$ | 411,414,144 | 3 |
| $17E_0/3$ | 322,232,223 | 3 |
| $14E_0/3$ | 321,312,231,213,123,132 | 6 |
| $4E_0$ | 222 | 1 |
| $11E_0/3$ | 311,131,113 | 3 |
| $3E_0$ | 221,212,122 | 3 |
| $2E_0$ | 211,121,112 | 3 |
| $E_0$ | 111 | 1 |
| 0 | | |

FIGURE 1.   Energies, quantum numbers and degeneracies of the states of a particle in a cubical box of side $L$. The energy $E_0$ of the ground state is $3\hbar^2\pi^2/2mL^2$.

It is also of interest to consider periodic boundary conditions for the cubical box, that is, boundary conditions which require that $\psi_E$ assume the same value on any pair of opposite walls of the box. For this case,

$$\psi_E = \frac{1}{\sqrt{V}} \exp\left[2i\pi(n_1 x + n_2 y + n_3 z)/L\right],$$

where

$$E = E_{n_1 n_2 n_3} = \frac{2\pi^2\hbar^2}{mL^2}(n_1{}^2 + n_2{}^2 + n_3{}^2), \qquad n_i = 0, \pm1, \pm2, \ldots. \qquad (17)$$

By computing the total number of states $N(E)$ of energy less than or equal to $E$, we can find the density of states $\rho(E)$, just as we did in one dimension. It is not hard to show in this way that

$$\rho(E) = \frac{(2m)^{3/2}}{4\pi^2\hbar^3} V\sqrt{E}, \qquad (18)$$

which is a very useful and important result.

---

**Exercise 1.**

(a)   Derive equation (18).

(b)   Construct the analog of Figure 1 for the states of equation (17).

---

As a second example, we discuss the three-dimensional isotropic harmonic oscillator, described by the potential

$$V(\mathbf{r}) = \tfrac{1}{2}\, m\omega^2 r^2 = \tfrac{1}{2}\, m\omega^2(x^2 + y^2 + z^2).$$

This potential is of the form of equation (16), and the stationary state solutions of Schrödinger's equation can be expressed as

$$\psi_{n_1 n_2 n_3} = \psi_{n_1}(x)\psi_{n_2}(y)\psi_{n_3}(z)$$

$$E_{n_1 n_2 n_3} = (n_1 + n_2 + n_3 + \tfrac{3}{2})\,\hbar\omega \qquad (19)$$

$$n_i = 0, 1, 2, \ldots,$$

where $\psi_{n_i}(x_i)$ is the one-dimensional harmonic oscillator stationary state function. The ground state, with energy $3\hbar\omega/2$, is not degenerate, but all the remaining states are, and increasingly so as their energy increases. For example, the first excited state has energy $5\hbar\omega/2$ and is three-fold degenerate (any one of the $n_i$ can be unity, the other two are zero), the second excited state has energy $7/2\hbar\omega$ and is six-fold degenerate (any $n_i$ can be two, the others zero, or any two can be unity, the remaining one zero). It can be shown that the state of energy $(n + 3/2)\hbar\omega$ is $[(n + 1)(n + 2)/2]$-fold degenerate, this being the number of ways $n$ can be expressed as the sum of three non-negative integers.

---

**Exercise 2.**   Prove the three-dimensional harmonic oscillator state of energy $(n + 3/2)\hbar\omega$ is $[(n + 1)(n + 2)/2]$-fold degenerate.

---

## 3. CENTRAL POTENTIALS; ANGULAR MOMENTUM STATES

We now consider the important case of motion in a spherically symmetric potential, such as the Coulomb potential between point charges, or the gravitational potential between point masses. Such a potential depends, of course, only upon the magnitude, and not upon the orientation, of the radius vector from a fixed point, which we shall take to be the origin. To describe motion in a potential of this kind, it is convenient to introduce spherical coordinates, $r$, $\theta$, $\phi$, which are defined as follows. Consider a point $P$ with coordinates $x, y, z$ in some right-handed rectangular system. The displacement vector of $P$ relative to the origin is then

$$\mathbf{r} = \hat{e}_x x + \hat{e}_y y + \hat{e}_z z,$$

where $\hat{e}_x$, $\hat{e}_y$ and $\hat{e}_z$ are unit vectors along the $x$, $y$ and $z$ axes respectively. As illustrated in Figure 2, $\theta$ is then defined as the angle between $\mathbf{r}$ and the $z$-axis, while $\phi$ is defined as the angle between the projection of $\mathbf{r}$ in the $x$-$y$ plane and the $x$-axis measured clockwise when looking toward positive $z$. The range of $\phi$ is taken as zero to $2\pi$, and the range of $\theta$ is taken as zero to $\pi$. The radial coordinate $r$ is, of course, the magnitude of $\mathbf{r}$, and its range is zero to infinity. From Figure 2, the relationship between the

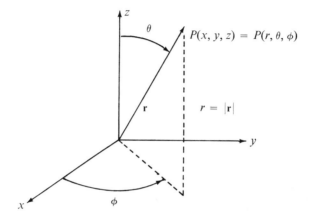

FIGURE 2.   Spherical coordinate system.

spherical and rectangular coordinates of $P$ is seen to be

$$x = r \sin\theta \cos\phi \qquad r = \sqrt{x^2 + y^2 + z^2}$$
$$y = r \sin\theta \sin\phi \qquad \tan\phi = y/x \qquad (20)$$
$$z = r \cos\theta \qquad \cos\theta = z/\sqrt{x^2 + y^2 + z^2}.$$

The coordinate system thus defined is clearly orthogonal, and the volume element can be shown to be

$$d^3\mathbf{r} = r^2 \, dr \sin\theta \, d\theta \, d\phi.$$

The element of area on a unit sphere, or element of *solid angle*, is commonly denoted by $d\Omega$ and is given by

$$d\Omega = \sin\theta \, d\theta \, d\phi, \qquad (21)$$

whence also

$$d^3\mathbf{r} = r^2 \, dr \, d\Omega. \qquad (22)$$

We now seek to write Schrödinger's equation

$$-\frac{\hbar^2}{2m}\nabla^2\psi_E + V(r)\psi_E = E\psi_E \tag{23}$$

in spherical coordinates. This requires that we express the Laplacian operator $\nabla^2$ in these coordinates. Now, from equation (20), we have

$$\frac{\partial}{\partial x} = \frac{\partial r}{\partial x}\frac{\partial}{\partial r} + \frac{\partial\theta}{\partial x}\frac{\partial}{\partial\theta} + \frac{\partial\phi}{\partial x}\frac{\partial}{\partial\phi} = \frac{x}{r}\frac{\partial}{\partial r} + \frac{xz}{r^3\sin\theta}\frac{\partial}{\partial\theta} - \frac{y}{x^2}\cos^2\phi\frac{\partial}{\partial\phi}$$

$$= \sin\theta\cos\phi\frac{\partial}{\partial r} + \frac{\cos\theta\cos\phi}{r}\frac{\partial}{\partial\theta} - \frac{\sin\phi}{r\sin\theta}\frac{\partial}{\partial\phi},$$

with similar expressions for $\partial/\partial y$ and $\partial/\partial z$. Recalling that

$$\nabla^2 = \frac{\partial^2}{\partial x^2} + \frac{\partial^2}{\partial y^2} + \frac{\partial^2}{\partial z^2},$$

we eventually find, upon putting all this together, that equation (23) becomes

$$-\frac{\hbar^2}{2m}\left[\frac{1}{r^2}\frac{\partial}{\partial r}\left(r^2\frac{\partial\psi_E}{\partial r}\right) + \frac{1}{r^2\sin\theta}\frac{\partial}{\partial\theta}\left(\sin\theta\frac{\partial\psi_E}{\partial\theta}\right) + \frac{1}{r^2\sin^2\theta}\frac{\partial^2\psi_E}{\partial\phi^2}\right]$$

$$+ V\psi_E = E\psi_E, \tag{24}$$

which is the desired expression of Schrödinger's equation in spherical coordinates.[2] In spite of its complicated appearance, this equation is separable, as we now proceed to show.

We first separate the angular and radial coordinates by writing

$$\psi_E(r, \theta, \phi) = R(r)Y(\theta, \phi). \tag{25}$$

After multiplying through by $-2mr^2/\hbar^2$, we then obtain in the usual way

$$\frac{1}{Y}\left[\frac{1}{\sin\theta}\frac{\partial}{\partial\theta}\left(\sin\theta\frac{\partial Y}{\partial\theta}\right) + \frac{1}{\sin^2\theta}\frac{\partial^2 Y}{\partial\phi^2}\right] = -\beta \tag{26}$$

$$\frac{1}{R}\left[\frac{d}{dr}\left(r^2\frac{dR}{dr}\right) + r^2\frac{2m}{\hbar^2}[E - V(r)]R\right] = \beta, \tag{27}$$

where $\beta$ is the separation constant. The first of these equations can now in turn be separated by writing

$$Y(\theta, \phi) = \Theta(\theta)\Phi(\phi).$$

Multiplying through by $\sin^2\theta$, we then obtain, again in the usual way,

$$\frac{1}{\Phi}\frac{d^2\Phi}{d\phi^2} = -\alpha^2 \tag{28}$$

[2] For a derivation of expressions for the Laplacian operator in spherical and other curvilinear coordinate systems, see Reference [7].

and

$$\frac{1}{\Theta} \left[ \sin \theta \frac{d}{d\theta} \left( \sin \theta \frac{d\Theta}{d\theta} \right) + \beta \sin^2 \theta \, \Theta \right] = \alpha^2, \tag{29}$$

where $\alpha^2$ is a second separation constant. *Observe the striking fact that the angular equations are independent of the potential $V(r)$ and of the energy $E$.* The angular functions are thus universal functions which appear for *any central potential.* Now it turns out, as we show in a moment, that equation (26) defines states of *definite angular momentum.* If we temporarily accept this assertion, and if we recall that angular momentum is a constant of the motion for a central potential, this behavior is not too surprising. It simply expresses the fact that the states in a central potential involve a universal set of functions characterizing the equally universal angular momentum states of the system.

We now show that equation (26) indeed defines states of definite angular momentum, as asserted. Recall that, with respect to some fixed point which we take to be the origin, the classical angular momentum vector **L** is given by

$$\mathbf{L} = \mathbf{r} \times \mathbf{p}. \tag{30}$$

Quantum mechanically, **L** is taken to be the same function of the quantum mechanical dynamical variables, and hence is a vector operator. In configuration space its rectangular components are, explicitly,

$$L_x = (yp_z - zp_y) = \frac{\hbar}{i} \left( y \frac{\partial}{\partial z} - z \frac{\partial}{\partial y} \right) \tag{31}$$

$$L_y = (zp_x - xp_z) = \frac{\hbar}{i} \left( z \frac{\partial}{\partial x} - x \frac{\partial}{\partial z} \right) \tag{32}$$

$$L_z = (xp_y - yp_x) = \frac{\hbar}{i} \left( x \frac{\partial}{\partial y} - y \frac{\partial}{\partial x} \right). \tag{33}$$

It is not hard to show that in spherical coordinates these rectangular components become

$$L_x = -\frac{\hbar}{i} \left( \sin \phi \frac{\partial}{\partial \theta} + \mathrm{ctn}\, \theta \cos \phi \frac{\partial}{\partial \phi} \right) \tag{34}$$

$$L_y = \frac{\hbar}{i} \left( \cos \phi \frac{\partial}{\partial \theta} - \mathrm{ctn}\, \theta \sin \phi \frac{\partial}{\partial \phi} \right) \tag{35}$$

$$L_z = \frac{\hbar}{i} \frac{\partial}{\partial \phi} \tag{36}$$

and hence that, after more algebra,

$$L^2 \equiv L_x{}^2 + L_y{}^2 + L_z{}^2 = -\hbar^2 \left[ \frac{1}{\sin\theta} \frac{\partial}{\partial\theta} \sin\theta \frac{\partial}{\partial\theta} + \frac{1}{\sin^2\theta} \frac{\partial^2}{\partial\phi^2} \right]. \quad (37)$$

We thus see that equation (26) is equivalent to

$$L^2 Y = \beta\, \hbar^2 Y \qquad (38)$$

and hence that, as was to be proved, $Y$ defines a state in which the *magnitude* of the angular momentum has a definite value, namely $\sqrt{\beta}\hbar$. We defer a detailed discussion of angular momentum and its properties to Chapter X. Before going on, however, we remark briefly on the *orientation* of the angular momentum vector. Note that, since $L_z = (\hbar/i)\partial/\partial\phi$, equation (28) is equivalent to

$$L_z{}^2\, \Phi = \alpha^2\hbar^2\Phi. \qquad (39)$$

This means that $Y = \Theta\Phi$ is also an eigenfunction of $L_z$, with eigenvalue $\alpha\hbar$. Thus the angular momentum states $Y$ are states in which the magnitude of the angular momentum vector and its projection on the z-axis are both fixed, with $\beta$ determining the magnitude and $\alpha$ the projection.[3]

Observe that the form of the Laplacian in spherical coordinates, and therefore of the kinetic energy operator, has been established by the preceding analysis to be

$$\frac{p^2}{2m} = -\frac{\hbar^2}{2m}\nabla^2 = -\frac{\hbar^2}{2m}\left[ \frac{1}{r^2}\frac{\partial}{\partial r}\left(r^2\frac{\partial}{\delta r}\right)\right] + \frac{L^2}{2mr^2}, \qquad (40)$$

where $L^2$ is given by equation (37). We now give a direct and instructive alternative derivation of this important result, a derivation in which the angular momentum enters from the outset instead of appearing in rather mysterious fashion at the end. We start with the classical vector identity

$$(\mathbf{A} \times \mathbf{B}) \cdot (\mathbf{C} \times \mathbf{D}) = (\mathbf{A} \cdot \mathbf{C})(\mathbf{B} \cdot \mathbf{D}) - (\mathbf{A} \cdot \mathbf{D})(\mathbf{B} \cdot \mathbf{C}), \qquad (41)$$

which is readily verified by expressing each of the four vectors in rectangular components and writing each side out explicitly.

We use this vector identity by making the identification

$$\mathbf{A} = \mathbf{C} = \mathbf{r},$$
$$\mathbf{B} = \mathbf{D} = \mathbf{p}. \qquad (42)$$

For the classical case, this at once gives

$$L^2 = r^2 p^2 - (\mathbf{r} \cdot \mathbf{p})^2, \qquad (43)$$

which, when multiplied through by $(1/2mr^2)$, already resembles the

---

[3] It turns out, as we shall see, that the orientation of the angular momentum vector cannot be specified more precisely on the quantum mechanical level, and that this lack of precision is nothing more than a manifestation of the uncertainty principle.

sought-after equation (40). Equation (42) can be applied in the quantum mechanical case as well, provided that the *ordering* of the non-commuting vectors **r** and **p** is *maintained* in every term. Choosing the order **ABCD**, so that the left side is just $(\mathbf{r} \times \mathbf{p})^2 = L^2$, we thus can write this vector identity as

$$L^2 = \sum_{i,j} x_i p_j x_i p_j - \sum_{i,j} x_i p_j x_j p_i,$$

where the terms on the right have been written out in rectangular component form, with every factor appearing in its proper order. Using the commutation relation

$$p_j x_i = x_i p_j + \frac{\hbar}{i}\, \delta_{ij},$$

the first summation gives

$$\sum_{i,j} x_i p_j x_i p_j = \sum_{i,j} x_i^2 p_j^2 + \frac{\hbar}{i} \sum_{i,j} x_i p_j\, \delta_{ij}$$

$$= r^2 p^2 + \frac{\hbar}{i}\, \mathbf{r} \cdot \mathbf{p},$$

while the second gives

$$\sum_{i,j} x_i p_j x_j p_i = \sum_{i,j} x_i x_j p_j p_i + \frac{3\hbar}{i}\, \mathbf{r} \cdot \mathbf{p}$$

$$= \sum_{i,j} x_j p_j x_i p_i + \frac{2\hbar}{i}\, \mathbf{r} \cdot \mathbf{p}$$

$$= (\mathbf{r} \cdot \mathbf{p})^2 + \frac{2\hbar}{i}\, \mathbf{r} \cdot \mathbf{p}.$$

Hence

$$L^2 = r^2 p^2 - (\mathbf{r} \cdot \mathbf{p})^2 - \frac{\hbar}{i}\, (\mathbf{r} \cdot \mathbf{p}), \tag{44}$$

which differs from the classical result only by the term proportional to $\hbar$. As the final step, observe that

$$(\mathbf{r} \cdot \mathbf{p})^2 + \frac{\hbar}{i}\, (\mathbf{r} \cdot \mathbf{p}) = -\hbar^2 \left[ \left( r\frac{\partial}{\partial r} \right)^2 + r\frac{\partial}{\partial r} \right]$$

$$= -\hbar^2 \frac{\partial}{\partial r} \left( r^2 \frac{\partial}{\partial r} \right),$$

whence equation (44) can be rewritten, upon multiplying from the left by $1/r^2$ and solving for $p^2$, in the form

$$p^2 = -\frac{\hbar^2}{r^2} \frac{\partial}{\partial r} \left( r^2 \frac{\partial}{\partial r} \right) + \frac{L^2}{r^2}, \tag{45}$$

which is recognized as equation (40) except for the common factor of $1/2m$. From this derivation, the first term in equation (45) is recognized as the quantum analog of the square of the radial momentum, or, upon division by $2m$, as the kinetic energy associated with the radial motion.

We now turn to the specific angular functions defined by the differential equations (28) and (29). Unfortunately, the latter of these functions is rather complicated. However, its detailed structure is not particularly relevant for our present purposes, and we shall therefore simply write down the answer.[4] It turns out that single-valued well-behaved solutions of equations (28) and (29) are obtained only if $\beta$ and $\alpha$ are such that

$$\beta = l(l+1), \quad l = 0, 1, 2, \ldots$$
$$\alpha = m \quad , \quad m = 0, \pm 1, \pm 2, \ldots, \pm l. \tag{46}$$

Note that $l$ must be a non-negative integer and that, for a given $l$, $m$ can take on only those integral values which range from $-l$ to $l$.

The normalized solution $Y(\theta, \phi) = \Theta(\theta)\Phi(\phi)$, which we denote by $Y_l^m(\theta, \phi)$ for a particular value of $l$ and of $m$, is usually expressed as

$$Y_l^m(\theta, \phi) = (-1)^{\frac{m+|m|}{2}} \sqrt{\frac{2l+1}{4\pi} \frac{(l-|m|)!}{(l+|m|)!}} P_l^m(\theta) \, e^{im\phi}, \tag{47}$$

where the function $P_l^m(\theta) \equiv P_l^{-m}(\theta)$ is known as an *associated Legendre function*. It is a product of $\sin^{|m|} \theta$ and a polynomial of degree $(l-|m|)$ in $\cos \theta$, in even powers if $(l-|m|)$ is even, in odd powers if $(l-|m|)$ is odd. For reference, the first few $Y_l^m$ are given in Table I.

| | $l = 0$ | $l = 1$ | $l = 2$ |
|---|---|---|---|
| $m = 0$ | $Y_0^0 = \dfrac{1}{\sqrt{4\pi}}$ | $Y_1^0 = \sqrt{\dfrac{3}{4\pi}} \cos \theta$ | $Y_2^0 = \sqrt{\dfrac{5}{4\pi}} \dfrac{3\cos^2 \theta - 1}{2}$ |
| $m = \pm 1$ | — | $Y_1^{\pm 1} = \mp \sqrt{\dfrac{3}{8\pi}} \sin \theta \, e^{\pm i\phi}$ | $Y_2^{\pm 1} = \mp \sqrt{\dfrac{5}{24\pi}} \, 3 \sin \theta \cos \theta \, e^{\pm i\phi}$ |
| $m = \pm 2$ | — | — | $Y_2^{\pm 2} = \sqrt{\dfrac{5}{96\pi}} \, 3 \sin^2 \theta \, e^{\pm 2i\phi}$ |

TABLE I.   Normalized angular momentum eigenfunctions.

The normalized angular momentum eigenfunctions are known mathematically as *spherical harmonics*. They form a complete orthonormal set on a unit sphere. By orthonormality for these functions, we mean that

---

[4] A complete derivation is given in Chapter X. See also References [1] through [5].

$$\int Y_l^{m*}(\theta, \phi) Y_{l'}^{m'}(\theta, \phi) \, d\Omega$$

$$= \int_0^\pi \int_0^{2\pi} Y_l^{m*} Y_{l'}^{m'} \sin \theta \, d\theta \, d\phi = \delta_{ll'} \delta_{mm'}, \quad (48)$$

as may readily be verified for any of the particular examples in Table I.

Although, as we have said, we shall defer a derivation of any of the above features until Chapter X, where we will use a factorization method to construct the solutions, it is instructive to at least indicate how equations (28) and (29) can be handled directly. The first is trivial; it yields at once $\Phi(\phi) = e^{i\alpha\phi}$. Since $\phi = 0$ and $\phi = 2\pi$ refer to the same point in space, $\alpha$ must be an integer, say $m$, if $\Phi$ is to be a single-valued function. Equation (29) is more complicated, but it can be solved by a power series method in which $\Theta$ is expressed in the form

$$\Theta = \sin^{|m|} \theta \sum_p c_p \cos^p \theta.$$

It turns out that this series diverges at $\theta = 0$ and $\pi$, unless it is forced to terminate. The condition that it do so for given $m$ is that $\beta = l(l+1)$ with $(l - |m|)$ restricted to nonnegative integral values. The resulting polynomials are the associated Legendre polynomials, and the conditions on $\alpha$ and $\beta$ are those of equation (46).

Thus far we have considered only the angular dependence of the stationary state solutions of Schrödinger's equation. We must next take up equation (27) for the radial function. Expressing $\beta$ in terms of $l$, and rearranging slightly, this equation becomes

$$-\frac{\hbar^2}{2m} \frac{1}{r^2} \frac{d}{dr} \left( r^2 \frac{dR_{El}}{dr} \right) + \left[ V(r) + \frac{\hbar^2 l(l+1)}{2mr^2} \right] R_{El} = ER_{El}, \quad (49)$$

where we have now introduced the subscripts $E$ and $l$ to denote the dependence of $R$ on these quantities. Of course, the entire stationary state wave function is expressed as the product

$$\psi_{Elm} = R_{El}(r) Y_l^m(\theta, \phi), \quad (50)$$

where we have again introduced appropriate subscripts to make explicit the fact that these are simultaneously states of definite energy $(E)$, angular momentum $(l)$ and z-component of angular momentum $(m)$.

Before going on to consider the solutions of equation (49) for particular potentials, we point out some general properties of the stationary states. Our first remark has to do with the degeneracy of the states. Note that $R_{El}$ does not depend on $m$, but only on $l$. Hence we obtain an eigenfunction corresponding to the same energy for each permissible value of $m$ for a given $l$. Since $m$ can take on any integral value running from

$-l$ to $l$, there are $(2l + 1)$ such values, and thus *the states in a central potential are intrinsically $(2l + 1)$-fold degenerate*. As we have seen, $m$ measures the projection of $L$ on the $z$-axis and hence it is essentially determined by the orientation of the angular momentum vector. The degeneracy in question is a consequence of the fact that the Hamiltonian is independent of this orientation when the potential is spherically symmetric.

Our second remark has to do with the parity of the states. The parity operator, it is recalled, changes the sign of all coordinates. Thus, by definition, for arbitrary $f$,

$$Pf(x, y, x) = f(-x, -y, -z).$$

Reference to equation (20), or to Figure 1, then shows that in spherical coordinates

$$Pf(r, \theta, \phi) = f(r, \pi - \theta, \phi \pm \pi).$$

In particular, therefore,

$$P\psi_{Elm}(r, \theta, \phi) = R_{El}(r) \ Y_l{}^m(\pi - \theta, \phi \pm \pi).$$

Recalling that $P_l{}^m(\theta)$ is an even or odd function of $\cos\theta$, depending upon whether $(l - |m|)$ is even or odd, we see that $P_l{}^m(\theta)$ has parity $(-1)^{l-|m|}$, while the remaining factor, $e^{im\phi}$, in $Y_l{}^m$, has parity $(-1)^m$. Hence

$$P\psi_{Elm}(r, \theta, \phi) = (-1)^l \psi_{Elm}(r, \theta, \phi), \qquad (51)$$

and the states thus have definite parity. The parity is even or odd, depending only upon whether $l$ is even or odd, and not at all upon $m$.

Finally, we remark on the relation between the radial equation (49) and the equation, familiar by now, for the stationary states in one rectilinear dimension. If we think of the combination $V(r)$ plus the centrifugal terms $l(l + 1)\hbar^2/2mr^2$ as equivalent to an effective potential

$$\tilde{V}(r) \equiv V(r) + \frac{l(l + 1)\hbar^2}{2mr^2}, \qquad (52)$$

then the radial equation is already seen to bear a close resemblance to one-dimensional motion. Just how close this connection is becomes apparent if we write the radial part of the wave function $R_{El}$ in the form

$$R_{El}(r) = \frac{u_{El}(r)}{r} \qquad (53)$$

Substitution into equation (49) then gives, almost at once,

$$-\frac{\hbar^2}{2m}\frac{d^2u_{El}}{dr^2} + \left[ V(r) + \frac{\hbar^2 l(l + 1)}{2mr^2} \right] u_{El} = Eu_{El}, \qquad (54)$$

which is seen to be rather simpler than equation (49) and, more important, identical in form to the stationary state Schrödinger equation in one rectilinear dimension for motion in the effective potential $\tilde{V}$. However, equation (54) has meaning only for positive values of the coordinate $r$. Further, if $R_{El}(r)$ is bounded at the origin, then, according to equation (53),

$$u_{El}(r = 0) = 0. \tag{55}$$

From this we see that the solutions of the radial equations are exactly the same as the *odd state* solutions in the one-dimensional problem of motion in the symmetrical potential $\tilde{V} = V(|x|) + \hbar^2 l(l+1)/x^2$, since these odd states automatically vanish at the origin. *The even one-dimensional states do not satisfy equation (55) and hence do not appear in the spectrum.* All of the one-dimensional techniques we have learned are thus seen to be applicable to the study of motion in three dimensions. Note, moreover, that for given $l$, the radial states are unique; there is one and only one simultaneous radial eigenfunction of $E$ and $l$, for continuum states as well as for bound states. However, eigenstates of *any* given energy in the continuum can always be found for *every* value of $l$; and the continuum states in three dimensions are thus infinitely degenerate corresponding to the infinite set of possible $l$ values. Furthermore, it may happen that even in the discrete part of the spectrum states of different $l$ occur which have the same energy. The degeneracy thus introduced, which is an addition to the intrinsic $(2l+1)$-fold degeneracy discussed earlier for each state of given $l$, is commonly called an *accidental degeneracy*. This nomenclature is sometimes inappropriate, since, in fact, such degeneracies are not always accidental but instead may be a consequence of additional symmetries in the Hamiltonian beyond the spherical symmetry we have assumed for $V(r)$. We shall shortly present

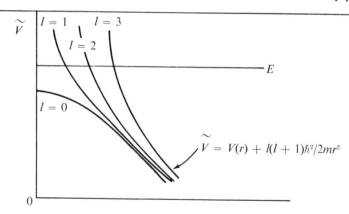

FIGURE 3.  Plot of the effective radial potential $\tilde{V} = V(r) + l(l+1)\hbar^2/2mr^2$ for the first few values of $l$, for a repulsive potential. Only continuum states of positive energy $E$ appear. For given $E$ one such state occurs for each value of $l$.

some examples which illustrate this behavior.

The above remarks are perhaps made clearer by reference to the figures. Thus Figure 3 shows the effect of the centrifugal potential when $V(r)$ is everywhere repulsive. As $l$ increases, the effective potential is seen to become increasingly repulsive and the spectrum thus consists entirely of continuum states of positive energy. For any given positive energy, one radial state exists for each value of $l$. In Figure 4, the more complicated and more interesting case of an attractive potential is presented. For positive energies the situation is the same as for repulsive potentials; the spectrum is continuous for every $l$. The spectrum of discrete, negative energy bound states depends, of course, upon the detailed behavior of $V(r)$. In the example shown, such states could exist for the particular $l$ values 0, 1 and 2, but evidently for $l \geq 3$ no bound states exist because $\tilde{V}$ is repulsive for every such state. The lowest bound state for a given $l$ has no radial nodes, the first excited state has one such node, and so on.

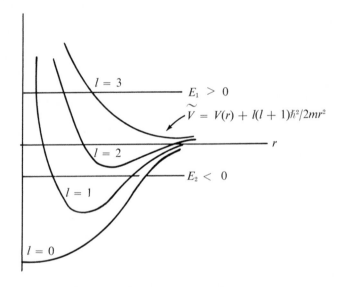

FIGURE 4. Plot of the effective radial potential $V = V(r) + l(l+1)\hbar^2/2mr^2$ for the first few values of $l$ for an attractive potential. For positive energy, such as $E_1$, the spectrum is continuous for every $l$. For those values of $l$ for which bound states of negative energy, such as $E_2$, exist, the spectrum is discrete.

This behavior is illustrated in Figure 5, where the spectrum is shown for $l = 0$ and $l = 1$. In the example chosen, which is that of a rather shallow short-range potential, it so happens that there are three states with $l = 0$ and two with $l = 1$. At the other extreme, when the potential is deep and long-range, as for the Coulomb potential, it turns out that there are infinitely many discrete bound states for *every* value of $l$, as we shall see.

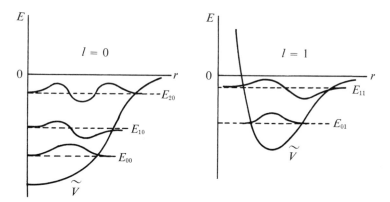

FIGURE 5. Discrete states for $l = 0$ and $l = 1$ in the attractive potential of Figure 4. In the example shown there are three bound states for $l = 0$, two for $l = 1$. The radial functions $u_{El} = rR_{El}$ are also shown for each state. If one of the allowed energies $E_{nl}$ for $l = 1$ happened to coincide with one of the energies for $l = 0$, this would be an example of an accidental degeneracy. As pointed out in the text, such degeneracies are sometimes a consequence of the symmetry properties of the Hamiltonian.

## 4. SOME EXAMPLES

We now consider a few special examples of motion in spherically symmetric potentials.

(a) *Spherically symmetric states ($l = 0$).* For spherically symmetric states, that is, states with $l = 0$, and hence with zero angular momentum, equation (54) reduces to

$$-\frac{\hbar^2}{2m} \frac{d^2 u_{E0}}{dr^2} + V(r)\, u_{E0} = E u_{E0},$$

where $u_{E0}(r)$ satisfies the boundary condition

$$u_{E0}(r = 0) = 0.$$

Thus the problem is exactly equivalent to that of finding the odd states characterizing one-dimensional motion in the symmetrical potential

$$V(|x|) = V(r).$$

The situation is particularly simple because of the absence of the invariably serious complications associated with the centrifugal potential.

As a first example, consider the states in a spherical square well potential. This example is quite important, because it happens to give a fair description of the very short range interaction between a neutron and proton in the deuteron. Such a potential is given by

$$V(r) = -V_0, \qquad r \leq a$$
$$V(r) = 0 \quad , \qquad r > a.$$

The corresponding one-dimensional potential is then

$$V(x) = -V_0, \qquad |x| \leq a$$
$$V(x) = 0 \quad , \qquad |x| > a,$$

that is, a symmetrical square well of width $2a$. We have already considered this problem in detail in Chapter VI, and among other things, we found that no bound state exists unless

$$V_0 \geq \left(\frac{\pi}{2}\right)^2 \frac{\hbar^2}{2ma^2}.$$

Now it turns out that the deuteron has only one bound state, and that this state is rather weakly bound. Hence $V_0$ exceeds this minimum value by just a little. The range of nuclear forces is known to be about 1.9 $\times 10^{-13}$ cm. Accepting this value for $a$, the depth of the potential can thus be estimated, and it turns out to be something like 40 Mev. Interestingly enough, this simple argument actually provided the first reliable value for the strength of nuclear forces.

(b) *Harmonic oscillator.* As a second example, consider the three-dimensional isotropic oscillator. Of course, we have already given a complete solution to this problem in rectangular coordinates, but it is instructive to re-examine the problem in spherical coordinates. The $l = 0$ states are at once simply the odd states of the one-dimensional oscillator, and hence have energies $3\hbar\omega/2$, $7\hbar\omega/2$, $11\hbar\omega/2$, and so on. Recall now that the complete spectrum of the three-dimensional oscillator was found to be expressible as

$$E_n = (n + 3/2)\hbar\omega,$$

where the $n$th state was $[(n+2)(n+1)/2]$-fold degenerate. We thus see that the $n$th state contains among its $[(n+2)(n+1)/2]$ degenerate members, exactly one spherically symmetric state if $n$ is even, none if $n$ is odd. It is possible to show, although we shall not attempt to do so here, that states of higher angular momentum appear in the spectrum in the following way. For $l = 1$, the allowed energies are $5/2\ \hbar\omega$, $9/2\ \hbar\omega$, . . .; for $l = 2$ the allowed energies are $7\ \hbar\omega/2$, $11\ \hbar\omega/2$, . . .; and so on.

In general, for angular momentum $l$, the allowed energies are $(l + 3/2)\hbar\omega$, $(l + 2 + 3/2)\hbar\omega$, $(l + 4 + 3/2)\hbar\omega$, .... Otherwise stated, the degenerate members of the $n$th energy state are those states of angular momentum $l = n, n - 2, n - 4, \ldots$, and so on down to $l = 0$, for even $n$, or to $l = 1$ for odd $n$.

The spectrum, classified according to angular momentum, is shown in Figure 6. The degeneracy of each state in this scheme can be obtained upon noting that a state of angular momentum $l$ is $(2l + 1)$-fold degenerate. Thus the ground state, which is $l = 0$, is not degenerate, the first excited state which contains only an $l = 1$ state is three-fold degenerate, the second excited state contains a nondegenerate $l = 0$ state and a five-fold-degenerate $l = 2$ state so that the total degeneracy is six-fold, and so on, in agreement with our earlier rule. The pervasive occurrence of degeneracies between states of different $l$ values is an excellent example of "accidental" degeneracies which are not accidental at all. These degeneracies occur because of the special structure of the potential for the isotropic harmonic oscillator which makes Schrödinger's equation separable in both rectangular and spherical coordinates.

---

**Exercise 3.** Verify that the oscillator states corresponding to energy $E_1 = 5\hbar\omega/2$ are indeed $l = 1$ states, as we claimed above. Do this by showing that

$$\psi_{100} \pm i\psi_{010} = f(r)Y_1^{\pm 1}(\theta, \phi)$$

$$\psi_{001} = f(r)Y_1^0(\theta, \phi),$$

where $\psi_{n_1 n_2 n_3}(x, y, z)$ is defined by equation (19). Identify $f(r)$ and by substitution verify that it actually satisfies the correct radial Schrödinger's equation.

---

| | | | | | Degeneracy |
|---|---|---|---|---|---|
| $E_4 = 11/2\,\hbar\omega$ | — | | — | | — | 15 |
| $E_3 = 9/2\,\hbar\omega$ | | — | | — | | 10 |
| $E_2 = 7/2\,\hbar\omega$ | — | | — | | | 6 |
| $E_1 = 5/2\,\hbar\omega$ | | — | | | | 3 |
| $E_0 = 3/2\,\hbar\omega$ | — | | | | | 1 |
| | $l = 0$ | $l = 1$ | $l = 2$ | $l = 3$ | $l = 4$ | |

FIGURE 6. States of the three-dimensional harmonic oscillator classified according to angular momentum. The degeneracy of each state is also indicated.

(c) *Motion of a free particle.* We have already discussed the states of a free particle in rectangular coordinates. The states so obtained were simultaneous eigenfunctions of the Hamiltonian and of the linear momentum. In spherical coordinates we obtain states which are, instead, simultaneous eigenfunctions of $H$ and of the angular momentum, that is, they have definite values of $E$, $l$ and $m$. The radial functions are solutions of equation (49) with $V(r)$ set equal to zero. They thus satisfy

$$\frac{1}{r^2}\frac{d}{dr}\left(r^2\frac{dR_{El}}{dr}\right) + \left[k^2 - \frac{l(l+1)}{r^2}\right]R_{El} = 0, \tag{56}$$

where

$$k^2 = 2mE/\hbar^2. \tag{57}$$

Alternatively, these solutions may be obtained from equation (54), which, for a free particle, can be written in the form

$$\frac{d^2u_{El}}{dr^2} + \left[k^2 - \frac{l(l+1)}{r^2}\right]u_{El} = 0, \tag{58}$$

where, it is recalled, $u_{El} = rR_{El}$.

Consider first the case $l = 0$. From equation (58), recalling also that $u_{El}(r)$ must vanish at the origin, we have at once

$$u_{E0} \sim \sin kr$$

and hence

$$R_{E0} \sim \frac{\sin kr}{r}.$$

For $l$ different from zero the situation is more complicated. However, it turns out that the solutions of equation (56) are a well-studied set of functions which can be compactly defined as follows:

$$j_l(kr) \equiv (-1)^l \left(\frac{r}{k}\right)^l \left(\frac{1}{r}\frac{d}{dr}\right)^l \frac{\sin kr}{kr}. \tag{59}$$

The function $j_l(kr)$ is called a *spherical Bessel function* of order $l$. In terms of these functions, and up to a multiplicative constant, the radial free particle states are given by

$$R_{El}(r) = j_l(kr), \qquad k = \sqrt{2mE}/\hbar, \tag{60}$$

which clearly reduce to the correct solution for $l = 0$. The proof that the functions defined by equation (59) are, for general $l$ values, actually solutions of equation (46) is not difficult, but we shall not bother to carry it out.[5]

From equation (59), the first few spherical Bessel functions are readily found to be

$$j_0(kr) = \frac{\sin kr}{kr}$$

$$j_1(kr) = \frac{\sin kr}{(kr)^2} - \frac{\cos kr}{kr}$$

$$j_2(kr) = \left[ \frac{3}{(kr)^3} - \frac{1}{kr} \right] \sin kr - \frac{3}{(kr)^2} \cos kr,$$

(61)

and these are sufficient to illustrate the general structure of these functions as polynomials in $(1/kr)$ multiplying trigonometric functions. From equation (59), it is not difficult to obtain the explicit behavior of $j_l(kr)$ when $r$ is very small and also when $r$ is very large. In the former case, expanding $(\sin kr)/kr$ in a power series in $kr$, it is seen that the first contributing term in this series is that in $(kr)^{2l}$, and we thus find,

$$r \to 0, \qquad j_l(kr) \simeq \frac{2^l \, l!}{(2l+1)!} \, (kr)^l.$$

(62)

In the latter case, the dominant term is that inversely proportional to

---

[5] A simple proof by induction can be constructed along the following lines. First note that equation (59) is completely equivalent to the recursion relation,

$$j_{l+1}(kr) = -\left(\frac{r}{k}\right)^{l+1} \left(\frac{1}{r}\frac{d}{dr}\right) \left[\left(\frac{k}{r}\right)^l j_l(kr)\right],$$

(59a)

which expresses $j_{l+1}$ in terms of $j_l$. The equivalence is established by starting with the known expression for $j_0$ and applying the recursion relation $l$ times. To demonstrate that equation (59) defines the solutions of equation (56), it thus suffices to show that if $j_{l_0}$ is a solution for some $l = l_0$, then the right side of equation (59a) is necessarily a solution for $l = l_0 + 1$. For this purpose, introduce

$$S_l(kr) \equiv \left(\frac{k}{r}\right)^l j_l(kr),$$

in terms of which equation (59a) becomes

$$S_{l+1} = \frac{1}{r}\frac{dS_l}{dr}.$$

(59b)

It is then readily demonstrated that the right side of equation (59b) indeed satisfies the same differential equation as the left side, and the desired result follows.

For orientation, we remark in passing that equation (56) is a form of *Bessel's equation.* Specifically, it can be shown that

$$j_l(kr) = \sqrt{\frac{\pi}{2kr}} \, J_{l+1/2} \, (kr),$$

where $J_\nu(kr)$ is an ordinary (cylindrical) Bessel function of order $\nu$. For a discussion of the properties of Bessel functions see References [1] through [5].

$r$, and we eventually find,

$$r \to \infty, \qquad j_l(kr) \simeq \frac{\sin (kr - l\pi/2)}{kr}. \tag{63}$$

The solutions are thus simple spherical waves at large distances from the origin. Near the origin, however, the centrifugal barrier dominates and the wave function is seen to become smaller and smaller in that region as $l$ increases.

---

**Exercise 4.**  Derive equations (62) and (63) from equation (59).

---

As a simple example involving free particle states of definite angular momentum, we briefly consider the states of a particle confined in a spherical container of radius $a$. The wave function must vanish at the walls of the sphere and hence the spectrum of allowed energies is determined by the transcendental equations,

$$j_l(ka) = 0, \qquad l = 0, 1, 2, \ldots .$$

For $l = 0$, we see from equation (61) that this reduces to the requirement that $\sin ka = 0$, which means that $ka$ must simply be an integral multiple of $\pi$. For $l = 1$, the situation is not quite so simple, since we must solve $\tan ka = ka$, which can only be done numerically. Evidently, the equations rapidly become more and more complicated as $l$ increases, and accordingly we shall not discuss them further.

One highly interesting and important aspect of our results remains to be discussed. We found in Section 1 that the stationary states of a free particle could be written as ordinary de Broglie waves corresponding to definite, but arbitrarily oriented, vector momentum $\mathbf{p}$, as expressed by equation (13). Writing

$$\mathbf{p} = \hbar k \hat{n}_\mathrm{p},$$

where $\hat{n}_\mathrm{p}$ is a unit vector along p, equation (13) becomes

$$\psi_\mathrm{p} = e^{ik\hat{n}_\mathrm{p} \cdot \mathbf{r}}. \tag{64}$$

On the other hand, we have just seen that, for any $l$ and $m$,

$$\psi_{Elm} = j_l(kr) Y_l^m(\theta, \phi) \tag{65}$$

is also a free particle state function corresponding to the same energy. These two representations are *complementary* in that the former describes stationary states of well-defined linear momentum but poorly defined angular momentum, while the latter conversely describes states of well-defined angular momentum but poorly defined linear momentum. Classically, of course, free particle states are such that both are precisely defined. This is not so in quantum mechanics – not by accident but as a direct consequence of the easily verifiable noncommutativity of the linear and angular momentum operators.

The representations of equations (64) and (65) are both complete in that an arbitrary free particle state of energy $E$ can be expressed as a superposition of either. In particular, then, each must be expressible in terms of the other. Using standard properties of the spherical harmonics and spherical Bessel functions, it can be shown[6] after some effort that

$$\psi_\mathrm{p}(\mathbf{r}) = 4\pi \sum_{l,m} i^l Y_l^{m*} (\theta_\mathrm{p}, \phi_\mathrm{p}) \psi_{Elm}(r, \theta, \phi) ,\qquad (66)$$

where $\theta_\mathrm{p}$ and $\phi_\mathrm{p}$ define the angular orientation of $\mathbf{p}$ in the same way that $\theta$ and $\phi$ define the angular orientation of $\mathbf{r}$. Conversely, in view of the orthonormality of the spherical harmonics,

$$\psi_{Elm}(r, \theta, \phi) = \frac{1}{4\pi i^l} \int Y_l^m (\theta_\mathrm{p}, \phi_\mathrm{p}) \psi_\mathrm{p}(\mathbf{r}) \, d\Omega_\mathrm{p} ,\qquad (67)$$

where $d\Omega_\mathrm{p} = \sin \theta_\mathrm{p} \, d\theta_\mathrm{p} \, d\phi_\mathrm{p}$ is the element of solid angle about the unit vector $\hat{n}_\mathrm{p}$ defining the direction of $\mathbf{p}$. For a given energy, equation (66) is seen to express a state of linear momentum $\mathbf{p}$ as a superposition over all angular momentum states, while equation (67) expresses a state of definite angular momentum as a superposition over all orientations of the linear momentum.

An important special case of equation [66] is that in which $\hat{n}_\mathrm{p}$ lies along the $z$-axis, the polar axis of the spherical coordinate system, so that $\theta_\mathrm{p} = 0$. Since

$$Y_l^m(0, \phi) = \sqrt{\frac{2l+1}{4\pi}} \, \delta_{m0} ,$$

we obtain, upon inserting the explicit forms of $\psi_\mathrm{p}$ and $\psi_{Elm}$,

$$\psi_\mathrm{p} = e^{ikr \cos \theta} = \sum_{l=0}^{\infty} i^l (2l+1) \, j_l(kr) \, P_l (\cos \theta) ,\qquad (68)$$

where we have also used the fact that, according to equation (47),

$$Y_l^{m=0} (\theta, \phi) = \sqrt{\frac{2l+1}{4\pi}} \, P_l (\cos \theta) ,$$

with $P_l (\cos \theta)$ denoting the ordinary Legendre polynomial. Now, because the linear momentum has been chosen to lie along the $z$-axis, the $z$-component of angular momentum must be zero, and so we have found; the superposition of equation (51) involves only states with $m$ equal to zero.

---

[6] For a derivation of equation (66) see, for example, Reference [22], Chapter IX, Section 9.

## 5. THE HYDROGENIC ATOM

We now obtain the stationary states of a system consisting of a single electron and an atomic nucleus in purely electrostatic interaction. Denoting the magnitude of the electronic charge by $e$, and considering a nucleus of atomic number $Z$, and hence of charge $Ze$, the electrostatic potential is taken to be

$$V(r) = -\frac{Ze^2}{r}.$$

For $Z = 1$ this system is the hydrogen atom, for $Z = 2$ it is singly-ionized helium, for $Z = 3$ it is doubly-ionized lithium, and so on. Such systems are called hydrogenic because of their obvious similarity to the hydrogen atom. Note that we have written the Coulomb potential appropriate to a *point charge* nucleus, since this is an excellent approximation on the atomic scale of distances.

We must now solve the radial equation, for a state of given angular momentum $l$,

$$-\frac{\hbar^2}{2m}\frac{d^2u_{El}}{dr^2} + \left[-\frac{Ze^2}{r} + \frac{l(l+1)\hbar^2}{2mr^2}\right]u_{El} = Eu_{El}. \qquad (69)$$

Recall that $m$ is the reduced mass of the system and thus is given by

$$m = \frac{m_e m_n}{m_e + m_n},$$

where $m_e$ is the mass of the electron and $m_n$ that of the nucleus. Of course $m$ differs from $m_e$ by less than one part in a thousand, but the measurements of atomic spectra are sufficiently precise that the reduced mass effect is actually detectable.

It is convenient to introduce a dimensionless coordinate $y$ and a dimensionless binding energy $W$ by writing

$$r = \frac{\hbar^2}{me^2}y$$

$$E = -\frac{me^4}{2\hbar^2}W, \qquad (70)$$

whence equation (52) takes on the much simpler appearance

$$\frac{d^2u_{El}}{dy^2} - \left[-\frac{2Z}{y} + \frac{l(l+1)}{y^2} + W\right]u_{El} = 0, \qquad (71)$$

and it is this equation we must now solve. We shall do so by the power series method. Note first, however, that for sufficiently large values of $y$ the square bracket in equation (71) differs only slightly from $W$. Hence, asymptotically the behavior of $u_{El}$ is dominated by the exponential factor $e^{\pm\sqrt{W}y}$. Physical admissibility demands the negative sign and hence we write

$$u_{El} = v_{El}(y)\ e^{-\sqrt{W}y}. \tag{72}$$

Substitution into equation (71) then yields

$$\frac{d^2v_{El}}{dy^2} - 2\sqrt{W}\ \frac{dv_{El}}{dy} - \left[\frac{l(l+1)}{y^2} - \frac{2Z}{y}\right]v_{El} = 0. \tag{73}$$

Next we determine the behavior of $v_{El}$ near the origin. For this purpose, write

$$v_{El} \sim y^s.$$

Substitution into equation (73) then gives

$$s(s-1)y^{s-2} - 2\sqrt{W}\ sy^{s-1} - l(l+1)y^{s-2} + 2Zy^{s-1} = 0$$

or

$$s(s-1) - l(l+1) + 2y(Z - \sqrt{W}) = 0.$$

The last term is negligible for sufficiently small $y$ and hence $s$ must be such that

$$s(s-1) = l(l+1)$$

or

$$s = l+1, \qquad -l.$$

Since $v_{El}$ must be bounded at the origin, only the former is permitted and we thus see that

$$y \to 0, \qquad v_{El} \sim y^{l+1}.$$

Consequently, we now seek a power series solution of the form

$$v_{El} = y^{l+1} \sum_{q=0}^{\infty} c_q y^q. \tag{74}$$

Substitution into equation (73) then gives

$$\sum_{q=0}^{\infty} c_q(q+l+1)(q+l)y^{q+l-1}$$

$$- 2\sqrt{W} \sum_{q=0}^{\infty} c_q(q+l+1)y^{q+l}$$

$$- \sum_{q=0}^{\infty} c_q l(l+1)y^{q+l-1} + 2Z \sum_{q=0}^{\infty} c_q y^{q+l} = 0$$

or, collecting terms, we can write

$$\sum_{q=0}^{\infty} c_q \, q[q + (2l + 1)]y^{q+l-1} = 2 \sum_{q=0}^{\infty} c_q[\sqrt{W}(q + l + 1) - Z]y^{q+l}.$$

Equating coefficients of like powers of $y$, we then at once obtain

$$c_{q+1} = 2 \frac{\sqrt{W} \, (q + l + 1) - Z}{(q + 1)(q + 2l + 2)} \, c_q. \qquad (75)$$

This recursion formula permits successive determination of all the expansion coefficients in terms of $c_0$, which is merely an arbitrary multiplicative constant, and thus provides us with an explicit power series solution of Schrödinger's equation. Further, this solution is guaranteed to be well-behaved at the origin. We must still, however, examine its behavior at infinity. This behavior is determined by the properties of the series for large $q$. From equation (75) we see that

$$q \to \infty, \qquad \frac{c_{q+1}}{c_q} \sim \frac{2\sqrt{W}}{q}.$$

Now this ratio is exactly the same as that obtained for the successive terms in the expansion of the exponential function, and hence, for $y \to \infty$, $v_{El}$ is dominated by the exponential factor $e^{2\sqrt{W}y}$. This means that $u_{El} = v_{El}e^{-\sqrt{W}y}$ diverges at infinity like $e^{\sqrt{W}y}$, and the general solution is not physically admissible.

This behavior is no surprise, since, as we showed earlier, $u_{El}$ behaves at infinity like $e^{\pm\sqrt{W}y}$. Any general solution, therefore, consists of a linear combination of increasing and decreasing exponentials and the increasing exponential necessarily dominates, as we have found. This difficulty can be avoided only if the series terminates, which means that $c_{n'}$, say, and therefore all subsequent $c_q$, must vanish. This will be so if $W$ is such that

$$\sqrt{W} = Z/(n' + l),$$

that is, if the binding energy of the atom is such that

$$W_{n'l} = \frac{Z^2}{(n' + l)^2}, \qquad n' = 1, 2, 3, \ldots, \qquad (76)$$

which then defines the spectrum of discrete bound states. Note that $n'$ cannot take on the value zero since the wave function is identically zero if $c_0$ vanishes.

For a given $n'$ and $l$, according to equation (74), $v_{El}$ is thus a definite polynomial of degree $n' - 1$ multiplied by the factor $y^{l+1}$. These polynomials are known as *associated Laguerre polynomials*,[7] and are denoted by the symbol $L_k{}^q(z)$. They can be compactly expressed as follows.

[7] The properties of these polynomials are discussed in References [1] through [5].

First introduce the ordinary Laguerre polynomial $L_k(z)$, defined by

$$L_k(z) = e^z \left(\frac{d}{dz}\right)^k (z^k e^{-z}).$$  (77)

Then the associated Laguerre polynomials are given by

$$L_k^q(z) = \left(\frac{d}{dz}\right)^q L_k(z).$$  (78)

From these expressions, $L_k(z)$ is seen to be a polynomial of degree $k$ in $z$, $L_k^q$ a polynomial of degree $k - q$ in $z$. An alternative expression for these polynomials can be obtained in terms of an appropriate generating function; specifically it can be shown that

$$\frac{e^{-zs/(1-s)}}{1-s} = \sum_{k=0}^{\infty} L_k(z) \frac{s^k}{k!},$$

whence, from equation (78),

$$\left(-\frac{s}{1-s}\right)^q \frac{e^{-zs/(1-s)}}{1-s} = \sum_{k=q}^{\infty} L_k^q(z) \frac{s^k}{k!}.$$  (79)

Finally, it can also be shown that

$$\int_0^{\infty} e^{-z} z^{q+1} [L_k^q(z)]^2 \, dz = (2k + 1 - q) \frac{(k!)^3}{(k-q)!}.$$  (80)

In terms of these polynomials, it then turns out that

$$v_{El} = y^{l+1} L_{n'+2l}^{2l+1} (2y\sqrt{W_{n'l}}),$$

and it is not too hard to verify, by direct substitution, that this expression is indeed a solution of equation (68). Next, from equation (72), the radial wave function is given by

$$R_{El} = \frac{u_{El}}{r} = y^l L_{n'+2l}^{2l+1} (2y\sqrt{W_{n'l}}) e^{-\sqrt{W_{n'l}}y}.$$  (81)

Using equation (80), the normalized stationary states are found to be

$$\psi_{Elm} = c_{n'l} \left[\frac{2Zr}{(n'+l)a_0}\right]^l L_{n'+2l}^{2l+1} \left(\frac{2Zr}{(n'+l)a_0}\right) e^{-Zr/(n'+l)a_0} Y_l^m(\theta, \phi),$$  (82)

where

$$c_{n'l} = \left[\frac{2Z(n'+l)}{a_0}\right]^{3/2} \left[\frac{(n'-1)!}{2(n'+l)[(n'+2l)!]^3}\right]^{1/2}$$  (83)

and where, in re-expressing $y$ in terms of $r$, we have introduced the *Bohr radius* $a_0$, defined by

$$a_0 = \frac{\hbar^2}{me^2} \simeq 0.53 \times 10^{-8} \text{ cm}. \tag{84}$$

The energy $E$ of these states, according to equations (70) and (76), is

$$E = -\frac{Z^2 e^2}{2a_0} \frac{1}{(n'+l)^2}, \qquad \begin{cases} n' = 1, 2, \dots \\ l = 0, 1, 2, \dots \end{cases} \tag{85}$$

Observe now that the energy depends only upon the sum of the radial quantum number $n'$ and of the angular momentum quantum number $l$. It is convenient and customary to introduce at this stage the *principal quantum number* $n$, defined by

$$n = n' + l, \tag{86}$$

in terms of which

$$E = E_n = -\frac{Z^2 e^2}{2a_0 n^2}, \tag{87}$$

which is seen to be the familiar Bohr formula.[8] The state function can also be expressed in terms of the principal quantum number, of course, and we obtain at once

$$\psi_{nlm} = D_{nl} \left( \frac{2Zr}{na_0} \right)^l L_{n+l}^{2l+1} \left( \frac{2Zr}{na_0} \right) e^{-Zr/na_0} Y_l^m(\theta, \phi), \tag{88}$$

where

$$D_{nl} = \left( \frac{2Zn}{a_0} \right)^{3/2} \left[ \frac{(n-l-1)!}{2n[(n+l)!]^3} \right]^{1/2}. \tag{89}$$

It is important to keep in mind that, according to equation (86), $n$ can take on, for given $l$, only the values $l+1, l+2, \dots$, while for given $n$, $l$ can take on only the values $n-1, n-2, \dots, 0$. The first few normalized stationary state functions are given below:

$$\psi_{100} = 2 \left( \frac{Z}{a_0} \right)^{3/2} e^{-Zr/a_0} Y_0^0; \qquad E_1 = -\frac{Z^2 e^2}{2a_0}$$

$$\psi_{200} = \left( \frac{Z}{2a_0} \right)^{3/2} \left( 2 - \frac{Zr}{a_0} \right) e^{-Zr/2a_0} Y_0^0; \qquad E_2 = -\frac{Z^2 e^2}{2a_0^2} \cdot \frac{1}{4} \tag{90}$$

$$\psi_{21m} = \left( \frac{Z}{2a_0} \right)^{3/2} \frac{Zr}{a_0 \sqrt{3}} e^{-Zr/2a_0} Y_1^m; \qquad E_2 = -\frac{Z^2 e^2}{2a_0^2} \cdot \frac{1}{4}.$$

.
.
.

[8] Observe that the energy is proportional to $Z^2$, even though the strength of the interaction is proportional to $Z$. The explanation is that the mean radius of the atom for a state of given $n$ is inversely proportional to the strength of interaction and hence inversely proportional also to $Z$. Because the energy, being electrostatic in origin, is proportional to the strength *divided* by the radius, the $Z^2$ dependence follows.

The degeneracy of the hydrogenic states can be found in the following way. For a given principal quantum number $n$, and thus for a given bound state energy $E_n$, $l$ can take on all values ranging from zero to $n - 1$, as we have seen. For each $l$ value, there is a $(2l + 1)$-fold degeneracy and hence $N$, the total number of states with energy $E_n$, is

$$N = \sum_{l=0}^{n-1} (2l + 1) = n^2.$$

Again we have an excellent example of pervasive "accidental" degeneracies which are not accidental. They arise in this case because Schrödinger's equation for a Coulomb potential happens to be separable in parabolic as well as spherical coordinates.

The spectrum of the hydrogenic atom is shown in Figure 7. As a matter of nomenclature, dating back to the early days of spectroscopy, states with $l = 0$ are called $s$-states; with $l = 1$, $p$-states; with $l = 2$, $d$-states; with $l = 3$, $f$-states; with $l = 4$, $g$-states; and so on in alphabetical order for states of still higher $l$. From the figure it is readily seen how the states crowd together as $E$ approaches zero, in contrast to the behavior of the states in a square well. This is a consequence of the long range of the Coulomb interaction, as is the fact that there are infinitely many discrete bound states for each value of $l$.

We have now obtained a complete and exact solution for the bound states of a (nonrelativistic) hydrogenic atom. Unfortunately, these states

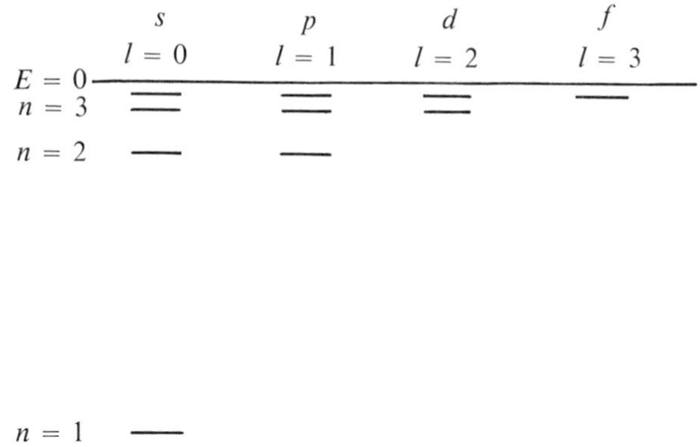

FIGURE 7.  The hydrogenic spectrum.

have turned out to be rather complicated in general, and their qualitative behavior is not very readily apparent. The simplest states are those of maximum angular momentum and of maximum $z$-component of angular momentum, for a given $n$, that is, the states $\psi_{n, l=n-1, m=n-1}$. The state functions in this case have the simple form, omitting normalization factors,

$$\psi_{n,n-1,n-1} = r^{n-1} e^{-Zr/na_0} \sin^{n-1} \theta.$$

For large quantum numbers $n$, these states are quite well localized in angle and in radius. In particular, the wave function has a relatively sharp maximum about the equatorial plane and about the Bohr radius $n^2 a_0/Z$. Thus such states actually exhibit the behavior predicted by the original Bohr theory, in the sense that they are centered about the circular orbits of that theory, and rather precisely so for sufficiently large quantum numbers.

---

**Exercise 5.** Verify the above assertions for the state $\psi_{n,n-1,n-1}$.

---

**Problem 1.**

(a)   Find the propagator $K(\mathbf{r}, \mathbf{r}'; t - t_0)$ for a free particle in three dimensions.

(b)   Consider a wave packet which at $t = 0$ is a Gaussian,

$$\psi(\mathbf{r}, t = 0) = A \, e^{i p_0 \cdot (r - r_0)/\hbar} \, e^{-(r - r_0)^2/2L^2}.$$

Find $A$ if the wave packet is normalized.

(c)   Show that $\psi(\mathbf{r}, t = 0)$ is a minimum uncertainty wave packet.

(d)   Find $\psi(\mathbf{r}, t)$.

(e)   Find $\phi(\mathbf{p}, t = 0), \phi(\mathbf{p}, t)$.

(f)   Find $\langle x \rangle$ at $t = 0$ and for any $t > 0$.

(g)   Find $\langle \mathbf{p} \rangle, \langle p^2 \rangle$ at $t = 0$ and for any $t > 0$.

Note: All integrals can be factored into products of one-dimensional integrals, familiar from our earlier work. The problem can, however, be more easily worked directly in three dimensions. Either procedure is acceptable.

**Problem 2.** Using the results of Chapter VI for a one-dimensional oscillator,

(a)   Find the propagator for a three-dimensional oscillator.

(b)   Discuss the motion of an arbitrary wave packet in a three-dimensional oscillator.

**Problem 3.**

(a)   Calculate the polarizability of a three-dimensional isotropic harmonic oscillator.

(b)   What is the degeneracy of the states of a three-dimensional oscillator in a uniform external field $\mathbf{E} = \mathcal{E} \, \mathbf{e}_x$?

**Problem 4.** To good approximation, the nucleus can be regarded as a *uniformly* charged sphere of radius $R_0 \ll a_0$.

(a)  What is the electrostatic potential between an electron and a nucleus so regarded?

(b)  Find an expression for the first-order correction to the ground state energy of a hydrogenic atom arising from the finite nuclear size. What is the order of magnitude of this correction, assuming that $R_0 \simeq (2Z)^{1/3} \times 10^{-13}$ cm? How does it depend upon $Z$?

(c)  The same for a $\mu$-mesonic atom, that is, a system consisting of a nucleus and a negative $\mu$-meson. (The mass of the $\mu$-meson is approximately 207 electron masses.) For what value of $Z$, if any, does first-order perturbation theory become inadequate? [The earliest accurate estimates of nuclear sizes were obtained by analysis of a $\mu$-mesonic atomic spectra.]

**Problem 5.** Consider an anisotropic harmonic oscillator described by the potential

$$V(x, y, z) = \tfrac{1}{2} m\omega_1^2 \, (x^2 + y^2) + \tfrac{1}{2} m\omega_2^2 z^2 .$$

(a)  Find the stationary states using rectangular coordinates. What are the degeneracies of the states, assuming $\omega_1$ and $\omega_2$ are incommensurate?

(b)  Can the stationary states be eigenstates of $L^2$? of $L_z$? Explain in each case.

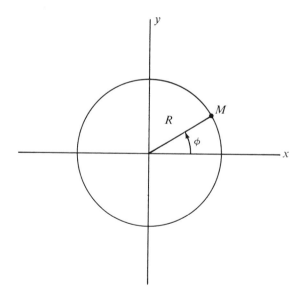

FIGURE 8.  Motion of a particle on a circular track.

**Problem 6.** A particle of mass $M$ is constrained to move on a frictionless, vertical circular track of radius $R$. Neglecting gravity, the Hamiltonian of the system is $H = L_z^2/2MR^2$. Since $L_z = (\hbar/i)(\partial/\partial\phi)$, Schrödinger's equation is

$$-\frac{\hbar}{2MR^2}\frac{\partial^2\psi}{\partial\phi^2} = -\frac{\hbar}{i}\frac{\partial\psi}{\partial t}; \quad \psi = \psi(\phi, t),$$

where, as shown in Figure 8, $\phi$ is the angular coordinate of the particle.

(a)   Solve Schrödinger's equation to find the allowed energies $\mathcal{E}_m$ and the energy eigenstates $\psi_m$. What are the degeneracies of these states, if any?

(b)   Suppose the gravitational field is now included. The Hamiltionian is then

$$H = L_z^2/2MR^2 + MgR \sin \phi.$$

Treating the gravitational term as a perturbation, show that the first-order correction to the energies $\mathcal{E}_m$ vanishes. Calculate the second-order correction to the energy.

(c)   Suppose the gravitational term is much too large to be treated generally as a perturbation. (For what combination of the parameters will that be true?) Estimate the ground state energy using the Rayleigh–Ritz method and then using the WKB approximation.

(d)   Discuss the transition between states bounded in $\phi$ and states unbounded in $\phi$, using the WKB approximation. For what states, if any, will the perturbation results of part (b) apply?

**Problem 7.** Suppose two identical spinless particles are moving on the circular track described in Problem 6. Neglect gravity.

(a)   Assuming the particles do not interact with each other, solve Schrödinger's equation to find the allowed energies and the energy eigenstates. What are the degeneracies of these states, if any?

(b)   Suppose the particles interact weakly according to the potential $V(\phi_1, \phi_2) = V_0[1 + \cos(\phi_1 - \phi_2)]$. Use perturbation theory to find the corrections to the unperturbed ground state energy.

**Problem 8.** A particle of mass $M$ is constrained to move on the surface of a sphere of radius $R$, but is otherwise free.

(a)   Solve Schrödinger's equation for the allowed energies and the energy eigenstates.

(b)   Classically, the orbit of such a particle lies in a plane and is thus kinematically equivalent to motion on a circular track, as in Problem 6, part (a). Show that, to good approximation, this equivalence is also attained quantum mechanically by considering, for large quantum numbers, states of maximum $L_z$.

**Problem 9.** A hydrogen atom is placed in a uniform electric field directed along the $z$-axis. Neglecting spin, the relative motion of electron and proton is then described by the Hamiltonian

$$H = \frac{p^2}{2m} - \frac{e^2}{r} + e\,\mathcal{E}z.$$

(a) Is $L^2$ a constant of the motion? Is $L_z$? Do the states have definite parity? Explain briefly.

(b) Treating the term in the electric field as a perturbation, write down an expression for the second-order correction to the ground state energy (why second-order?). Include only non-vanishing matrix elements in your answer. Do not actually evaluate the integrals or perform the summations.

(c) Suggest some explicit trial function suitable for a variational calculation. Allow, as best you can, for the distortion of the atom caused by the electric field. Do not actually carry out any calculation, but justify your choice of trial function.

**Problem 10.** Consider the bound states of a particle in a spherically symmetric potential. Show that if there are no "accidental" degeneracies, $\langle x \rangle$ and $\langle p \rangle$ vanish for *any* stationary state. Why does the proof fail if there are accidental degeneracies? [Hint: What can you say about the parity of the stationary states?]

**Problem 11.** The nucleus $H^3$ (the triton) consisting of one proton and two neutrons is unstable. By beta emission it decays to $He^3$, consisting of two protons and one neutron. Assume that when this process takes place it does so instantaneously. Thus there occurs a sudden doubling of the Coulomb interaction between the atomic electron and the nucleus when tritium (the $H^3$ atom) decays by beta emission to $He^+$ (singly ionized $He^3$). If the tritium atom is in its ground state when it decays, what is the probability that the $He^+$ ion in its ground state immediately after the decay? In the $2s$-state? In the $2p$-state? In any state other than an $s$-state?

**Problem 12.** The weakness of gravitational compared to electrostatic interactions is dramatically illustrated by considering a system of two neutrons under the sole influence of their mutual gravitational attraction.

(a) Write expressions for the bound state energies and for the "Bohr radius" for such a system.

(b) Estimate, to the nearest power of ten, the numerical value of the ground state energy (in electron volts) and of the Bohr radius (in cm, in light years).

**Problem 13.** Three identical non-interacting particles are described by the following Hamiltonian,

$$H = \sum_{i=1}^{3} H_{\mathrm{sp}} (r_i); \qquad H_{\mathrm{sp}} (r) = \frac{p^2}{2m} + \frac{1}{2} m\omega^2 r^2.$$

(a)   Find the eigenfunctions and energies of the system, *without regard to symmetrization.*

(b)   Give the degeneracy of the lowest three states, including exchange degeneracy.

(c)   The same for the physically realizable states of spinless particles.

(d)   What is the ground state energy for spin one-half particles? What is the degeneracy of the ground state?

**Problem 14.**   A two-particle system is described by the Hamiltonian

$$H = \frac{p_1^2}{2m_1} + \frac{p_2^2}{2m_2} + \frac{1}{2} m_1\omega^2 r_1^2 + \frac{1}{2} m_2\omega^2 r_2^2 + V_0\, e^{-(r_1-r_2)^2/a^2}.$$

Treating the Gaussian interaction term as a perturbation and transforming to center of mass coordinates,

(a)   Find the energy of the ground state and first excited state to first order.

(b)   Use equation (VII–39) to find an upper bound to the second-order correction to the ground state energy.

(c)   Use a Gaussian of variationally determined width as a trial function in the Rayleigh–Ritz method to estimate the ground state energy.

**Problem 15.**   The same as Problem 14, but for identical spinless particles.

**Problem 16.**   A two-particle system is described by the Hamiltonian

$$H = \frac{p_1^2}{2m_1} + \frac{p_2^2}{2m_2} + \frac{1}{2} m_1\omega^2 r_1^2 + \frac{1}{2} m_2\omega^2 r_2^2 + \frac{1}{2} k(\mathbf{r}_1 - \mathbf{r}_2)^2.$$

(a)   Find the exact solutions by transforming to center-of-mass coordinates.

(b)   Sketch the spectrum in the weak and strong coupling limits, $k \ll \mu\omega^2$ and $k \gg \mu\omega^2$, respectively, where $\mu$ is the reduced mass.

(c)   The same for identical spinless particles.

**Problem 17.**   Consider the hydrogen atom in its ground state. Suppose by some magic that the Coulomb interaction is suddenly turned off at $t = 0$ and the electron moves off as a free particle. Treating the proton as infinitely massive for simplicity,

(a)   Find the probability $\rho(\mathbf{p}, t)d^3\mathbf{p}$ that a measurement of the electron's momentum at any time $t > 0$ will yield a value in the volume element $d^3\mathbf{p}$ at $\mathbf{p}$. How does the result depend upon $t$? Upon the direction of $\mathbf{p}$? Explain.

(b)  What is the probability that a measurement of the electron's energy will yield a value between $E$ and $E + dE$?

(c)  Find the probability that a measurement of the electron's position at any time $t > 0$ will yield a value in the volume element $d^3\mathbf{r}$ at $r$. [Hint: Use the free particle propagator.] Without working out all the details, show qualitatively how this probability changes with time, starting at $t = 0$. Explain briefly.

(d)  In contrast to the distributions in parts (a), (b) and (c), a measurement of the electron's angular momentum must always yield a unique and precise value. What is this unique value?

**Problem 18.**  Obtain an expression for $d/dt \langle \psi | \mathbf{p} \cdot \mathbf{r} | \psi \rangle$ and then, by considering a stationary state $\psi_E$, prove the virial theorem,

$$\langle \psi_E | T | \psi_E \rangle = \langle \psi_E | \tfrac{1}{2} \, \mathbf{r} \cdot \nabla V | \psi_E \rangle,$$

where $T$ is the kinetic energy, $V$ the potential energy. Use the virial theorem to show that $E = 2 \langle \psi_E | V | \psi_E \rangle$ for the harmonic oscillator and $E = \tfrac{1}{2} \langle \psi_E | V | \psi_E \rangle$ for the hydrogen atom.

**Problem 19.**  The neutron and proton interact through a strong short-range force. Reasonable approximations to such a force are:

(i)  Yukawa type:  $V(r) = -\dfrac{V_0 R}{r} e^{-r/R}$

(ii)  Exponential:  $V(r) = -V_0 \, e^{-r/R}$

(iii)  Square well:  $V(r) = \begin{cases} -V_0, & r < R \\ 0, & r > R. \end{cases}$

(a)  Use the Rayleigh–Ritz method to find an expression for the ground state binding energy $\epsilon$ for each of the three cases. Use $e^{-r/2R}$ as a trial function.[9]

(b)  Taking $\epsilon = 2.2$ MeV and $R = 2 \times 10^{-13}$ cm, find $V_0$ in each case. Draw the three interactions on the same graph.

(c)  As a check on the accuracy of the Rayleigh–Ritz method, solve the square well case exactly, using graphical methods to find $\epsilon$, for the particular value of $V_0$ found in part (b).

---

[9] One can do considerably better by introducing a trial function containing a variational parameter, such as $e^{-\alpha r/2R}$. Unfortunately, however, the resulting equations are complicated and must be solved numerically in all three cases. We remark that case (ii) can actually be solved *exactly* in terms of Bessel functions. See Reference [29], pp. 218–220. The variational method is discussed in these pages as well.

# X

# *Angular momentum and spin*

## 1. ORBITAL ANGULAR MOMENTUM OPERATORS AND COMMUTATION RELATIONS

If at some instant a structureless particle passes with momentum **p** through a point whose displacement from an arbitrarily chosen origin is **r**, then its angular momentum **L** with respect to that origin is

$$\mathbf{L} = \mathbf{r} \times \mathbf{p} \tag{1}$$

or, in component form,

$$
\begin{aligned}
L_x &= yp_z - zp_y \\
L_y &= zp_x - xp_z \\
L_z &= xp_y - yp_x.
\end{aligned}
\tag{2}
$$

The corresponding quantum mechanical dynamical variable, called the *orbital angular momentum operator*, is given by these relations with, of course, **r** and **p** interpreted as quantum mechanical dynamical variables. Since the commutator of $x_i$ and $p_j$ vanishes for $i \neq j$, it is easily verified that **L** is Hermitian.

---

**Exercise 1.** Prove that **L** is Hermitian.

---

The commutation relations between the rectangular components of **L** are obtained as follows. We have, for example,

$$
\begin{aligned}
(L_x, L_y) &= ([yp_z - zp_y], [zp_x - xp_z]) \\
&= (yp_z, zp_x) - (yp_z, xp_z) - (zp_y, zp_x) + (zp_y, xp_z).
\end{aligned}
$$

Consider the first term,

$$(yp_z, zp_x) = yp_zzp_x - zp_xyp_z.$$

Because $y$ and $p_x$ commute with each other and with $z$ and $p_z$, this becomes

$$(yp_z, zp_x) = yp_x(p_z, z) = \frac{\hbar}{i} yp_x.$$

Similarly, the last term becomes

$$(zp_y, xp_z) = xp_y(z, p_z) = -\frac{\hbar}{i} xp_y.$$

On the other hand, since $y$, $x$ and $p_z$ mutually commute, the second term vanishes, and because $z$, $p_y$ and $p_x$ mutually commute, so does the third. Thus we obtain

$$(L_x, L_y) = i\hbar (xp_y - yp_x)$$

or, recognizing the term in parentheses as $L_z$, simply

$$(L_x, L_y) = i\hbar L_z. \tag{3}$$

From the structure of $\mathbf{L}$, the remaining commutation relations are obtained at once by cyclic permutation of the coordinates, and we thus have

$$(L_y, L_z) = i\hbar L_x \tag{4}$$

and

$$(L_z, L_x) = i\hbar L_y. \tag{5}$$

These three relations are equivalent to the single *vector commutation relation*

$$\mathbf{L} \times \mathbf{L} = i\hbar \mathbf{L}, \tag{6}$$

as is easily verified by writing out the rectangular components. The operator nature of $\mathbf{L}$ is made quite explicit by equation (6), since the vector product of a purely numerical vector with itself vanishes.

It is recalled that simultaneous eigenfunctions or a collection of operators exist only if the operators in question mutually commute. Hence, the noncommutativity of the rectangular components of angular momentum at once means that *a complete set of states cannot be defined for which each component of the angular momentum vector has a precise and definite value*. Instead, only a single component of $\mathbf{L}$ can be precisely specified; the remaining two components perpendicular to it are necessarily uncertain.[1] Noting that the orientation of the coordinate axes is open to choice, we see that we may choose angular momentum states in such a way that the projection of **L** on any arbitrary axis has a definite value. Normally this *axis of quantization* (for angular momentum), as it is called, is taken to be the $z$-axis, in which case $L_z$ has a definite value, but $L_x$ and $L_y$ do not.

Although we have shown that the *orientation* of the angular momentum vector cannot be completely specified on the quantum level, the question of its *magnitude* has not been discussed. We now take up this question. To do so, we introduce the square of the angular momentum operator,

$$L^2 = \mathbf{L} \cdot \mathbf{L} = L_x^2 + L_y^2 + L_z^2. \tag{7}$$

We must now examine the commutation properties of $L^2$ with respect to its rectangular components. Let us look first at $(L_x, L^2)$. Since the commutator of $L_x$ with itself vanishes, we have

$$(L_x, L^2) = (L_x, L_y^2) + (L_x, L_z^2).$$

Now, it is easily verified by expanding both sides that

$$(L_x, L_y^2) = (L_x, L_y) L_y + L_y(L_x, L_y)$$

and hence, from equation (3),

$$(L_x, L_y^2) = i\hbar (L_z L_y + L_y L_z).$$

Similarly, using equation (5),

$$(L_x, L_z^2) = (L_x, L_z)L_z + L_z(L_x, L_z) = -i\hbar (L_y L_z + L_z L_y).$$

Combining these results we thus see that

$$(L_x, L^2) = 0.$$

In the same way, we find at once

$$(L_y, L^2) = (L_z, L^2) = 0$$

or, in brief,

$$(\mathbf{L}, L^2) = 0. \tag{8}$$

It then follows that $L^2$ and any *one* of its rectangular components, say $L_z$, can be simultaneously specified and, further, that $L^2$ and $L_z$ form a complete set of commuting operators for the specification of angular momentum states. These states are characterized by the magnitude of

---

[1] Observe, however, that states with angular momentum identically zero are not ruled out, even though all three components of **L** have definite values for such states. The commutation relations are not violated, because each side of equation (6) vanishes when operating on a state of zero angular momentum.

the angular momentum and by its projection on the arbitrarily oriented
$z$-axis, both of which have definite values.

Before going on to explicitly construct the angular momentum eigenfunctions, we briefly consider the properties of angular momentum for a
system of particles. For such a system, the total angular momentum is

$$\mathbf{L} = \sum_i \mathbf{L}_i, \tag{9}$$

where $\mathbf{L}_i$ is the angular momentum of the $i$th particle, and is given by

$$\mathbf{L}_i = \mathbf{r}_i \times \mathbf{p}_i.$$

Evidently, since dynamical variables referring to different particles
commute, we have

$$(\mathbf{L}_i, \mathbf{L}_j) = 0, \qquad i \neq j,$$

and it follows after some elementary algebra that

$$\mathbf{L} \times \mathbf{L} = i\hbar\, \mathbf{L},$$

just as for a single particle. Our conclusions about the general character
of angular momentum states for a single particle thus apply, without
modification, to the angular momentum of a system of particles.

---

**Exercise 2.** Verify the vector commutation relation for the total angular
momentum of a system of particles.

---

An important feature of many-particle systems is that the angular
momentum associated with the center-of-mass motion is *not* the total
angular momentum of the system. This is in contrast to linear momentum; the total linear momentum of a system and the linear momentum
of its center of mass are, of course, identical. As an example, we specifically consider only the simplest case, that of a two-particle system.
Transforming to center-of-mass and relative coordinates, we have for
the center-of-mass angular momentum

$$\mathbf{L}_{cm} = \mathbf{R} \times \mathbf{P}, \tag{10}$$

and for the angular momentum of the relative motion

$$\mathbf{L}_r = \mathbf{r} \times \mathbf{p}, \tag{11}$$

where

$$\mathbf{r} = \mathbf{r}_1 - \mathbf{r}_2, \qquad \mathbf{p} = \mu\!\left(\frac{\mathbf{p}_1}{m_1} - \frac{\mathbf{p}_2}{m_2}\right)$$

$$\mathbf{R} = \frac{m_1\mathbf{r}_1 + m_2\mathbf{r}_2}{M}, \qquad \mathbf{P} = \mathbf{p}_1 + \mathbf{p}_2.$$

It is then easily verified, by direct substitution, that the total angular momentum of the system is the sum of the relative and center-of-mass angular momentum vectors. Further, in view of the fact that the center-of-mass and relative coordinates satisfy the usual (canonical) commutation relations, it follows at once that

$$\mathbf{L}_{cm} \times \mathbf{L}_{cm} = i\hbar\,\mathbf{L}_{cm}$$

$$\mathbf{L}_r \times \mathbf{L}_r = i\hbar\,\mathbf{L}_r \tag{12}$$

$$(\mathbf{L}_r, \mathbf{L}_{cm}) = 0.$$

Consequently, the states of a two-particle system can be simultaneously classified with respect to the angular momentum of its center of mass and of its relative motion, and each is formally identical to the orbital angular momentum states of a single particle.

## 2. ANGULAR MOMENTUM EIGENFUNCTIONS AND EIGENVALUES

We now construct the simultaneous eigenfunctions, and the corresponding eigenvalues, of $L^2$ and $L_z$. We shall do so by a factorization method utilizing only the algebraic properties of the operators. The analysis is analogous to but somewhat more complicated than that encountered in our treatment of the harmonic oscillator. The operators which play a role similar to that of $a$ and $a\dagger$, turn out to be, as we shall see in a moment,[2]

$$L_+ = L_x + iL_y$$

$$L_- = L_x - iL_y. \tag{13}$$

Denoting the eigenvalue of $L^2$ by $\hbar^2\beta$ and of $L_z$ by $\hbar\alpha$ and denoting their simultaneous eigenfunctions, not necessarily normalized, by $Y_\beta{}^\alpha$, we have

$$L^2Y_\beta{}^\alpha = \hbar^2\beta Y_\beta{}^\alpha \tag{14}$$

$$L_zY_\beta{}^\alpha = \hbar\alpha Y_\beta{}^\alpha. \tag{15}$$

Because $L^2 - L_z{}^2 = L_x{}^2 + L_y{}^2$, and is thus a nonnegative operator, we must have

$$\alpha^2 \le \beta. \tag{16}$$

---

[2] For reasons which will become clear, $L_+$ is called the *raising operator* for angular momentum and $L_-$ is called the *lowering operator*. Note that $L_- = (L_+)\dagger$.

Now, $L_x$ and $L_y$ commute with $L^2$ and hence so do $L_+$ and $L_-$. Consequently,

$$L^2 L_\pm Y_\beta{}^\alpha = L_\pm L^2 Y_\beta{}^\alpha = \hbar^2 \beta L_\pm Y_\beta{}^\alpha, \tag{17}$$

that is, if $Y_\beta{}^\alpha$ is an eigenfunction of $L^2$ with eigenvalue $\hbar^2 \beta$, then so are the new functions $L_\pm Y_\beta{}^\alpha$.

Next, consider the commutator of $L_\pm$ and $L_z$. We have,

$$(L_\pm, L_z) = (L_x \pm iL_y, L_z)$$

$$= -i\hbar L_y \pm i(i\hbar L_x)$$

$$= \mp \hbar L_\pm$$

or, equally well,

$$L_z L_\pm = L_\pm (L_z \pm \hbar). \tag{18}$$

Operating on $Y_\beta{}^\alpha$ with this operator equation, we then obtain

$$L_z L_\pm Y_\beta{}^\alpha = L_\pm (L_z \pm \hbar) Y_\beta{}^\alpha$$

or, using equation (15),

$$L_z L_\pm Y_\beta{}^\alpha = \hbar (\alpha \pm 1) L_\pm Y_\beta{}^\alpha. \tag{19}$$

Together, equations (17) and (19) tell us that $L_\pm Y_\beta{}^\alpha$ is a simultaneous eigenfunction of $L^2$ and $L_z$ with eigenvalues $\hbar^2 \beta$ and $\hbar(\alpha \pm 1)$. Hence we write, according to our notation,

$$L_\pm Y_\beta{}^\alpha = Y_\beta{}^{\alpha \pm 1}, \tag{20}$$

where the normalization is again left unspecified.

Starting from a given state $Y_\beta{}^\alpha$, an eigenstate of $L_z$ with eigenvalue $\hbar\alpha$, we now see that by repeated operation with $L_+$ we can successively construct eigenstates of $L_z$ with eigenvalues $\hbar(\alpha + 1)$, $\hbar(\alpha + 2)$, and so on, each of which is also an eigenstate of $L^2$ with eigenvalue $\hbar^2 \beta$. Similarly, by repeated operation with $L_-$, we can construct a sequence of eigenstates of $L_z$ with eigenvalues $\hbar(\alpha - 1)$, $\hbar(\alpha - 2)$, and so on, and again each is an eigenstate of $L^2$ with the same eigenvalue $\hbar^2 \beta$. However, since $L^2 - L_z{}^2$ is a nonnegative operator, this ladder of states cannot continue indefinitely in either direction, but must terminate at each end. Call the eigenvalue of the top rung $\hbar\alpha_1$, and that of the bottom rung $(-\hbar\alpha_2)$. We must then have, according to equation (16),

$$\alpha_1{}^2 \leqslant \beta$$

$$\alpha_2{}^2 \leqslant \beta.$$

Further, since the ladder must necessarily contain an integral number of intervals between rungs, say $n$, we have

$$\alpha_1 + \alpha_2 = n; \qquad n = 0, 1, 2, \ldots \ldots \tag{21}$$

The termination of the ladder at its upper end requires that

$$L_+ Y_\beta^{\alpha_1} = 0 \tag{22}$$

and its termination at the lower end requires that

$$L_- Y_\beta^{-\alpha_2} = 0, \tag{23}$$

and we now use these conditions to determine the allowed values of $\alpha_1$, $\alpha_2$ and $\beta$.

To do so, we must first express $L^2$ in terms of $L_+$, $L_-$ and $L_z$. Now

$$L_+ L_- = (L_x + iL_y)(L_x - iL_y) = L_x{}^2 + L_y{}^2 - i(L_x, L_y)$$

$$= L_x{}^2 + L_y{}^2 + \hbar L_z,$$

and consequently

$$L^2 = L_x{}^2 + L_y{}^2 + L_z{}^2 = L_+ L_- - \hbar L_z + L_z{}^2. \tag{24}$$

Similarly,

$$L_- L_+ = L_x{}^2 + L_y{}^2 - \hbar L_z,$$

whence we obtain the alternative but equivalent expression

$$L^2 = L_- L_+ + \hbar L_z + L_z{}^2. \tag{25}$$

To proceed, operate on $Y_\beta^{\alpha_1}$ with $L^2$ in the form of equation (25), whence

$$L^2 Y_\beta^{\alpha_1} = L_- L_+ Y_\beta^{\alpha_1} + (\hbar L_z + L_z{}^2) Y_\beta^{\alpha_1}.$$

Now the first term on the right vanishes according to equation (22) and $Y_\beta^{\alpha_1}$ is a simultaneous eigenfunction of $L^2$ and of $L_z$ with eigenvalues $\hbar^2 \beta$ and $\hbar \alpha_1$, according to equations (14) and (15). Hence, canceling a common factor of $\hbar^2$, we must have

$$\beta = \alpha_1(\alpha_1 + 1). \tag{26}$$

In the same way, operating on $Y_\beta^{-\alpha_2}$ with $L^2$ in the form of equation (24), we find

$$\beta = \alpha_2(\alpha_2 + 1), \tag{27}$$

whence $\alpha_1(\alpha_1 + 1) = \alpha_2(\alpha_2 + 1)$. The only solution of this equation consistent with equation (21) is $\alpha_1 = \alpha_2$, and hence we have

$$\alpha_1 = \alpha_2 = \frac{n}{2} \equiv l, \tag{28}$$

where $l$ is integral or half-integral, depending on whether $n$ is even or odd. In either case,

$$\beta = l(l+1), \tag{29}$$

and the eigenvalues of $L_z$ range from $\hbar l$ to $-\hbar l$ in integral steps in $\hbar$. Denoting now the eigenvalues of $L_z$ by $\hbar m$ instead of by $\hbar \alpha$, we thus see that $m$ is integral or half-integral, depending on $l$, and takes on all values from $l$ to $-l$ in integral steps, as shown in Figure 1. It is easy to see that there are $(2l+1)$ eigenstates for $L_z$ for a given value of $l$.

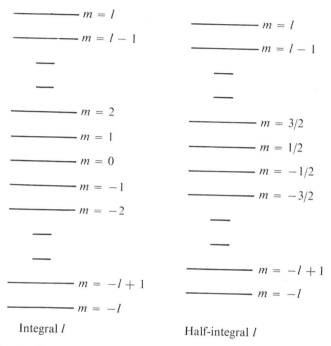

$$
\begin{aligned}
&m = l \\
&m = l-1 \\
&\phantom{m} \\
&\phantom{m} \\
&m = 2 \\
&m = 1 \\
&m = 0 \\
&m = -1 \\
&m = -2 \\
&\phantom{m} \\
&m = -l+1 \\
&m = -l
\end{aligned}
$$

Integral $l$

$$
\begin{aligned}
&m = l \\
&m = l-1 \\
&\phantom{m} \\
&m = 3/2 \\
&m = 1/2 \\
&m = -1/2 \\
&m = -3/2 \\
&\phantom{m} \\
&m = -l+1 \\
&m = -l
\end{aligned}
$$

Half-integral $l$

FIGURE 1.  Values of $m$ for states of given angular momentum $l$.

From now on we shall denote an angular momentum eigenstate with simultaneous eigenvalues $\hbar^2 l(l+1)$ and $\hbar m$ by $Y_l^m$, whence its eigenvalue equations take the form

$$
\begin{aligned}
L^2 Y_l^m &= \hbar^2 l(l+1) Y_l^m \\
L_z Y_l^m &= \hbar m Y_l^m.
\end{aligned} \tag{30}
$$

For brevity, such states are called states of angular momentum $l$ with $z$-component $m$. These states can be constructed quite simply by starting with the top state on the ladder, $Y_l^l$, and then operating successively with the lowering operator $L_-$. Thus, we write specifically,

$$Y_l^m = \frac{c_l^m}{c_l^l} (L_-)^{l-m} Y_l^l, \tag{31}$$

where the $c_l{}^m$ are normalizing constants and where, it is recalled, $Y_l{}^l$ is defined by

$$L_+ Y_l{}^l = 0. \tag{32}$$

Of course, the angular momentum states can equally well be obtained by starting with the bottom state $Y_l{}^{-l}$ and operating successively with $L_+$. In either case, using the already established properties of the raising and lowering operators, these states can be explicitly constructed and their relevant properties can be determined. Leaving the details to Exercise 3, below, we simply state the results:[3]

$$\left.\begin{aligned}
Y_l{}^m &= \hbar^{m-l} \sqrt{\frac{(l+m)!}{(2l)!\,(l-m)!}} \; (L_-)^{l-m}\, Y_l{}^l \\[2mm]
Y_l{}^m &= \hbar^{-m-l} \sqrt{\frac{(l-m)!}{(2l)!\,(l+m)!}} \; (L_+)^{l+m}\, Y_l{}^{-l}
\end{aligned}\right\} \tag{33a}$$

$$L_\pm Y_l{}^m = \hbar \sqrt{l(l+1) - m(m \pm 1)} \; Y_l{}^{m\pm 1}. \tag{33b}$$

---

**Exercise 3.**

   (a)   Show that equation (33b) follows from equation (33a) and hence that these two equations are equivalent.

   (b)   Derive equation (33b) by directly evaluating $\langle L_\pm Y_l{}^m | L_\pm Y_l{}^m \rangle$, assuming the $Y_l{}^m$ to be normalized. Hint: Start by establishing that

$$\langle L_\pm Y_l{}^m | L_\pm Y_l{}^m \rangle = \langle Y_l{}^m | L_\mp L_\pm Y_l{}^m \rangle,$$

and then use equations (24) and (25).

   (c)   Verify that the $Y_l{}^m$ are orthogonal.

---

It is important to observe that all our results to this stage have followed simply and directly as consequences of the angular momentum commutation relation, equation (6). We have used the specific realization of $L$ as the dynamical variable corresponding to orbital angular momentum, and thus defined by equation (1), only in establishing this commutation relation. Our results accordingly hold for any operator satisfying equation (6), whether or not it represents *orbital* angular momentum.

   Even with this generality in mind, there is one unexpected feature in our results, namely the appearance of half-integral angular momentum

---

[3] See Reference [22], especially Chapter XIII and Appendix B, for a particularly complete and careful presentation.

states as possible states in the spectrum. Recall that, in our treatment of orbital angular momentum states for a particle moving in a central potential, we deduced that these states should have integral angular momentum from the requirement that the state function be single-valued. Actually, this requirement was too stringently applied since, in fact, only physically observable quantities need be single-valued. Now such quantities are always expressed in terms of expectation values and are thus second-degree functionals of the state function of the system. This means that there can be no *a priori* objection to a state function merely because it can assume two different signs at a given point in space, for the square of such a function is single-valued.

To put it another way, no physical significance can be attached to the *absolute* sign of the state function. Recall also that the question of single-valuedness arose specifically in connection with the eigenfunctions $e^{im\phi}$ of $L_z$, and also with the fact that $\phi = 0$ and $\phi = 2\pi$ refer to the same point in space. We thus see that half-integral angular momentum, and therefore half-integral $m$, corresponds precisely to the previously ignored possibility of an ambiguity of sign. Note, however, that the *relative* sign of any two state functions *is* physically significant, since interference terms depend on the relative phase. As a consequence, the single-valuedness requirement would be violated if some of the states of a given system could have integral angular momentum and some half-integral. To make this explicit, suppose $\psi_1$ is an integral angular momentum state, $\psi_2$ a half-integral angular momentum state, and consider the superposition state

$$\psi(r, \theta, \phi) = \psi_1(r, \theta, \phi) + \psi_2(r, \theta, \phi).$$

Then

$$\psi(r, \theta, \phi + 2\pi) = \psi_1(r, \theta, \phi) - \psi_2(r, \theta, \phi),$$

whence $|\psi|^2$ is not single-valued and such combinations are accordingly forbidden.

On the basis of this general and purely formal discussion, we thus conclude that, in principle, the states of a given system can have integral or half-integral angular momentum, but only one or the other exclusively, and never a mixture. In particular, this means that while the orbital angular momentum states could indeed have integral angular momentum, as we earlier assumed, the alternative possibility of half-integral angular momentum exists and must be examined. Which is correct is then a matter for experiment to determine. The hydrogen atom can readily be used for this test, and it turns out that the spectrum computed assuming half-integral values of orbital angular momentum is not in agreement with experiment. *Half-integral values of orbital angular momentum are thus ruled out.*[4] As we shall see shortly, this is not the case for the *intrinsic* angular momentum or *spin* of a particle. Both possibilities for the spin are, in fact, observed in nature.

Because we shall be concerned with more than one kind of angular momentum, as the discussion above implies, we now introduce an appropriate and more or less standard notation to permit us to distinguish among them. We shall continue to denote the *orbital* angular momentum operator by **L** and its eigenstates by $Y_l^m$. The *spin* angular momentum operator will be denoted by **S** and its eigenstates by $\chi_s^m$, which is to say that $\chi_s^m$ is defined by

$$S^2\chi_s^m = \hbar^2 s(s+1)\chi_s^m$$
$$S_z\chi_s^m = \hbar m\chi_s^m. \tag{34}$$

Finally, we shall use the symbol **J** as a *generic* symbol for the angular momentum operator, referring to either orbital or spin angular momentum, as the case may be. Its eigenstates will be denoted by $Y_j^m$, whence

$$J^2 Y_j^m = \hbar^2 j(j+1)Y_j^m$$
$$J_z Y_j^m = \hbar m Y_j^m. \tag{35}$$

For ease in writing, in all three cases we have used $m$ as the quantum number associated with the $z$-component of angular momentum. Whenever it becomes necessary to make a distinction, we shall simply introduce appropriate subscripts and write $m_l$ or $m_s$ or $m_j$, depending on circumstances.

The point of these notational matters is that *all* angular momentum operators satisfy the same vector commutation relations, namely, equation (6) for **L**,

$$\mathbf{S} \times \mathbf{S} = ih\,\mathbf{S} \tag{36}$$

for **S**, and, for **J**,

$$\mathbf{J} \times \mathbf{J} = ih\,\mathbf{J}. \tag{37}$$

Nonetheless, there are some distinct differences between orbital and spin angular momentum, as is made clear by the fact that only the latter can assume half-integral values. We shall use **J** to write those general relations which are valid for either. In other words, *all expressions written in terms of* **J** *apply equally to both orbital and spin angular momentum*. On the other hand, expressions written in terms of **L** or **S**,

---

[4] Arguments other than the simple empirical one we have given are available for ruling out half-integral orbital angular momentum. In particular, difficulties are encountered with the probability flux for such states, according to J. M. Blatt and V. F. Weisskopf, *Theoretical Nuclear Physics,* Wiley (1952), Appendix A.

depending on the context, either will be particular cases of the general relations or will refer to some special property of one which is not shared by the other. An example of the latter is the representation of orbital angular momentum in terms of spherical harmonics; no such representation exists for spin. Equations (6) and (36) are examples of the former; both are particular cases of the general angular momentum commutation rule, equation (37).

We shall also ultimately be concerned with the *total angular momentum* of composite systems. In general, the total angular momentum is compounded from both spin and orbital contributions and hence exhibits only the properties they share in common. It was precisely to exhibit these common properties that **J** was introduced, ánd hence *we shall also denote the total angular momentum of a general system by* **J**.

As we have already emphasized, *all* of our results to this stage have been derived using only the vector commutation relation for angular momentum, and they thus hold for any kind of angular momentum. For future reference, as well as to make their generality explicit, we now rewrite at least the principal results in terms of **J**:

$$\left. \begin{aligned} Y_j{}^m &= \hbar^{m-j} \sqrt{\frac{(j+m)!}{(2j)!\,(j-m)!}} \; (J_-)^{j-m} \, Y_j{}^j \\[2mm] Y_j{}^m &= \hbar^{-m-j} \sqrt{\frac{(j-m)!}{(2j)!\,(j+m)!}} \; (J_+)^{j+m} \, Y_j{}^{-j} \end{aligned} \right\} \tag{38a}$$

$$J_\pm Y_j{}^m = \hbar \sqrt{j(j+1) - m(m \pm 1)} \; Y_j{}^{m\pm 1}, \tag{38b}$$

where, of course,

$$J_\pm = J_x \pm i J_y \tag{39a}$$

and

$$\begin{aligned} J_- J_+ &= J^2 - \hbar J_z - J_z{}^2 \\ J_+ J_- &= J^2 + \hbar J_z - J_z{}^2. \end{aligned} \tag{39b}$$

One interesting feature of the angular momentum operator is that its projection on the $z$-axis is always less than its absolute magnitude; thus, as we have mentioned before, its orientation is not precisely defined. It is illuminating to discuss this behavior in terms of the uncertainty principle. Recall that, according to equation (V-49), we have, for any pair of Hermitian operators $A$ and $B$,

$$(\Delta A)^2 (\Delta B)^2 \geq \tfrac{1}{4} \, |\langle (A, B) \rangle|^2.$$

We now use this relation to examine the effects of the noncommutativity of the rectangular components of **J**. First observe, however, that for any state $Y_j{}^m$

$$\langle Y_j^m | J_\pm | Y_j^m \rangle \sim \langle Y_j^m | Y_j^{m\pm 1} \rangle = 0,$$

and hence, since $J_x$ and $J_y$ are linear combinations of $J_+$ and $J_-$,

$$\langle Y_j^m | J_x | Y_j^m \rangle = \langle Y_j^m | J_y | Y_j^m \rangle = 0.$$

Consequently, for such a state,

$$(\Delta J_x)^2 = \langle J_x^2 \rangle$$

and

$$(\Delta J_y)^2 = \langle J_y^2 \rangle,$$

whence

$$(\Delta J_x)^2 + (\Delta J_y)^2 = \langle J_x^2 + J_y^2 \rangle = \langle J^2 - J_z^2 \rangle.$$

On the other hand, according to the uncertainty principle,

$$(\Delta J_x)^2 (\Delta J_y)^2 \geq \frac{1}{4} |\langle ( J_x, J_y) \rangle|^2 = \frac{\hbar^2}{4} \langle J_z \rangle^2.$$

Now the orientation of the $x$- and $y$-axes is arbitrary, and hence we infer that

$$(\Delta J_x)^2 = (\Delta J_y)^2,$$

whence the first relation becomes

$$(\Delta J_x)^2 = \frac{1}{2} \langle J^2 - J_z^2 \rangle,$$

while the second becomes

$$(\Delta J_x)^2 \geq \frac{\hbar}{2} |\langle J_z \rangle|.$$

Comparing these expressions, then, we must have

$$\langle J^2 - J_z^2 \rangle \geq \hbar |\langle J_z \rangle|,$$

so that $\langle J^2 \rangle$ must always be greater than $\langle J_z^2 \rangle$. Specifically,

$$\langle J^2 \rangle \geq \langle J_z^2 \rangle + \hbar |\langle J_z \rangle| = |\langle J_z \rangle| (|\langle J_z \rangle| + \hbar),$$

and hence, for a state of given $J^2$, say $\hbar^2 j(j+1)$, we have

$$\hbar^2 j(j+1) \geq |\langle J_z \rangle| (|\langle J_z \rangle| + \hbar).$$

It then follows that the maximum possible value of $\langle J_z \rangle$ is correctly $\hbar j$, and not $\hbar \sqrt{j(j+1)}$ as it would be if the angular momentum were oriented precisely along the $z$-axis.

These quantum mechanical features of the angular momentum can be given an oversimplified but helpful geometrical interpretation. For the state $Y_j^m$, the angular momentum **J** can be visualized as a vector of length $\sqrt{j(j+1)}\,\hbar$ lying on the surface of a cone with altitude $m\hbar$ centered about the $z$-axis, as illustrated in Figure 2. In this picture,

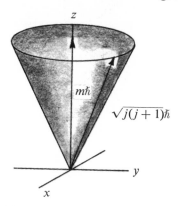

FIGURE 2. Geometrical interpretation of the properties of the angular momentum for the state $Y_j^m$.

all orientations of **J** on the surface of the cone are to be regarded as equally likely. Hence, as should be the case,

$$\langle J_x \rangle = \langle J_y \rangle = 0.$$

Further, by the Pythagorean theorem,

$$\langle J_x^2 \rangle = \langle J_y^2 \rangle = (\sqrt{j(j+1)}\hbar)^2 - (m\hbar)^2 = [j(j+1) - m^2]\,\hbar^2,$$

which is also the correct result, as we have, in effect, just demonstrated above. States of different $m$ for a given $j$ then correspond to cones of different altitude and angular opening. The angular momentum cone can never close completely, its smallest aperture coming for $m = j$, when the angular opening is $\cos^{-1}(j/\sqrt{j(j+1)})$. The precise orientation of classical angular momentum vectors is recovered, however, in the classical limit $j \gg 1$, as it must be.

These features are further illustrated in Figure 3, where the angular momentum cones are drawn to scale for the particular angular momentum states $j = 1$ and $j = 2$.

It still remains for us to exhibit the orbital angular momentum states in configuration space, and thereby to establish the relationship between our present results and those given in Chapter IX. We start with equation (32). Referring back to equations (34) and (35) of Chapter IX, we see that, in configuration space,

$$L_+ = \hbar\, e^{i\phi} \left[ \frac{\partial}{\partial\theta} + i \operatorname{ctn}\theta\, \frac{\partial}{\partial\phi} \right]$$

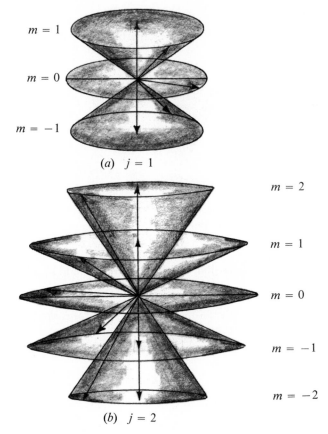

$m = 1$

$m = 0$

$m = -1$

(a)  $j = 1$

$m = 2$

$m = 1$

$m = 0$

$m = -1$

$m = -2$

(b)  $j = 2$

FIGURE 3.   Geometrical representation of angular momentum states for $j = 1$ and $j = 2$.

and

$$L_- = -\hbar\, e^{-i\phi}\left[\frac{\partial}{\partial\theta} - i\, \text{ctn}\, \theta\frac{\partial}{\partial\phi}\right].$$

Writing $Y_l^l(\theta, \phi) = f_l(\theta)\, e^{il\phi}$, equation (32) then becomes

$$\frac{df_l}{d\theta} = (l\, \text{ctn}\, \theta)\, f_l,$$

whence, as is easily verified,

$$f_l(\theta) \sim (\sin\,\theta)^l,$$

and hence

$$Y_l^l = c_l^l\,(\sin\,\theta)^l\, e^{il\phi}.$$

It follows just as readily that $Y_l^{-l}$ has exactly the same form. Evaluation of the normalization integral then leads, finally, to the result

$$Y_l^{\pm l}(\theta, \phi) = (\mp)^l \sqrt{\frac{(2l+1)!}{4\pi}} \frac{\sin^l \theta}{2^l l!} e^{\pm il\phi}.$$

Observe that the phase of the normalization constant is *not* determined by the normalization condition. Our choice of this arbitrary phase, which is that of most authors, is already embodied in equations (33). For further discussion see reference [22].

---

**Exercise 4.** By performing the angular integration, verify that $Y_l^{\pm l}$ as given above is indeed normalized to unity.

---

States with $m \neq \pm l$ can be obtained by successive operation with $L_\mp$ on $Y_l^{\pm l}$ and, after some manipulation, the results can be summarized in the form

$$Y_l^m(\theta, \phi) = (-1)^l \sqrt{\frac{(2l+1)!\,(l+m)!}{4\pi(2l)!\,(l-m)!}} \frac{e^{im\phi} \sin^{-m} \theta}{2^l l!} \frac{d^{l-m}}{d(\cos \theta)^{l-m}}$$
$$\times (1 - \cos^2 \theta)^l$$

or, equivalently,

$$Y_l^m(\theta, \phi) = (-1)^{m+l} \sqrt{\frac{(2l+1)!\,(l-m)!}{4\pi(2l)!\,(l+m)!}} \frac{e^{im\phi} \sin^m \theta}{2^l l!} \frac{d^{l+m}}{d(\cos \theta)^{l+m}}$$
$$\times (1 - \cos^2 \theta)^l.$$

---

**Exercise 5.** By carrying out the indicated differentiations, find $Y_1^0$, $Y_1^{\pm 1}$, $Y_2^{\pm 2}$, $Y_2^{\pm 1}$, $Y_2^0$, and compare with Table I, Chapter IX.

---

## 3. ROTATION AND TRANSLATION OPERATORS

We now establish an interesting and informative relationship between rotational transformations of the space coordinates and the orbital angular momentum operator. Consider an infinitesimal rotation of the coordinates through an angle $\delta\phi$ about the $z$-axis. Denote the operator which induces this transformation by $\delta R_z$; that is, $\delta R_z$, acting on any scalar function $f(r, \theta, \phi)$, is *defined* by

$$\delta R_z f(r, \theta, \phi) \equiv f(r, \theta, \phi + \delta\phi). \tag{40}$$

Because $\delta\phi$ is infinitesimal, we expand the right-hand side in a Taylor series,

$$f(r, \theta, \phi + \delta\phi) = f(r, \theta, \phi) + \delta\phi \frac{\partial f}{\partial \phi} + \cdots,$$

or, retaining only first-order terms,

$$f(r, \theta, \phi + \delta\phi) = \left(1 + \delta\phi \frac{\partial}{\partial \phi}\right) f(r, \theta, \phi). \tag{41}$$

Thus upon comparison of equations (40) and (41), we see that

$$\delta R_z = 1 + \delta\phi \frac{\partial}{\partial \phi}.$$

Now

$$L_z = \frac{\hbar}{i} \frac{\partial}{\partial \phi}$$

and thus

$$\delta R_z = 1 + \frac{i\delta\phi}{\hbar} L_z, \tag{42}$$

which establishes a deep and fundamental connection between space rotations and the angular momentum operators.

Next, we use this result to generate a rotation about the $z$-axis through some finite angle, say $\beta$. We do this by repeatedly operating with $\delta R_z$. Denote such a finite rotation operator by $R_z(\beta)$, which, operating upon an arbitrary scalar function $f$, is explicitly defined by

$$R_z(\beta) f(r, \theta, \phi) \equiv f(r, \theta, \phi + \beta).$$

Now

$$R_z(n\delta\phi) = (\delta R_z)^n = \left(1 + i\frac{\delta\phi}{\hbar} L_z\right)^n$$

and hence, writing $n\delta\phi = \beta$,

$$R_z(\beta) = \lim_{\delta\phi \to 0} \left(1 + i\frac{\delta\phi}{\hbar} L_z\right)^{\beta/\delta\phi}.$$

Proceeding to the limit, we thus obtain, from the definition of the exponential function,

$$R_z(\beta) = e^{i\beta L_z/\hbar} \equiv \sum_n \frac{1}{n!} \left(\frac{i\beta L_z}{\hbar}\right)^n.$$

That this result is correct may be directly verified by operating with $R_z(\beta)$

in power series form on an arbitrary function $f(r, \theta, \phi)$ and then comparing the result with the Taylor series representation of $f(r, \theta, \phi + \beta)$. More generally, the operator $R_{\hat{n}}(\beta)$, which induces a rotation through an angle $\beta$ about an axis oriented along an arbitrary unit vector $\mathbf{n}$, is seen to be

$$R_{\hat{n}}(\beta) = e^{i\beta\hat{n}\cdot\mathbf{L}/\hbar}, \tag{43}$$

while the corresponding infinitesimal rotation operator $\delta R_{\hat{n}}$ is

$$\delta R_{\hat{n}} = 1 + i\delta\phi\hat{n} \cdot \mathbf{L}/\hbar.$$

From these results we see that if, for any system, the Hamiltonian $H$ commutes with the angular momentum operator $\mathbf{L}$ then $H$ is necessarily invariant under an arbitrary rotation of the coordinate axes. From this point of view, then, conservation of angular momentum, which is a consequence of the commutativity of $\mathbf{L}$ and $H$, implies that there is no preferred set of coordinate axes for the system in question. Hence the dynamical requirement that the total angular momentum be conserved for any isolated system is equivalent to the much deeper and purely geometrical requirement that *space be intrinsically isotropic*.

The relation between space rotations and angular momentum also provides us with a geometrical interpretation of the noncommutativity of different components of angular momentum. Explicitly, we see that this property is a direct and immediate consequence of the noncommutativity of finite rotations in space about different axes. For example, it is easily verified that rotation about the $x$-axis followed by rotation about the $y$-axis gives an entirely different result from that obtained when these rotations are carried out in the opposite order.

Another important result of these considerations is that they provide a basis for a better understanding of the intrinsic $(2l + 1)$-fold degeneracy of states of angular momentum $l$ in a central potential. This degeneracy occurs whenever the Hamiltonian is rotationally invariant, and it is merely a reflection of the fact that the choice of $z$-axis is entirely arbitrary, in which circumstance no physical significance attaches to the eigenvalues of $L_z$. We remark, however, that if for the system in question, the isotropy of space is destroyed in some way, then this degeneracy is removed. An important example is the Zeeman splitting of atomic states in the presence of an external magnetic field. The direction of the magnetic field singles out a particular spatial axis, and the actual splitting is produced by the interaction of the external field with the magnetic moments generated by the orbital motion of the charged constituents of the atomic system and with whatever intrinsic magnetic moments these constituents may have.

As an interesting addendum to these observations, we note that if space is homogeneous as well as isotropic, that is, if there is no preferred origin of coordinates, then we expect invariance with respect to translations of the coordinates. To make this explicit, introduce the infinitesimal translation operator $T_{\delta r}$, defined by

$$T_{\delta r} f(\mathbf{r}) \equiv f(\mathbf{r} + \delta \mathbf{r}). \tag{44}$$

Thus $T_{\delta r}$ is seen to translate the coordinates through $\delta \mathbf{r}$. Since $\delta \mathbf{r}$ is infinitesimal, we expand $f(\mathbf{r} + \delta \mathbf{r})$ in a Taylor series,

$$f(\mathbf{r} + \delta \mathbf{r}) = f(\mathbf{r}) + \delta \mathbf{r} \cdot \nabla f(\mathbf{r})$$

or, using $\mathbf{p} = \hbar \nabla / i$,

$$f(\mathbf{r} + \delta \mathbf{r}) = (1 + i \delta \mathbf{r} \cdot \mathbf{p}/\hbar) f(\mathbf{r}). \tag{45}$$

Comparing equations (44) and (45), we then see that

$$T_{\delta r} = 1 + i \delta \mathbf{r} \cdot \mathbf{p}/\hbar, \tag{46}$$

which thus relates space translations to the linear momentum. By arguments analogous to those used for rotations, it follows that the operator $T_a$ which induces translation through a finite distance $\mathbf{a}$ is given by

$$T_a = e^{i \mathbf{a} \cdot \mathbf{p}/\hbar} = \sum_n \frac{1}{n!} (i \mathbf{a} \cdot \mathbf{p}/\hbar)^n. \tag{47}$$

The requirement that the total linear momentum of an isolated system be conserved is now seen to be equivalent to the geometrical requirement that *space be intrinsically homogeneous*. Further, the mutual commutativity of different components of the linear momentum is seen to follow from the indifference of the final results of a sequence of translations to the order in which they are carried out. A net translation can always be expressed as a sum of any set of its components taken in any order.

## 4. SPIN: THE PAULI OPERATORS

To the present we have been primarily concerned with the properties of orbital angular momentum, that is, with angular momentum which depends only upon the state of motion of a particle and not at all upon the specific characteristics of the particle itself. We now take up a second kind of angular momentum which occurs in nature and which is an intrinsic attribute of certain elementary particles. This intrinsic angular momentum, or *spin* as it is called, satisfies the usual angular momentum commutation rules, but it differs sharply from orbital angular momentum in the following ways:

(a)  Spin is a specific property of specific types of particles and is independent of their state of motion. Accordingly, the spin of a particle may be regarded as an internal degree of freedom which is in some sense associated with the internal structure of these particles.

(b)   The spin angular momentum can take on half-integral as well as integral values, but exclusively one or the other for a given type of particle.

(c)   For a given type of particle, the spin has a fixed and immutable magnitude. For example, the common constituents of matter, the electron, neutron and proton, all have spin one-half in units of $\hbar$, so do neutrinos and muons, and so do the anti-particles of all of these. Photons have spin one. Some particles, such as pions and kaons, carry no intrinsic angular momentum and thus have spin zero.

(d)   Spin angular momentum is a purely quantum mechanical quantity. In the sense of the correspondence principle, there is no classical limit for spin. Otherwise stated, no classical description can be supplied for the internal degrees of freedom associated with spin.

Although spin is an internal property of the elementary particles, it must be coupled in some way to the external world if it is to have any physical significance. This coupling occurs in several domains. The *strong interactions,* which are those which predominate in nuclear physics, are very much dependent upon spin. So are the *weak interactions,* which describe beta decay processes. On the more familiar level of *electromagnetic interactions,* any charged particle carrying spin also carries an associated magnetic moment, with the magnetic moment vector being proportional to the spin vector.

Whatever the nature of these couplings, there are certain general consequences of the existence of spin. For one thing, the connection between spin and statistics discussed in Chapter VIII plays a dominant role in the thermodynamical–statistical properties of matter. For a second, at the quantum level, spin angular momentum enters crucially into the profoundly significant angular momentum conservation laws, just as crucially as does orbital angular momentum. Historically, however, the existence of spin was not inferred from such general features, but rather from the special effects of the electron's magnetic moment upon atomic states. The simplest of these effects involves the splitting of atomic states in an external magnetic field. Since the degeneracy of a state of total angular momentum $j$ is $(2j + 1)$, it follows that the value of $j$ for a given state is unambiguously determined by the number of states into which it is split.

A detailed and systematic analysis of these Zeeman splittings, as they are called, was shown by Goudsmit and Uhlenbeck in 1925 to imply that the electron carries an intrinsic angular momentum, of fixed magnitude one-half in units of $\hbar$, which must be added to the orbital angular momentum to give the total angular momentum $j$. Thus, for example, an $s$-state $(l = 0)$ of a hydrogenic atom is split into two components, since $j = 1/2$ for such a state. A $p$-state $(l = 1)$ can combine with the spin in two ways. In one, the orbital and spin angular momentum are parallel $(j = 3/2)$, in the other anti-parallel $(j = 1/2)$. The former thus splits into four components, the latter into two, giving a total of six components instead of three, as there would be if the electron were spinless. And so it goes for the states of higher angular momentum; for a given $l$ there occur two states, $j = l + 1/2$ and $j = l - 1/2$, and the total number of states in the spectrum is seen to be doubled. Note that all of this refers only to a hydrogenic atom, since the description is far more complicated for an atom containing more than one electron.

We now want to develop a formalism for handling spin. We shall restrict our attention to the most important case, that of spin one-half. As we have stated, the existence of electron spin produces a doubling of the single-particle electronic states corresponding to the two, and only two, possible orientations of the spin one-half angular momentum vector with respect to some arbitrarily chosen axis. We shall denote this axis as the $z$-axis. We now want to write a general expression for an arbitrary single particle spin dependent wave function $\psi(r, \text{spin})$. This can only be a linear combination of the two possible spin states, that in which the $z$-component of spin is $+\hbar/2$ and that in which it is $-\hbar/2$. We thus write the superposition

$$\psi(\mathbf{r}, \text{spin}) = \psi_+(\mathbf{r})\chi_+ + \psi_-(\mathbf{r})\chi_-, \qquad (48)$$

where $\psi_\pm(\mathbf{r})$ are the space dependent parts of the state function and where $\chi_+$ and $\chi_-$ are eigenfunctions of the spin angular momentum corresponding, respectively, to spin up or down.[5] These functions thus carry all the information about spin. The probability density for finding the particle at $\mathbf{r}$ with spin up is $\psi_+^*(\mathbf{r})\psi_+(\mathbf{r})$, and with spin down it is $\psi_-^*(\mathbf{r})\psi_-(\mathbf{r})$, whence the probability density for finding the particle at $\mathbf{r}$, independently of its spin, is $\psi_+^*\psi_+ + \psi_-^*\psi_-$. It thus follows that the normalization integral for such a function is to be understood as

$$\langle \psi(\mathbf{r}, \text{spin}) | \psi(\mathbf{r}, \text{spin}) \rangle = \langle \psi_+(\mathbf{r}) | \psi_+(\mathbf{r}) \rangle + \langle \psi_-(\mathbf{r}) | \psi_-(\mathbf{r}) \rangle, \quad (49)$$

where each of the terms on the right has its conventional meaning as a configuration space integral. Equation (49) follows from equation (48), provided we understand that

$$\langle \psi_1(\mathbf{r})\chi_\pm | \psi_2(\mathbf{r})\chi_\pm \rangle = \langle \psi_1(\mathbf{r}) | \psi_2(\mathbf{r}) \rangle \langle \chi_\pm | \chi_\pm \rangle \qquad (50)$$

[5] In the notation of equation (34), these functions are simply $\chi_{1/2}{}^{1/2}$ and $\chi_{1/2}{}^{-1/2}$, respectively. To save writing, we have abbreviated them in a way which has become standard in the literature.

and that the $\chi_\pm$ are orthonormal,

$$\langle \chi_+|\chi_+ \rangle = \langle \chi_-|\chi_- \rangle = 1$$
$$\langle \chi_+|\chi_- \rangle = \langle \chi_-|\chi_+ \rangle = 0. \tag{51}$$

It may be helpful to compare this way of writing the state function with the way in which an ordinary vector **A**, say, is written in component form,

$$\mathbf{A} = A_x \hat{\mathbf{e}}_x + A_y \hat{\mathbf{e}}_y + A_z \hat{\mathbf{e}}_z,$$

where $\hat{\mathbf{e}}_x$, $\hat{\mathbf{e}}_y$ and $\hat{\mathbf{e}}_z$ are orthonormal unit vectors. In correspondence with equation (49), we have

$$A^2 = A_x^2 + A_y^2 + A_z^2.$$

A spin dependent state function may thus be regarded as a *two-component function*, one component for each possible spin orientation, with the spin states $\chi_\pm$ playing the role of unit vectors.

We next develop the properties of the spin states and of the spin angular momentum operator, **S**, which acts upon these states. Of course, **S** must satisfy the usual commutation rules for angular momentum, equation (36),

$$\mathbf{S} \times \mathbf{S} = i\hbar \mathbf{S},$$

and, from their definition as eigenstates corresponding to $z$-components of spin $+ \hbar/2$ and $- \hbar/2$, respectively, $\chi_+$ and $\chi_-$ must satisfy the relations

$$S_z \chi_+ = \frac{\hbar}{2} \chi_+$$
$$S_z \chi_- = -\frac{\hbar}{2} \chi_-. \tag{52}$$

Further, because both are states of total spin $\hbar/2$, we must have

$$S^2 \chi_\pm = \tfrac{1}{2} \left( \tfrac{1}{2} + 1 \right) \hbar^2 \chi_\pm = \tfrac{3}{4} \hbar^2 \chi_\pm.$$

Reference to equation (48) then shows that, for a perfectly *arbitrary* state function, $\psi(r, \text{spin})$,

$$S^2 \psi(r, \text{spin}) = \tfrac{3}{4} \hbar^2 \psi(r, \text{spin}),$$

which is to say that, in contrast to the case for orbital angular momentum, $S^2$ is a purely *numerical* operator,

$$S^2 = \tfrac{3}{4} \hbar^2. \tag{53}$$

Note also that

$$S_z^2 \chi_\pm = \frac{\hbar^2}{4} \chi_\pm,$$

and hence $S_z^2$ is also a purely numerical operator by the same argument. Now $S_x^2$, $S_y^2$ and $S_z^2$ are entirely equivalent dynamical variables, and they must thus all share this last property, whence

$$S_x^2 = S_y^2 = S_z^2 = \frac{\hbar^2}{4}, \tag{54}$$

which is clearly consistent with equation (53).

Equations (53) and (54) mean that no differential operator characterization exists, in any representation, for the spin angular momentum. This poses no problem, of course, since *an operator is completely defined by the results it yields when it operates upon an arbitrary state.* These results have already been given for $S^2$ and $S_z$, and it remains only to similarly specify $S_x$ and $S_y$. Instead of working with these latter operators directly, it is more convenient to work with the raising and lowering operators $S_+$ and $S_-$, which are defined, as usual, by

$$S_\pm = S_x \pm iS_y. \tag{55}$$

Applying the completely general relation, equation (38b), we at once obtain for the special case under consideration, $j = \frac{1}{2}$ and $m = \pm \frac{1}{2}$,

$$
\begin{aligned}
S_+\chi_+ &= 0 & S_+\chi_- &= \hbar\chi_+ \\
S_-\chi_+ &= \hbar\chi_- & S_-\chi_- &= 0
\end{aligned}
\tag{56}
$$

and, inverting equation (55),

$$
\begin{aligned}
S_x\chi_+ &= \frac{\hbar}{2}\chi_- & S_x\chi_- &= \frac{\hbar}{2}\chi_+ \\
S_y\chi_+ &= \frac{i\hbar}{2}\chi_- & S_y\chi_- &= -\frac{i\hbar}{2}\chi_+ .
\end{aligned}
\tag{57}
$$

It is easily verified that $S_x^2$ and $S_y^2$ are numerical operators satisfying equation (54) and that the proper commutation relations are satisfied.

---

**Exercise 6.**

(a)   Obtain equation (57) starting with equation (38b).

(b)   Taking $S_x$, $S_y$ and $S_z$ to be *defined* by equations (52) and (57), show that equation (54) and the vector commutation relation, equation (36), are both satisfied. Do this by letting the spin operators act upon the perfectly arbitrary spin state of equation (48).

---

The algebra of the spin one-half operators is quite unusual, as we have already seen, primarily as a consequence of equation (54). We now develop this algebra a little further. From equation (56), it follows at once that $S_\pm^2 = 0$. Hence, using equation (55),

$$0 = (S_x \pm iS_y)^2 = S_x^2 - S_y^2 \pm i(S_xS_y + S_yS_x)$$

or, since $S_x^2 = S_y^2 = \hbar^2/4$,

$$S_xS_y + S_yS_x = 0.$$

The same relation must hold between any different pair of components, of course, since all components of **S** are dynamically equivalent. This kind of expression, which is like the commutator except for a *plus* sign instead of a minus sign, is called an *anticommutator*, and the anticommutator bracket is defined by, for any pair of operators,

$$(A, B)_+ \equiv AB + BA = (B, A)_+. \tag{58}$$

Denoting the *x, y* and *z* components of *S* by numerical subscripts, we can thus write the *anticommutation* relations for spin one-half,

$$(S_i, S_j)_+ = \frac{\hbar^2}{2} \delta_{ij}. \tag{59}$$

Using this important result, the commutation relations can now be somewhat simplified. We have, with *i, j* and *k* in cyclic order,

$$S_iS_j - S_jS_i = i\hbar S_k$$

and hence, since $S_i$ and $S_j$ anticommute,

$$S_iS_j = \frac{i\hbar}{2} S_k, \tag{60}$$

or alternatively, multiplying by $S_k$ from either left or right,

$$S_iS_jS_k = S_kS_iS_j = i\hbar^3/8, \tag{61}$$

where, we repeat for emphasis, *i, j* and *k* are to be taken in cyclic order.

Equation (60) is particularly useful for the following reason. Consider some completely arbitrary spin dependent operator. Suppose it to involve a term in the *n*th power of the components of **S**, in some order. Equations (54) and (60) then assure that such a term can *always* be reduced either to a *spin independent* term or to term *linear* in the spin operators. To see how this works, consider the following examples:

$$(1) \quad S_xS_yS_xS_zS_y = S_xS_yS_x \left(-\frac{i\hbar}{2}\right) S_x$$

$$= S_xS_y \left(-\frac{i\hbar^3}{8}\right)$$

$$= \frac{\hbar^4}{16} S_z.$$

In the first line we replaced $S_z S_y$ by $(-i\hbar/2)S_x$, in the second, $S_x^2$ by $\hbar^2/4$ and in the third, $S_x S_y$ by $(i\hbar/2)S_z$.

$$(2) \quad S_x S_z S_x S_y S_x = S_x \left(\frac{i\hbar^3}{8}\right) S_x$$

$$= \frac{i\hbar^5}{32}.$$

In the first line we replaced $S_z S_x S_y$ by $i\hbar^3/8$, according to equation (61), and in the second, $S_x^2$ by $\hbar^2/4$. Since any power of the spin operators can be reduced in the above way, we see that the *most general possible* spin dependent operator $A$ must be expressible as a linear function of the spin, that is, as

$$A = A_0 + A_1 S_x + A_2 S_y + A_3 S_z, \tag{62}$$

where the $A_i$ are arbitrary spin *independent* operators of the type we have worked with all along.

We are now in a position to show that the spin operator is indeed completely specified by equations (52) and (57). We do this by computing the result of operating with the *arbitrary* operator $A$, equation (62), on an *arbitrary* state $\psi$, equation (48). We have at once

$$A\psi = A_0\,(\psi_+\chi_+ + \psi_-\chi_-) + \frac{\hbar}{2}\,A_1(\psi_+\chi_- + \psi_-\chi_+) + \frac{i\hbar}{2}\,A_2\,(\psi_+\chi_- - \psi_-\chi_+)$$

$$+\frac{\hbar}{2}\,A_3(\psi_+\chi_+ - \psi_-\chi_-)$$

or, collecting terms,

$$A\psi = \left[\left(A_0 + \frac{\hbar}{2}\,A_3\right)\psi_+ + \frac{\hbar}{2}\,(A_1 - iA_2)\psi_-\right]\chi_+$$

$$+\left[\left(A_0 - \frac{\hbar}{2}\,A_3\right)\psi_- + \frac{\hbar}{2}\,(A_1 + iA_2)\psi_+\right]\chi_-. \tag{63}$$

Thus, for example, the expectation value of $A$ is

$$\langle\psi|A|\psi\rangle = \langle\psi_+|A_0 + \frac{\hbar}{2}\,A_3|\psi_+\rangle + \frac{\hbar}{2}\,\langle\psi_+|A_1 - iA_2|\psi_-\rangle$$

$$+\langle\psi_-|A_0 - \frac{\hbar}{2}\,A_3|\psi_-\rangle + \frac{\hbar}{2}\,\langle\psi_-|A_1 + iA_2|\psi_+\rangle. \tag{64}$$

We now give some specific applications of these results.

(1)   Let the operator $A$ of equation (62) be given by

$$A = \hat{\mathbf{n}} \cdot \mathbf{S},$$

where $\hat{\mathbf{n}}$ is an arbitrary unit vector with rectangular components $n_x$,

$n_y$ and $n_z$. The expectation value of the component of the spin angular momentum along the $\hat{\mathbf{n}}$-axis is then, according to equation (64),

$$\langle \psi | \mathbf{n} \cdot \mathbf{S} | \psi \rangle = n_z \frac{\hbar}{2} \left[ \langle \psi_+ | \psi_+ \rangle - \langle \psi_- | \psi_- \rangle \right]$$

$$+ \frac{\hbar}{2} (n_x + i n_y) \langle \psi_- | \psi_+ \rangle + \frac{\hbar}{2} (n_x - i n_y) \langle \psi_+ | \psi_- \rangle. \qquad (65)$$

Observe that the last term is the complex conjugate of the second, so that the result is properly real. The special case in which *either* $\psi_+$ or $\psi_-$ is zero, so that the state $\psi$ is a state of spin up or down with respect to the $z$-axis, respectively, gives the expected result

$$\langle \psi_\pm | \mathbf{n} \cdot \mathbf{S} | \psi_\pm \rangle = \pm n_z \, \hbar/2.$$

(2)   As a similar but more important and more complicated example, let

$$A = \mathbf{L} \cdot \mathbf{S},$$

where $\mathbf{L}$ is the orbital angular momentum operator. We then obtain

$$\langle \psi | \mathbf{L} \cdot \mathbf{S} | \psi \rangle = \frac{\hbar}{2} \langle \psi_+ | L_z | \psi_+ \rangle - \frac{\hbar}{2} \langle \psi_- | L_z | \psi_- \rangle$$

$$+ \frac{\hbar}{2} \langle \psi_- | L_+ | \psi_+ \rangle + \frac{\hbar}{2} \langle \psi_+ | L_- | \psi_- \rangle. \qquad (66)$$

If $\psi_+$ is a state of $z$-component of orbital angular momentum $m\hbar$, we see that the last two terms give a contribution only if $\psi_-$ contains states with $z$-component of angular momentum $(m + 1)\hbar$.

(3)   As a final example, consider a state of the special, but important, form

$$\psi(r, \text{spin}) = \phi(\mathbf{r})(\alpha \chi_+ + \beta \chi_-), \qquad (67)$$

$$|\alpha|^2 + |\beta|^2 = 1, \qquad \langle \phi | \phi \rangle = 1.$$

We then obtain, for the general operator $A$,

$$\langle \psi | A | \psi \rangle = \left\langle \phi \left| \left\{ A_0 + (\alpha \beta^* + \alpha^* \beta) \frac{\hbar}{2} A_1 + i(\alpha \beta^* - \alpha^* \beta) \frac{\hbar}{2} A_2 \right. \right. \right.$$

$$\left. \left. \left. + (|\alpha|^2 - |\beta|^2) \frac{\hbar}{2} A_3 \right\} \right| \phi \right\rangle. \qquad (68)$$

The special operator of the first example, $A = \hat{\mathbf{n}} \cdot \mathbf{S}$, gives the simple result

$$\langle \psi | \hat{\mathbf{n}} \cdot \mathbf{S} | \psi \rangle = \{ n_x (\alpha \beta^* + \alpha^* \beta) + i n_y (\alpha \beta^* - \alpha^* \beta)$$

$$+ n_z (|\alpha|^2 - |\beta|^2) \} \frac{\hbar}{2}. \qquad (69)$$

Further applications are left to the problems.

Although it is clearly unnecessary to do so, as the above examples demonstrate, it is frequently helpful to give an *explicit* realization of the spin operators. This is readily done using a matrix representation, noting that, because there are just two spin states, only two-by-two matrices are required. From the definition of the matrix elements of an operator, we thus write, for the $i$th component of $\mathbf{S}$,

$$S_i = \begin{pmatrix} \langle \chi_+ | S_i | \chi_+ \rangle & \langle \chi_+ | S_i | \chi_- \rangle \\ \langle \chi_- | S_i | \chi_+ \rangle & \langle \chi_- | S_i | \chi_- \rangle \end{pmatrix}.$$

The matrix elements are easily calculated using equations (52) and (57). For example,

$$\langle \chi_+ | S_x | \chi_+ \rangle = 0 = \langle \chi_- | S_x | \chi_- \rangle$$

$$\langle \chi_+ | S_x | \chi_- \rangle = \frac{\hbar}{2} = \langle \chi_- | S_x | \chi_+ \rangle,$$

and similarly for $S_y$ and $S_z$; whence we find

$$S_x = \frac{\hbar}{2} \begin{pmatrix} 0 & 1 \\ 1 & 0 \end{pmatrix}$$

$$S_y = \frac{\hbar}{2} \begin{pmatrix} 0 & -i \\ i & 0 \end{pmatrix} \tag{70}$$

$$S_z = \frac{\hbar}{2} \begin{pmatrix} 1 & 0 \\ 0 & -1 \end{pmatrix}$$

Using the laws of matrix multiplication, it is not hard to verify that the components of $\mathbf{S}$ satisfy the various relations derived earlier.

---

**Exercise 7.** Use the laws of matrix multiplication to show that the $S_x$, $S_y$ and $S_z$ defined by equation (70) satisfy equations (54), (59) and (60).

---

Although we have worked with $\mathbf{S}$ itself throughout, it is customary and convenient to eliminate the all-pervasive factors of $\hbar/2$ which appear in our analysis. We thus introduce the dimensionless *Pauli operator* $\boldsymbol{\sigma} = \sigma_x \hat{\mathbf{e}}_x + \sigma_y \hat{\mathbf{e}}_y + \sigma_z \hat{\mathbf{e}}_z$ by writing

$$\mathbf{S} = \frac{\hbar}{2} \boldsymbol{\sigma}. \tag{71}$$

Taking over our previous results, we have at once

$$\sigma^2 = 3$$

$$\sigma_x{}^2 = \sigma_y{}^2 = \sigma_z{}^2 = 1$$

$$\sigma \times \sigma = 2i\sigma$$

$$(\sigma_i, \sigma_j)_+ = 2\delta_{ij} \tag{72}$$

$$\sigma_i \sigma_j = i\sigma_k$$

$$\sigma_i \sigma_j \sigma_k = i,$$

while the matrix representation of $\sigma$ is

$$\sigma_x = \begin{pmatrix} 0 & 1 \\ 1 & 0 \end{pmatrix} \qquad \sigma_y = \begin{pmatrix} 0 & -i \\ i & 0 \end{pmatrix} \qquad \sigma_z = \begin{pmatrix} 1 & 0 \\ 0 & -1 \end{pmatrix}. \tag{73}$$

It is easily seen that $\sigma_x$, $\sigma_y$, $\sigma_z$ and the unit matrix form a complete set of two-by-two matrices in the sense that any arbitrary two-by-two matrix can be expressed in terms of them (why?). This statement is an alternative and more transparent version of our earlier statement that an arbitrary spin dependent operator can always be expressed as a linear function of the spin. (Why are these two statements equivalent?)

This matrix representation for the spin operators suggests a similar representation for the two-component state functions of the theory. Specifically, the spin functions $\chi_+$ and $\chi_-$ can be represented by *column matrices*[6] defined by

$$\chi_+ \equiv \begin{pmatrix} 1 \\ 0 \end{pmatrix}, \qquad \chi_- \equiv \begin{pmatrix} 0 \\ 1 \end{pmatrix}.$$

That these definitions are consistent follows upon verification of the fact that equations (52) and (57) hold, as they must, when regarded as purely matrix equations. The details are left to the exercises.

Next, we introduce the *adjoints* of these matrices, these being the *row matrices*

$$\chi_+{}^\dagger = (1 \quad 0), \qquad \chi_-{}^\dagger = (0 \quad 1),$$

in agreement with the general definition of the adjoint of a matrix, equation (VII–49). Observing that, according to the usual rules of matrix multiplication,

$$\chi_+{}^\dagger \chi_+ = \chi_-{}^\dagger \chi_- = 1$$

and

$$\chi_+{}^\dagger \chi_- = \chi_-{}^\dagger \chi_+ = 0,$$

---

[6] Such column matrices are also often called *column vectors*, or simply vectors.

we see that these relations are precisely equivalent to the Dirac bracket expressions defined in equation (51). Specifically, upon comparison, we have

$$\langle \chi_\pm | \chi_\pm \rangle \equiv \chi_\pm^\dagger \chi_\pm.$$

We have thus provided a picturization of Dirac brackets for spin states, lacking in our earlier definition, which is rather useful and convenient even though it contains no new information.

A general column matrix is now defined as an arbitrary linear combination of $\chi_+$ and $\chi_-$ and hence, for example, the general spin dependent state of equation (48) is expressed in matrix language as

$$\psi(r, \text{spin}) = \psi_+(\mathbf{r})\chi_+ + \psi_-(r)\chi_- = \begin{pmatrix} \psi_+ \\ \psi_- \end{pmatrix}.$$

Note, however, that Dirac bracket expressions involving *both* space and spin functions are still to be assigned the meaning of equation (50).[7]

---

**Exercise 8.** Verify that equations (52) and (57) hold when regarded as purely matrix equations.

---

Some final remarks are in order. We have introduced the spin in a purely *ad hoc* way as an empirical necessity, more or less as was done historically. The two-component theory was originated on just this empirical basis by Pauli. We want to mention, however, that all of these features were derived, without any *ad hoc* assumptions at all, by Dirac in 1930. Starting with a completely structureless electron, Dirac constructed a relativistic version of Schrödinger's equation which yielded the spin properties of the electron as one of its consequences. It also predicted the existence of the positron. We shall briefly discuss the Dirac equation in the next chapter to see how this comes about.

We have given a relatively complete discussion of spin one-half, and we know how to handle spinless particles, but what about other spins, for example unity? Since three orientations exist for unit spin, the state function describing a spin one particle must have three components. The algebra, although still straightforward, becomes much more complicated

---

[7] We remark that these ideas can equally well be applied to the representation of conventional state functions, $\psi(r)$. Consider $\psi$ to be expressed as a superposition of some complete set of orthonormal basis functions $\phi_m$. The coefficients in this superposition, say $c_m$, completely define $\psi$, which can thus be represented as an infinite-dimensional column matrix with $c_m$ as the $m$th element. Expressing operators as matrices in the same basis, any relations which hold in the conventional description also hold when regarded as purely matrix statements.

as a consequence, and we shall not attempt to develop it.

## 5. ADDITION OF ANGULAR MOMENTUM

Consider some isolated many-particle system. Its total angular momentum operator can be expressed as

$$\mathbf{J} = \sum_i \left( \mathbf{L}_i + \mathbf{S}_i \right), \tag{74}$$

where $\mathbf{L}_i$ is the orbital angular momentum and $\mathbf{S}_i$ the spin angular momentum, if any, of the $i$th particle. Since the angular momentum of an isolated system is conserved, the states of such a system can *always* be written as simultaneous eigenfunctions of $J^2$ and $J_z$, with eigenvalues $j(j+1)\hbar^2$ and $m\hbar$. However, it frequently happens that the system can be decomposed into subsystems which do not interact with each other to some approximation. To this approximation, we can discuss the system in terms of the angular momentum of each of these parts. Such ideas are quite familiar in classical physics. Thus the angular momentum of the solar system can be regarded as a composition of a number of quite distinct elements – the orbital angular momentum of each planet as it moves about the sun, the angular momenta of the various planetary moons as each moves about its planet and, finally, the angular momenta of all of these objects, and the sun, arising from the spinning motion of each about its axis. To first approximation all of these are uncoupled and separately conserved, and this provides an adequate description of the short-term behavior of the solar system. The long-term behavior requires a more precise treatment in which mutual interactions between the various angular momenta are taken into account. The individual angular momenta are no longer separately conserved, but only the total for the entire system.

We now seek a quantum mechanical description of the total angular momentum of a system as a composition of the angular momenta of some assembly of subsystems. We assume these to be sufficiently weakly inter acting that the effects of interactions can be handled by the methods of perturbation theory. We thus seek an appropriate set of unperturbed states, states of definite angular momentum for each subsystem and of definite total angular momentum for the whole. The process of combining angular momenta is completely trivial classically; one simply takes the vector sum in the usual way. Quantum mechanically, however, even this process is complicated, because none of the angular momentum vectors are precisely oriented. In effect, we must add together vectors, lying on cones, such as those illustrated in Figures 2 and 3, of varying angular openings, altitudes and orientations, to form a resultant which also lies on such a cone. We shall not attempt to give a complete answer, but will content ourselves with merely *enumerating* those states of total angular momentum which are actually achievable as a composition of states of definite angular momentum. Except for one or two special cases, we shall not explicitly construct these composite states.[8]

The question of enumerating the achievable composite angular momentum states, so simple and obvious it never enters one's head for classical systems, is not quite trivial quantum mechanically. As we shall shortly see, the answer, which is called the *vector addition theorem for angular momenta,* is the following: When a system of angular momentum $j_1$ is combined with a system of angular momentum $j_2$, the resulting total angular momentum $j$ has as its maximum possible value $j_1 + j_2$ and as its minimum $|j_1 - j_2|$. The other achievable values of $j$ lie at *integral* steps between these two extremes. The complete set of possibilities is thus $(j_1 + j_2)$, $(j_1 + j_2 - 1)$, $(j_1 + j_2 - 2)$, ..., $|j_1 - j_2|$.[9] Further, for each achievable value of $j$, the composite state with definite $z$-component $m$ is unique. However, these unique states are such that, in general, the $z$-components of $j_1$ and $j_2$ do *not* separately have definite values. This aspect is a direct manifestation of the quantum mechanical uncertainty in the orientation of angular momentum vectors, and it is precisely here that the complication of actually constructing the states arises.

To verify these assertions, let us now consider a composite system consisting of two non-interacting subsystems. Denote the angular momentum operator of the first by $\mathbf{J}_1$ and its angular momentum states by $\phi_{j_1 m_1}$. Similarly, let $\mathbf{J}_2$ and $\chi_{j_2 m_2}$ denote the same quantities for the second. According to these definitions we then have

$$J_1^2 \phi_{j_1 m_1} = j_1(j_1 + 1)\, \hbar^2 \phi_{j_1 m_1}$$
$$J_{1z} \phi_{j_1 m_1} = m_1 \hbar\, \phi_{j_1 m_1} \tag{75}$$

and

$$J_2^2 \chi_{j_2 m_2} = j_2(j_2 + 1)\, \hbar^2 \chi_{j_2 m_2}$$
$$J_{2z} \chi_{j_2 m_2} = m_2 \hbar\, \chi_{j_2 m_2}. \tag{76}$$

The states of the complete system can, of course, be expressed in terms of products of these functions. We now seek composite states which are simultaneous eigenfunctions of $J^2 \equiv (\mathbf{J}_1 + \mathbf{J}_2)^2$ and $J_z$, and

[8] For a complete discussion see Reference [22]. For a briefer treatment see References [24] and [25].

[9] In short, we have

$$(j_1 + j_2) \geqslant j \geqslant |j_1 - j_2|,$$

in analogy with the familiar *triangle rule* for classical vectors $\mathbf{A}$ and $\mathbf{B}$,

$$A + B \geqslant |\mathbf{A} + \mathbf{B}| \geqslant |A - B|.$$

also of $J_1{}^2$ and $J_2{}^2$, that is, we seek states in which the angular momenta $j_1$ and $j_2$ of the two subsystems add together to give a state of total angular momentum $j$ with $z$-component $m$. Denoting such a composite state function by $\psi_{jmj_1j_2}$, we have

$$\psi_{jmj_1j_2} = \sum_{m_1, m_2} C_{jmm_1m_2} \, \phi_{j_1m_1} \, \chi_{j_2m_2}, \tag{77}$$

since this is the most general superposition of product functions which is a simultaneous eigenfunction of $J_1{}^2$ and of $J_2{}^2$. The $C_{jmm_1m_2}$, called Clebsch–Gordan coefficients, can be determined from the requirement that $\psi_{jmj_1j_2}$ also be a simultaneous eigenfunction of $J^2$ and $J_z$. In order to verify the vector addition theorem, we now seek to enumerate the permissible values of $j$ and $m$.

With respect to $m$, the answer is immediate since $J_z = J_{1z} + J_{2z}$, whence we see by operating with $J_z$ on equation (77) that we must have

$$m = m_1 + m_2. \tag{78}$$

This means that the double sum of equation (77) reduces at once to a single sum. It also tells us that the maximum possible value of $m$, which is attained when $m_1$ and $m_2$ take their maximum values of $j_1$ and $j_2$, respectively, is simply $j_1 + j_2$.[10] This in turn tells us that the maximum possible value of $j$ is $j_{\max} = j_1 + j_2$. Consider next a state in which $m$ is $j_1 + j_2 - 1$. This state can be formed in two linearly independent ways, in one of which $m_1$ is $j_1$ and $m_2$ is $j_2 - 1$, while in the other $m_1$ is $j_1 - 1$ and $m_2$ is $j_2$.[11] One combination of these states must belong to the $j_{\max} = j_1 + j_2$ state already identified, but a second (orthogonal) combination also exists and it must therefore be associated with a state in which $j = j_1 + j_2 - 1$. Proceeding next to states with $m = j_1 + j_2 - 2$, we see that now three linearly independent states exist. Two of these must be associated with the total angular momentum states previously identified, while the third tells us that a state exists with $j = j_1 + j_2 - 2$. And so it goes, with $j$ decreasing in integral steps until all of the combinations are exhausted, which occurs when $j$ achieves its minimum value, $j_{\min} = |j_1 - j_2|$.

It is not difficult to verify that all possible states have indeed been

---

[10] Because there is only *one* such term in the superposition of equation (77), this particular state is at once *uniquely* determined to be

$$\psi_{(j_1+j_2),(j_1+j_2),j_1j_2} = \phi_{j_1j_1} \, \chi_{j_2j_2}. \tag{79}$$

It, and its companion state, $m = -(j_1 + j_2)$,

$$\psi_{(j_1+j_2),-(j_1+j_2),j_1j_2} = \phi_{j_1,-j_1} \, \chi_{j_2,-j_2} \tag{80}$$

are the only composite states which can always be *trivially* constructed.

[11] Specifically, these two linearly independent states are $\phi_{j_1,j_1-1} \, \chi_{j_2j_2}$ and $\phi_{j_1,j_1} \, \chi_{j_2,j_2-1}$.

included in this enumeration. The argument is the following. The degeneracy of the first subsystem is $(2j_1 + 1)$, that of the second is $(2j_2 + 1)$, and hence there must be $(2j_1 + 1)(2j_2 + 1)$ linearly independent states in *any* representation. We now also calculate the total number of states in the $(j, m)$ representation. The degeneracy of a state of total angular momentum $j$ is $2j + 1$ and hence, assuming $j_1 \geq j_2$ for definiteness, we must have, if all states are to be accounted for,

$$(2j_1 + 1)(2j_2 + 1) = \sum_{j=j_1-j_2}^{j=j_1+j_2} (2j + 1).$$

The sum can be evaluated by writing it in reverse order, adding it to the original and dividing by two. The first term of each sum taken together is $2(2j_1 + 1)$, and so is the sum of every corresponding pair of terms. There are $(2j_2 + 1)$ terms in all and hence the desired result follows.

We have thus verified the previously stated rules, according to which angular momentum is to be added together. As claimed, these rules are equivalent to the rules for the addition of ordinary vectors, but supplemented by the usual quantum conditions for angular momentum states. The same rules can also be applied to the addition of more than two angular momenta. To do so, add any two together, then add the result to a third, and so on.

There are many important examples involving the addition of angular momentum. In one class of these the Hamiltonian is approximately independent of spin, so that the total orbital angular momentum of all the particles and their total spin angular momentum form two non-interacting systems. The first of these is described in terms of states of definite **L**, the second by states of definite **S**. These are then coupled together by the weak spin dependent forces to form states of definite total angular momentum. This scheme is called either $L$–$S$ or Russell–Saunders coupling and is applicable to the atomic states in the first portion of the periodic table. At the other extreme is the class of problems in which the interaction between individual particles can be neglected, but not the spin dependence of the forces. In that case, each particle is described by a state of definite total angular momentum $j$ and the system as a whole by the sum of these single particle states. This scheme is called the $j$–$j$ coupling scheme and it is applicable to atomic states in the latter portion of the periodic table and to nuclear states in the shell model approximation.

We now work out an example of the addition of angular momenta. Specifically, we shall obtain the states of definite total spin for two spin one-half particles. This is the simplest possible example, sufficiently simple that all of the details can be presented. Let $\mathbf{S}_1$ denote the spin operator for the first particle and $\mathbf{S}_2$ that for the second. Let $\chi_{1\pm}$ and

$\chi_{2\pm}$ denote the corresponding spin states. According to equation (77), the most general spin state of the two-particle system, $\psi_{sm\frac{1}{2}\frac{1}{2}}$, can be written as the superposition of $(2 \cdot \frac{1}{2} + 1)(2 \cdot \frac{1}{2} + 1) = 4$ composite spin-states

$$\psi_{sm\frac{1}{2}\frac{1}{2}} = C_{sm++}\chi_{1+}\chi_{2+} + C_{sm+-}\chi_{1+}\chi_{2-} + C_{sm-+}\chi_{1-}\chi_{2+} + C_{sm--}\chi_{1-}\chi_{2-}$$

$$\equiv \chi_{sm}. \tag{81}$$

Because we are here talking about pure spin states, we have denoted the spin angular momentum eigenvalues by $s$ rather than by $j$ and we have abbreviated the composite spin states by $\chi_{sm}$. Also, as a notational convenience, the subscripts $m_1$ and $m_2$ in the summand of equation (77) have been replaced by plus and minus signs. Finally, the superposition sum has been written out explicitly since it contains only four terms.

We now seek those four *particular* linear combinations which are simultaneous eigenstates of the total spin and of its $z$-component. The total spin operator is, of course,

$$\mathbf{S} = \mathbf{S}_1 + \mathbf{S}_2 = \frac{\hbar}{2}(\sigma_1 + \sigma_2) \equiv \frac{\hbar}{2}\sigma, \tag{82}$$

and the states we seek are thus those for which

$$S^2\chi_{sm} = s(s+1)\hbar^2\chi_{sm}$$

$$S_z\chi_{sm} = m\hbar\chi_{sm} \tag{83}$$

or, in terms of the more convenient Pauli operators,

$$\sigma^2\chi_{sm} = 4s(s+1)\chi_{sm}$$

$$\sigma_z\chi_{sm} = 2m\chi_{sm}. \tag{84}$$

These states are readily identified with the aid of the vector addition theorem. According to that theorem, two spin one-half particles can combine only in such a way as to form a system of total spin unity or total spin zero. Now the states of maximum possible angular momentum and maximum $|m|$ are always trivially constructed, being given by equations (79) and (80). For the present case these are the states $s = 1$, $m = \pm 1$, and we thus have at once for these states,

$$\chi_{11} = \chi_{1+}\chi_{2+}$$

$$\chi_{1,-1} = \chi_{1-}\chi_{2-}. \tag{85}$$

The remaining two states of the system both have $m = 0$, and each is thus some linear combination of $\chi_{1+}\chi_{2-}$ and $\chi_{1-}\chi_{2+}$. What linear combination of these two states corresponds to $s = 1$, $m = 0$, the missing member of the threefold set of $s = 1$ substates? Observing that each of the $s = 1$ states already identified in equation (85) is symmetrical with respect to the interchange of the spins of particles one and two, it follows that the state we seek must also be symmetrical. Hence, properly normalized, it is

$$\chi_{10} = \frac{1}{\sqrt{2}} \left( \chi_{1+}\chi_{2-} + \chi_{1-}\chi_{2+} \right). \tag{86}$$

Evidently the remaining state of the system, that with both $j$ and $m = 0$, must be a similar linear combination, but orthogonal to that of equation (86), and hence is the *antisymmetric* state,

$$\chi_{00} = \frac{1}{\sqrt{2}} \left( \chi_{1+}\chi_{2-} - \chi_{1-}\chi_{2+} \right). \tag{87}$$

---

**Exercise 9.** Verify the assertion that the $j = 1, m = 0$ state must be symmetrical because the $j = 1$, $m = \pm 1$ states are. Do this by considering a rotation of the axis of quantization and showing that the relevant rotation operator is symmetric in the spins of the two particles.

---

That the states we have found are indeed simultaneous eigenfunctions of $S^2$ and $S_z$ (or equivalently of $\sigma^2$ and $\sigma_z$) is not hard to verify. Specifically, according to equation (84), we must show that, for the $s = 1$ states,

$$\left. \begin{array}{l} \sigma^2 \chi_{1m} = 8\chi_{1m} \\ \sigma_z \chi_{1m} = 2m\chi_{1m} \end{array} \right\} \quad m = 1, 0, -1 \tag{88}$$

and, for the $s = 0$ state,

$$\sigma^2 \chi_{00} = 0$$
$$\sigma_z \chi_{00} = 0. \tag{89}$$

The equations in $\sigma_z$ are transparently correct, but those in $\sigma^2$ are not entirely trivial. They can be simplified by observing that

$$\sigma^2 \equiv (\sigma_1 + \sigma_2)^2 = \sigma_1^2 + \sigma_2^2 + 2\sigma_1 \cdot \sigma_2$$

or, because $\sigma_1^2 = \sigma_2^2 = 3$,

$$\sigma^2 = 6 + 2\sigma_1 \cdot \sigma_2. \tag{90}$$

Comparison with equations (88) and (89) then shows that $\sigma_1 \cdot \sigma_2$ must yield unity when it operates on a state with $s = 1$ and minus three when it operates on a state of $s = 0$. That this is actually the case now follows readily, but we leave the details to the problems.

To recapitulate, with the help of the vector addition theorem we have explicitly constructed the so-called *triplet* spin state, with $s = 1$ and $m = 1, 0, -1$, and the *singlet* spin state with $s = m = 0$. The former was found to be symmetric under exchange of spins, the latter anti-symmetric.[12] The normalized state functions and their properties are summarized in Table I.

| Triplet | Singlet |
|---|---|
| $s = 1$ | $s = 0$ |
| $\sigma^2 = 8, \sigma_1 \cdot \sigma_2 = 1$ | $\sigma^2 = 0, \sigma_1 \cdot \sigma_2 = -3$ |
| $m = 1 : \chi_{1,1} = \chi_{1+}\chi_{2+}$ | |
| $m = 0 : \chi_{1,0} = \dfrac{1}{\sqrt{2}} (\chi_{1+}\chi_{2-} + \chi_{1-}\chi_{2+})$ | $m = 0 : \chi_{0,0}$ |
| | $\quad = \dfrac{1}{\sqrt{2}} (\chi_{1+}\chi_{2-} - \chi_{1-}\chi_{2+})$ |
| $m = -1 : \chi_{1,-1} = \chi_{1-}\chi_{2-}$ | |
| $\chi_{1,m}$ is *symmetric* under exchange | $\chi_{0,0}$ is *antisymmetric* under exchange |

TABLE I.   Normalized spin states of two spin one-half particles.

As a second and very important example, we consider the addition of spin and orbital angular momentum for the case of spin one-half. Leaving the details to Exercise 10, below, we merely state the results. For a given orbital angular momentum $l \neq 0$, there are two states of total angular momentum $j = l \pm \frac{1}{2}$, in agreement with the vector addition theorem. These states, which we denote by $\psi_{jm_jl}$ are defined by

$$J^2\psi_{jm_jl} = \hbar^2 j(j + 1)\psi_{jm_jl}$$

$$J_z\psi_{jm_jl} = \hbar m_j\psi_{jm_jl} \qquad\qquad (89)$$

$$L^2\psi_{jm_jl} = \hbar^2 l(l + 1)\psi_{jm_jl}.$$

They are given in terms of the eigenfunctions $\psi_{lm}$ of $L^2$ and $L_z$ and of the spin states $\chi_\pm$ by

$$j = l + \frac{1}{2}, \; m_j = m + \frac{1}{2}$$

$$\psi_{jm_jl} = \frac{1}{\sqrt{2l+1}} \left[ \sqrt{l+m+1}\; \psi_{lm}\chi_+ + \sqrt{l-m}\; \psi_{l,m+1}\chi_- \right] \qquad (90)$$

---

[12] This verifies our assertion in Chapter VIII that the totally antisymmetric states of two identical particles can be classified as either the product of an antisymmetric (singlet) spin state and a symmetric space state or as the product of a symmetric (triplet) spin state and an antisymmetric space state.

and by

$$j = l - \frac{1}{2}, \ m_j = m + \frac{1}{2}$$

$$\psi_{jm_jl} = \frac{1}{\sqrt{2l+1}} \left[ \sqrt{l-m} \ \psi_{lm}\chi_+ + \sqrt{l+m+1} \ \psi_{l,m+1}\chi_- \right]. \quad (91)$$

In equation (90), $m$ takes on all integral values between $-(l+1)$ and $l$, and in equation (91) all integral values between $-l$ and $l-1$.

The importance of this particular example is a consequence of the existence of the so-called *spin-orbit force*. For an electron moving in a central potential $V(r)$, this interaction is represented in the Hamiltonian by the term

$$H_{\text{spin-orbit}} = \frac{1}{2m^2c^2} \frac{1}{r} \frac{dV}{dr} \mathbf{L} \cdot \mathbf{S}, \quad (92)$$

where $L$ is the orbital angular momentum operator of the electron and $S$ is its spin operator. This term, which is relativistic in origin, arises as a consequence of the fact that the magnetic field produced by the motion of a charged particle interacts with its spin magnetic moment. A Hamiltonian containing such a term commutes with $J^2$, $J_z$ and $L^2$ but not with $L_z$. Hence its angular momentum states are just those of equations (90) and (91). This connection can be made quite explicit by observing that, since

$$J^2 = (\mathbf{L} + \mathbf{S})^2 = L^2 + 2\mathbf{L} \cdot \mathbf{S} + S^2,$$

we have for a spin one-half particle

$$\mathbf{L} \cdot \mathbf{S} = \frac{1}{2} \left( J^2 - L^2 - \frac{3}{4}\hbar^2 \right),$$

and the simultaneous eigenfunctions of $J^2$ and $L^2$ are therefore also eigenfunctions of $\mathbf{L} \cdot \mathbf{S}$. Specifically

$$(\mathbf{L} \cdot \mathbf{S}) \ \psi_{jm_jl} = \frac{1}{2} \left[ j(j+1) - l(l+1) - \frac{3}{4} \right] \hbar^2 \psi_{jm_jl}.$$

More specifically still, for $j = l + \frac{1}{2}$, $\mathbf{L} \cdot \mathbf{S}$ has the eigenvalue $l\hbar^2/2$, and for $j = l - \frac{1}{2}$ it has the eigenvalue $-(l+1)\hbar^2/2$.

---

**Exercise 10.** The $\psi_{jm_jl}$ of equations (90) and (91) are obviously eigenfunctions of $L^2$ and $J_z$, as claimed. Use equation (66) to verify that they are simultaneously eigenfunctions of $\mathbf{L} \cdot \mathbf{S}$, and therefore of $J^2$, also as claimed.

---

Consider now some system, such as the hydrogenic atom, described by the Hamiltonian

$$H = \frac{p^2}{2m} + V(r) + H_{\text{spin-orbit}}.$$

This Hamiltonian is separable into a product of radial and angular functions, as the substitution

$$\psi(\mathbf{r}, \text{spin}) = R_{jl} \, \psi_{jm_jl}$$

makes clear. Using the results obtained above for the eigenvalues of $\mathbf{L} \cdot \mathbf{S}$, the radial wave function $R_{jl}$ is seen to satisfy the equation

$$-\frac{\hbar^2}{2mr^2} \frac{d}{dr}\left(r^2 \frac{dR_{jl}}{dr}\right) + \left[V(r) + \frac{l(l+1)\hbar^2}{2mr^2} + \left\{\begin{array}{c} l \\ -l-1 \end{array}\right\} \frac{\hbar^2}{4m^2c^2 r} \frac{dV}{dr}\right] R_{jl}$$

$$= E_{jl} \, R_{jl}, \tag{93}$$

where the upper line in the braces refers to the state $j = l + \frac{1}{2}$, the lower to the state $j = l - \frac{1}{2}$. Of course the spin-orbit term is absent for $s$-states ($l = 0$), so that equation (93), which is exact, applies only for $l \neq 0$. The perturbation produced by this term is responsible for the *fine structure* of atomic states, while in the nuclear domain the existence of a strong spin-orbit interaction is an essential ingredient in the explanation of the observed shell structure of nuclei. In the atomic case, where this term is usually small, its contribution to the energy is given, to adequate approximation, by first-order perturbation theory. A one-electron state of given $l$ is split into a doublet, the energy shifts being

$$j = l + \frac{1}{2}: \quad \Delta E = \langle \phi_{El} | H_{\text{spin-orbit}} | \phi_{El} \rangle = \frac{\hbar^2}{4m^2c^2} \left\langle \frac{1}{r} \frac{dV}{dr} \right\rangle l$$

$$j = l - \frac{1}{2}: \quad \Delta E = -\frac{\hbar^2}{4mc^2} \left\langle \frac{1}{r} \frac{dV}{dr} \right\rangle (l+1).$$

Thus the energy separation of the two states is simply

$$\frac{\hbar^2}{4mc^2} \left\langle \frac{1}{r} \frac{dV}{dr} \right\rangle (2l+1).$$

The separation of the famous sodium D-lines is an example of the splitting produced by the spin-orbit interaction.

---

**Problem 1.**

(a) Compute the ground state energy of the hydrogen atom, assuming half-integral angular momentum. Compare the ionization energy obtained with the experimental value.

(b) What are the ground state eigenfunctions?

**Problem 2.** Find the ground state energy and eigenfunction for the three-dimensional isotropic harmonic oscillator, assuming half-integral orbital angular momentum.

**Problem 3.** Work out the commutation relations of $L_x$, $L_y$, $L_z$, $L^2$ with $p_x$, $p_y$, $p_z$, $p^2$, and with $x$, $y$, $z$, $r^2$.

**Problem 4.** Consider the motion of a particle in central potential $V(r)$. Let $\psi_{Elm}(r)$ be an eigenfunction of the Hamiltonian corresponding to total angular momentum $l$ and $z$-component $m$ in units of $\hbar$. Show that

$$\psi' \equiv e^{i\beta \hat{n} \cdot \mathbf{L}/\hbar} \, \psi_{Elm}(\mathbf{r})$$

is an eigenfunction of $H$ corresponding to the same energy $E$ and the same total angular momentum $l$, no matter what the value of $\beta$ or the orientation of $\hat{n}$. Is $\psi'$ also an eigenfunction of $L_z$? Explain.

**Problem 5.**
   (a)  Suppose $\hat{n}$ to be an arbitrarily oriented unit vector. Denote by $l$, $m$, $n$ its direction cosines with respect to the $x$, $y$, $z$ axes, respectively. Show that

$$\hat{n} \cdot \sigma = \begin{pmatrix} n & l - im \\ l + im & -n \end{pmatrix},$$

where $\sigma$ is the Pauli (vector) operator. Verify that $(\hat{n} \cdot \sigma)^2 = 1$, no matter what the orientation of $\hat{n}$.
   (b)  The most general spin one-half state is the superposition

$$\chi = a_+ \chi_+ + a_- \chi_- \qquad |a_+|^2 + |a_-|^2 = 1,$$

where $\chi_\pm$ are the eigenfunctions of $\sigma_z$ with eigenvalues $\pm 1$. Use the result of part (a) to find the values of $a_+$ and $a_-$ if $\chi$ is to be an eigenfunction of $\hat{n} \cdot \sigma$. [Hint: The eigenvalues of $(\hat{n} \cdot \sigma)$ are $\pm 1$. Why?]
   (c)  The result of part (b) gives the spin states referred to an arbitrary axis rather than to the $z$-axis. Suppose an electron has its spin oriented along the positive $x$-axis. What is its spin function? What is the probability that a measurement of its $z$-component of spin will yield the value $+ 1/2$?

**Problem 6.**
   (a)  Noting that $(L_z, \phi) = \hbar/i$ and that, because $0 \leqslant \phi \leqslant 2\pi$, $(\Delta\phi)^2$ is necessarily finite, for any state, how is it possible to have states of definite $L_z = m\hbar$ without violating the uncertainty principle, equation (V–49)? [Hint: Equation (V–49) holds for Hermitian operators only. For what class of functions $u(\phi)$ is $L_z$ Hermitian?]
   (b)  Consider the angular wave packet

$$u(\phi) = e^{im_0\phi} \sum_{s=-\infty}^{\infty} \exp\left[-(\phi - \phi_0 + 2s\pi)^2/2\gamma^2\right].$$

Show that for any *periodic* function $f(\phi)$

$$\int_0^{2\pi} f(\phi)u(\phi)\, d\phi = \int_{-\infty}^{\infty} f(\phi)\, e^{im_0\phi - (\phi - \phi_0)^2/2\gamma^2}\, d\phi.$$

(c) What is the probability $p_m$ that a measurement of $L_z$ for the wave packet $u(\phi)$ will yield $m\hbar$?

(d) By plotting $|u(\phi)|^2$ against $\phi$ and $p_m$ against $m$, discuss the mutual uncertainties in $\phi$ and in $L_z$.

**Problem 7.** Let $T_a$ denote the translation operator, $R_{\hat{n}}(\beta)$ the rotation operator, $P$ the parity operator, and $P_{ij}$ the exchange operator. Which, if any, of the following pairs commute:

(i) $T_a, T_b$; (ii) $R_{\hat{n}}(\beta), R_{\hat{n}}(\gamma)$; (iii) $R_{\hat{n}}(\beta), R_{\hat{n}'}(\beta)$;
(iv) $P, T_a$; (v) $P, P_{ij}$.

What must be the relation between $\hat{n}$ and $a$ if $T_a$ and $R_{\hat{n}}(\beta)$ commute?

**Problem 8.** Consider a system of two spinless identical particles. Show that the orbital angular momentum of their relative motion can only be even $(l = 0, 2, 4, \ldots)$.

**Problem 9.** Show by direct calculation that for the triplet spin states of two spin one-half particles

$$\sigma_1 \cdot \sigma_2\, \chi_{1m} = \chi_{1m}; \qquad m = 1, 0, -1,$$

while, for the singlet state,

$$\sigma_1 \cdot \sigma_2\, \chi_{00} = -3\chi_{00}.$$

**Problem 10.** Find $J^2\phi_{j_1m_1}\chi_{j_2m_2}$ where $\phi_{j_1m_1}$ and $\chi_{j_2m_2}$ are given by equations (75) and (76) and where $\mathbf{J} = \mathbf{J}_1 + \mathbf{J}_2$. Hint:

$$J^2 = J_1^2 + J_2^2 + 2\mathbf{J}_1 \cdot \mathbf{J}_2$$

$$\mathbf{J}_1 \cdot \mathbf{J}_2 = \tfrac{1}{2}(J_{1+}J_{2-} + J_{1-}J_{2+}) + J_{1z}J_{2z}.$$

**Problem 11.** The state function of an electron is given by

$$\psi = R(r)\left\{\sqrt{\tfrac{1}{3}}\, Y_1^0(\theta, \phi)\chi_+ + \sqrt{\tfrac{2}{3}}\, Y_1^1(\theta, \phi)\chi_-\right\}.$$

(a) Show directly that the $z$-component of the electron's total angular momentum is $1/2$ and that the electron has orbital angular momentum unity.

(b) What is the probability density for finding the electron with spin up at $r, \theta, \phi$? With spin down?

(c)  Show that the probability density for finding the electron at $r$, $\theta$, $\phi$, no matter what its spin, is spherically symmetric, that is, independent of $\theta$ and $\phi$.

**Problem 12.**  Write the most general configuration space state function consistent with the stated conditions for:

(a)  A particle in one dimension with definite linear momentum $p$.

(b)  A particle in one dimension with linear momentum $p$ of unspecified sign.

(c)  A particle in three dimensions with definite linear momentum vector **p**.

(d)  A particle in three dimensions with linear momentum of magnitude $p$ but unspecified direction.

(e)  A particle of definite angular momentum $l$ and $z$-component $m$.

(f)  A particle of definite angular momentum $l$ but with unspecified $z$-component.

(g)  A particle with unspecified total angular momentum but with definite $z$-component $m$.

**Problem 13.**  Write down the constants of the motion for each of the following cases (consider only the dynamical variables: energy, the components of linear momentum, the components of angular momentum, the square of the angular momentum, and the parity):

(a)  A free particle.

(b)  A particle in a central potential.

(c)  A particle in a cubical container.

(d)  A particle in a spherical container.

(e)  A particle in a cylindrical container with axis oriented along the $z$-axis.

(f)  A particle in a container of irregular shape.

(g)  A charged particle in a uniform electric field in the $z$-direction.

(h)  A charged particle in a time-varying but spatially uniform electric field in the $z$-direction.

**Problem 14.**

(a)  Show that an operator $A$ which commutes with $L_x$ and $L_y$ must also commute with $L^2$.

(b)  Suppose instead that $A$ commuted with $L_x^2$ and $L_y^2$. Could one draw similar conclusions about its commutator with $L^2$?

**Problem 15.**

(a)  Let $A$ be a spin-*independent* vector operator. Prove that

$$(\sigma \cdot \mathbf{A})^2 = A^2 + i\sigma \cdot (\mathbf{A} \times \mathbf{A}).$$

(b)  With $\hat{\mathbf{n}}$ an arbitrary unit vector and $\phi(\mathbf{r})$ an arbitrary spin-

*independent* function of position, prove that

$$e^{i\sigma \cdot \hat{n}\phi} = \cos \phi + i\sigma \cdot \hat{n} \sin \phi.$$

**Problem 16.** The most general spin one-half dependent operator being linear in the spin, reduce the following to linear functions:

(a)  $(1 + \sigma_x)^{1/2}$   (b)  $(1 + \sigma_x + i\sigma_y)^{1/2}$

(c)  $(\sigma_x + \sigma_y)^n$   (d)  $(1 + \sigma_x)^n$

(e)  $(\alpha\sigma_x + \beta\sigma_y)^n$

**Problem 17.** Taking the inverse of an operator $A$ to be defined by

$$A^{-1}A = AA^{-1} = 1,$$

reduce the following to linear form:

(a)  $\sigma_x^{-1}$   (b)  $(2 + \sigma_x)^{-1}$

(c)  $(1 + \sigma_x + i\sigma_y)^{-1}$   (d)  $(2 + \sigma_x)^{-1} (2 + \sigma_y)$

(e)  $(2 + \sigma_y)(2 + \sigma_x)^{-1}$   (f)  Does $(1 + \sigma_x)$ have an inverse?

**Problem 18.**

(a)   Show that, for arbitrary $f(\mathbf{r})$,

$$e^{i\mathbf{a} \cdot \mathbf{p}/\hbar} f(\mathbf{r}) = f(\mathbf{r} + \mathbf{a}).$$

(b)   Let $\psi(\mathbf{r}, t)$ be a solution of Schrödinger's equation for a particle moving in a potential $V(r)$. Show that $e^{i\mathbf{a} \cdot \mathbf{p}/\hbar} \psi(\mathbf{r}, t)$ is a solution of Schrödinger's equation for motion in the potential $V(\mathbf{r} + \mathbf{a})$.

**Problem 19.** Consider a transformation from a coordinate system $S$ to a coordinate system $S'$, corresponding to a simple change of origin

$$\mathbf{r}' = \mathbf{r} - \mathbf{a}.$$

*Note that $\mathbf{r}$ and $\mathbf{r}'$ are simply different coordinate labels for the same physical point in space, as shown in Figure 4.* Let the state function in $S$ be denoted by $\psi(\mathbf{r}, t)$ and in $S'$ by $\psi'(\mathbf{r}', t)$. Show that this transformation can be induced by the translation operator $T_a$ of equation (47) according to

$$\psi'(\mathbf{r}', t) = T_a \psi(\mathbf{r}', t)$$

and that, as must be the case (why?),

$$\langle \psi(\mathbf{r}, t) | \mathbf{r} | \psi(\mathbf{r}, t) \rangle = \langle \psi'(\mathbf{r}', t) | \mathbf{r}' + \mathbf{a} | \psi'(\mathbf{r}', t) \rangle.$$

Show, further, that $\mathbf{p}' = \mathbf{p}$, and that, as it must (why?),

$$|\phi'(\mathbf{p}, t)|^2 = |\phi(\mathbf{p}, t)|^2.$$

Is $\phi'(\mathbf{p}, t) = \phi(\mathbf{p}, t)$?

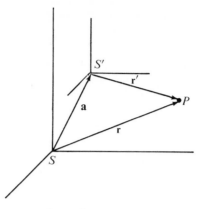

FIGURE 4. Coordinate transformation corresponding to a change of origin. The coordinates of $P$ are $\mathbf{r}$ in $S$ and $\mathbf{r}'$ in $S'$.

**Problem 20.** Consider a transformation in which an isolated physical system is displaced through a constant distance $\mathbf{a}$, the origin of coordinates being held fixed. A portion of the system originally at $\mathbf{r}'$ is thus translated to the point at $\mathbf{r}$ where

$$\mathbf{r} = \mathbf{r}' + \mathbf{a}.$$

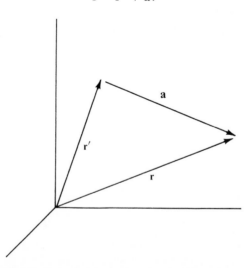

FIGURE 5. Transformation of $\mathbf{r}'$ into $\mathbf{r}$, corresponding to a uniform translation $\mathbf{a}$ of an isolated physical system with respect to a fixed origin.

Note that, as illustrated in Figure 5, and in contrast to the change of origin discussed in Problem 19, each point in space retains its unique labeling under this transformation. Let $\psi'(\mathbf{r}', t)$ denote the state function before this uniform translation is applied and $\psi(\mathbf{r}, t)$ the state function after. Show that all of the conclusions of Problem 19 hold, and thus that it is a matter of indifference whether the physical system is translated relative to the origin or the origin translated (in the opposite direction) relative to the system.

**Problem 21.** Show that the operator which induces a translation $\mathbf{p}_0$ in momentum space is given by

$$T_{\mathbf{p}_0} = e^{-i\mathbf{p}_0 \cdot \mathbf{r}/\hbar}. \tag{94}$$

**Problem 22.** Consider a *Galilean transformation*[13] between a coordinate system $S$ and a coordinate system $S'$ moving relative to each other with uniform velocity $\mathbf{v}$. Classically,

$$\mathbf{r}' = \mathbf{r} - \mathbf{v}t \tag{95a}$$

$$\mathbf{p}' = \mathbf{p} - m\mathbf{v}, \tag{95b}$$

but the *form* of the classical equations of motion is exactly the same when expressed in terms of either coordinates. Neither coordinate system is preferred, and the concept of absolute rest therefore has no meaning in a system governed by Newtonian mechanics. The same must be true quantum mechanically as well (why?). How can one show that this is indeed the case? Let the state function in $S$ be denoted by $\psi(\mathbf{r}, t)$ and in $S'$ by $\psi'(\mathbf{r}', t)$. Then, if $\psi$ is a solution of Schrödinger's equation

$$\left\{ -\frac{\hbar^2}{2m} \nabla_\mathbf{r}^2 + V(\mathbf{r}) \right\} \psi(\mathbf{r}, t) = -\frac{\hbar}{i} \frac{\partial \psi}{\partial t},$$

we must have the following conditions satisfied:

(i)   $\psi'(\mathbf{r}', t)$ must be a solution of Schrödinger's equation,

$$\left\{ -\frac{\hbar^2}{2m} \nabla_{\mathbf{r}'}^2 + V(\mathbf{r}' + \mathbf{v}t) \right\} \psi'(\mathbf{r}', t) = \left( -\frac{\hbar}{i} \frac{\partial}{\partial t} + c \right) \psi'(\mathbf{r}', t), \tag{96}$$

where we have allowed for the possibility of an additive constant in the energy, $c$, which cannot be ruled out because such a term is undetectable and hence has no physically significant consequences.

(ii)   The expectation value of any function of the coordinates must transform, in accord with equation (95a),

$$\langle \psi(\mathbf{r}, t) | f(\mathbf{r}) | \psi(\mathbf{r}, t) \rangle = \langle \psi'(\mathbf{r}', t) | f(\mathbf{r}' + \mathbf{v}t) | \psi'(\mathbf{r}', t) \rangle. \tag{97}$$

(iii)   The expectation value of any function of the momentum must

---

[13] A detailed treatment, from a different point of view than that developed here, is given in Reference [29], pp. 174–177. See also References [21], [22] and [28].

transform in accord with equation (95b),

$$\langle \psi(\mathbf{r}, t) | f(\mathbf{p}) | \psi(\mathbf{r}, t) \rangle = \langle \psi(\mathbf{r}', t) | f(\mathbf{p}' + m\mathbf{v}) | \psi(\mathbf{r}', t) \rangle. \qquad (98)$$

(a)   One might be tempted to assume that, in accord with equation (95a),

$$\psi'(\mathbf{r}', t) = \psi(\mathbf{r}' + \mathbf{v}t, t). \qquad (99)$$

Show, however, that this expression does *not* satisfy either condition (i) or condition (iii), equations (96) and (98).

(b)   To see what is wrong with equation (99), observe that it corresponds to

$$\psi'(\mathbf{r}', t) = e^{i\mathbf{p}\cdot\mathbf{v}t/\hbar} \, \psi(\mathbf{r}', t),$$

which is to say that $\psi'$ is generated only by the space translation part of the Galilean transformation, the accompanying momentum translation being omitted. Referring to equation (94), this suggests the combined transformation

$$\psi'(\mathbf{r}', t) = e^{-im\mathbf{v}\cdot\mathbf{r}'/\hbar} \, e^{i\mathbf{p}\cdot\mathbf{v}\,t/\hbar} \, \psi(\mathbf{r}', t)$$

$$= e^{-im\mathbf{v}\cdot\mathbf{r}'/\hbar} \, \psi(\mathbf{r}' + \mathbf{v}t, t). \qquad (100)$$

Show that this expression does indeed satisfy all three conditions and evaluate the additive constant in the energy.

(c)   Translations in momentum space and in configuration space are not commutative. Show, however, that if the pair of transformations leading to equation (100) had been carried out in the opposite order, the new function obtained would still satisfy all three conditions, only the additive constant in the energy being altered.

(d)   Neither (b) nor (c) is entirely satisfying, because each imposes an arbitrary and unnatural ordering of the two simultaneous and inseparable translations which make up the Galilean transformation of equation (95). To see how to solve this problem, consider the full transformation to be achieved by a sequence of infinitesimal transformations generated by velocity increments $\delta\mathbf{v}$. Show that infinitesimal transformations in momentum and configuration space are commutative, and then, by "integrating" these combined infinitesimal transformations, show that one obtains the desired simultaneous and symmetrical combined transformation

$$\psi'(\mathbf{r}', t) = e^{i\mathbf{v}\cdot(\mathbf{p}t - m\mathbf{r}')/\hbar} \, \psi(\mathbf{r}', t)$$

$$= e^{-i\mathbf{v}\cdot(m\mathbf{r}' - \mathbf{p}t)/\hbar} \, \psi(\mathbf{r}', t). \qquad (101)$$

(e)   To evaluate equation (101) use the following theorem, which we offer without proof:[16] If $A$ and $B$ are two operators, and if their commutator $(A,B)$ is some *number* $\alpha$, then $\exp[A + B] = \exp A \exp B \exp(-\frac{1}{2}\alpha)$. Using this result, show that

$$\psi'(\mathbf{r}', t) = \exp\left[-\frac{i}{\hbar}\left(m\mathbf{v} \cdot \mathbf{r}' + \frac{mv^2}{2}t\right)\right]\psi(\mathbf{r}' + \mathbf{v}t, t),\quad (102)$$

and that the additive constant in the energy now is zero.

(f)   Show that, for a system of particles, the Galilean transformation is just what we have found but with $m$ replaced by the total mass of the system, and with $\mathbf{r}$ and $\mathbf{p}$ replaced by the center-of-mass dynamical variables $\mathbf{R}$ and $\mathbf{P}$.

(g)   For an isolated system, where the center of mass moves as a free particle, examine the specific form assumed by the state function under the various transformations discussed above.

**Problem 23.**   A hydrogen atom in its ground state is moving with velocity $\mathbf{v}$ along the $z$-axis in the laboratory system of coordinates. If the proton in the atom is suddenly brought to rest, say by some kind of collision, what is the probability that the hydrogen atom will remain in its ground state? (For simplicity, take the electron mass to be negligible compared to the proton mass.) Use the results of Problem 22 to obtain the initial state of the system.

---

[16] The proof is given in Reference [29], p. 145.

# XI

## Some applications and further generalizations

### 1. THE HELIUM ATOM; THE PERIODIC TABLE

Studying the states of helium-like atoms provides an excellent example of the application of a variety of quantum mechanical ideas and techniques. By a helium-like atom, we mean a system consisting of two electrons and a nucleus of charge $Ze$ in mutual interaction. For simplicity, we shall neglect the motion of the nucleus, in effect treating it as if it were infinitely massive. At least at first, we shall also neglect all spin-dependent interactions. Our system thus reduces to a pair of electrons moving in the Coulomb potential of the nucleus and interacting electrostatically with each other. Taking the nucleus to lie at the origin of coordinates, the Hamiltonian is thus

$$H = \frac{p_1^2}{2m} + \frac{p_2^2}{2m} - \frac{Ze^2}{r_1} - \frac{Ze^2}{r_2} + \frac{e^2}{|\mathbf{r}_1 - \mathbf{r}_2|}, \tag{1}$$

where $\mathbf{r}_1$ and $\mathbf{r}_2$ are the coordinates of the two electrons, as indicated in Figure 1, while $\mathbf{p}_1$ and $\mathbf{p}_2$ denote their respective momenta. The last term in the expression for $H$, which describes the electron-electron interaction, is responsible for all the complications in the problem. Although

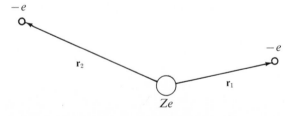

FIGURE 1.   Coordinate system for helium-like atom.

its relative effects clearly decrease steadily with increasing $Z$, this term is not particularly small in general. Nonetheless, we shall begin by treating it as a perturbation. Later we shall use the Rayleigh–Ritz variational method to obtain an improved result.

The unperturbed Hamiltonian, obtained by neglecting the electron-electron interaction, is separable and hence the unperturbed ground state wave function is simply a product of hydrogenic wave functions. Specifically, we have

$$\phi_0(r_1, r_2) = \phi_{100}(r_1)\phi_{100}(r_2), \tag{2}$$

where, with $a_0$ the Bohr radius,

$$\phi_{100}(r) = \frac{1}{\sqrt{\pi}}\left(\frac{Z}{a_0}\right)^{3/2} e^{-Zr/a_0}. \tag{3}$$

The unperturbed ground state energy is then just twice the hydrogenic ground state energy

$$\mathscr{E}_0 = -2 \cdot \frac{Z^2 e^2}{2a_0}, \tag{4}$$

while the first-order correction to the energy, $\Delta E_0$, is given by

$$\Delta E_0 = \left\langle \frac{e^2}{|\mathbf{r}_1 - \mathbf{r}_2|} \right\rangle = e^2 \int \int \frac{\phi_{100}^2(r_1)\phi_{100}^2(r_2)}{|\mathbf{r}_1 - \mathbf{r}_2|}\, d^3\mathbf{r}_1\, d^3\mathbf{r}_2. \tag{5}$$

This last is recognized as precisely the interaction energy of two superposed spherically symmetric charge clouds of charge density

$$\rho = e\, \phi_{100}^2(r),$$

and it is thus easy to give a reasonable estimate of the magnitude of $\Delta E_0$. Each charge cloud contains a total charge $e$ and each extends over a spatial region approximately $a_0/Z$ in radius, according to equation (3). Hence, up to a numerical factor of order unity, their mutual interaction energy, $\Delta E_0$, is simply $e^2/(a_0/Z) = Ze^2/a_0$. The important feature of this result is that it is *linear* in $Z$ and not quadratic, as is the unperturbed energy. This makes explicit the decreasing relative importance of the electron-electron interaction with increasing $Z$.

The actual value of the numerical factor in the expression for $\Delta E_0$ can readily be obtained using elementary electrostatic techniques. A somewhat more general procedure is to make use of the following expansion, familiar from potential theory, which we shall not prove,

$$\frac{1}{|\mathbf{r}_1 - \mathbf{r}_2|} = \begin{cases} \displaystyle\sum_{l=0}^{\infty} \frac{r_2^l}{r_1^{l+1}} P_l(\cos\theta), & r_1 \geqslant r_2 \\[2em] \displaystyle\sum_{l=0}^{\infty} \frac{r_1^l}{r_2^{l+1}} P_l(\cos\theta), & r_2 \geqslant r_1, \end{cases} \tag{6}$$

where $\theta$ is the angle between $\mathbf{r}_1$ and $\mathbf{r}_2$. Only the spherically symmetric term $(l = 0)$ is seen to contribute, in view of the orthogonality of the Legendre polynomials, and the resulting integral is not difficult to evaluate. By either method, the result turns out to be

$$\Delta E_0 = \frac{5}{8} \frac{Ze^2}{a_0},$$

so that the numerical factor by which our rough estimate must be multiplied is just 5/8. The first-order expression for the energy is then

$$E_0 = -\frac{Z^2 e^2}{a_0} + \frac{5}{8} \frac{Ze^2}{a_0}. \tag{7}$$

Both this result and the unperturbed energy are presented in Table I for the elements from helium $(Z = 2)$ to four times ionized carbon $(Z = 6)$. The observed energies are also tabulated. The first-order perturbation theory results are seen to be remarkably good, considering the crudeness of the calculation.

Next we consider a much improved, if still crude, calculation using a variational approach. A basic error in the first-order perturbation theory calculation arises because neither electron actually moves in the field of the bare nucleus, but each is shielded to some degree by the other. We can take this into account in a rough way by choosing as a trial function for each electron a hydrogenic wave function appropriate to a nucleus of charge $Z'e$, say, instead of $Ze$, and then determining $Z'$ variationally. Specifically, we write for the first electron

$$\phi_{\text{trial}} = \frac{1}{\sqrt{\pi}} \left( \frac{Z'}{a_0} \right)^{3/2} e^{-Z'r/a_0} \tag{8}$$

and similarly, for the second. We then find, after some straightforward but tedious calculations, that

$$E_0(Z') = -\frac{e^2}{a_0} \left( 2ZZ' - Z'^2 - \frac{5Z'}{8} \right),$$

which reduces, of course, to equation (7) for $Z' = Z$ (why?). Setting $dE_0/dZ'$ equal to zero, we obtain

$$Z' = Z - 5/16, \tag{9}$$

whence

$$E_0 = -\left( Z - \frac{5}{16} \right)^2 \frac{e^2}{a_0}, \tag{10}$$

which is seen to be lower, and hence more accurate, than the first-order perturbation theory result, equation (7). Note that the result is exactly what would be obtained if each electron moved independently in the field of a nucleus of charge $(Z - 5/16)e$. The variational results are also given in Table I and are seen to be quite close to the observed values.

| Element | Ground state energies $(eV)$ | | | |
|---------|------------------------------|------------|-------------|--------------|
|         | Unperturbed                  | First Order | Variational | Experimental |
| He         | 108.24 | 74.42  | 77.09  | 78.62  |
| Li$^+$     | 243.54 | 192.80 | 195.47 | 197.14 |
| Be$^{++}$  | 432.96 | 365.31 | 367.98 | 369.96 |
| B$^{+++}$  | 676.50 | 591.94 | 594.6  | 596.4  |
| C$^{++++}$ | 974.16 | 872.69 | 875.4  | 876.2  |

TABLE I.   Calculated and observed binding energies of helium-like atoms (after Pauling and Wilson, Reference [20]).

Now we come to the much harder problem of the excited states of helium-like atoms, which we shall discuss only in perturbation theory. From the unperturbed result we see at once that, except for very highly excited states, we need concern ourselves only with cases in which one electron remains in its lowest state (why?). A general unperturbed, symmetrized, not too highly excited state thus has the form

$$\Phi_{nlm}(\mathbf{r}_1, \mathbf{r}_2) = \frac{1}{\sqrt{2}} \left[ \phi_{100}(r_1)\phi_{nlm}(\mathbf{r}_2) \pm \phi_{100}(r_2)\phi_{nlm}(\mathbf{r}_1) \right], \quad (11)$$

where we now label the states by the quantum numbers of the excited electron only, since only these change from one state to the next. More important, since the ground state electron carries no orbital angular momentum, the total orbital angular momentum of the state $\Phi_{nlm}$ is $l$ and its $z$-component is $m$. Now for the spin-independent Hamiltonian we are considering, the total orbital angular momentum and its $z$-component are exact constants of the motion. Hence our classification of states will be exact within our spin-independent approximation, even if estimates of the energies are rather rough.

Note that states of different $l$ or of different $m$ are not coupled together by the perturbation and hence ordinary nondegenerate perturbation theory can be used directly to calculate the first-order correction to the energy of the states of equation (11). From our previous work in Chapter VIII, we know that the perturbed energies to this order have the form, independent of $m$,

$$E_{nl} = \mathscr{E}_{nl} + J_{nl} \pm K_{nl}, \quad (12)$$

where the plus sign appears for space symmetric states and the minus sign for space antisymmetric states. The Coulomb energy $J_{nl}$ is given by

$$J_{nl} = \int \int \phi_{100}^{2}(r_1)\phi_{nl0}^{2}(r_2) \frac{e^2}{|\mathbf{r}_1 - \mathbf{r}_2|} d^3r_1 \, d^3r_2$$

and $K_{nl}$, the exchange energy, by

$$K_{nl} = \int \int \phi_{100}(r_1)\phi_{nl0}(r_2) \frac{e^2}{|\mathbf{r}_1 - \mathbf{r}_2|} \phi_{100}(r_2)\phi_{nl0}(r_1) \, d^3r_1 \, d^3r_2.$$

These integrals can be evaluated, but we shall not bother to do so. Evidently each is just the familiar quantity $Ze^2/a_0$, up to numerical factors (which decrease with increasing $n$).

Finally, we must include the spin to complete our description. Since the electrons are identical spin one-half particles, they must satisfy the exclusion principle. Hence the space symmetric states must be multiplied by antisymmetric or singlet spin states, the space antisymmetric states by symmetric or triplet spin states. As one consequence we see that the spin of one electron or the other must be altered or "flipped" if a transition occurs between any state in the singlet series and any in the triplet series, as these sets of states are called. Such transitions almost never occur under normal conditions, and these series are practically independent. Historically, they were given quite separate names and origins, the singlet states being ascribed to something called *para-helium* and the triplet to something different called *ortho-helium*.

Although we have not attempted to evaluate the integrals for $J$ and for $K$, we note that $J$ is clearly positive and that $K$ must therefore be positive, too. This last follows, since the mean electrostatic interaction of the electrons in the space antisymmetric state must be smaller than in the space symmetric states because of the correlations imposed by symmetry. This will indeed be so, provided that $K$ has the same sign as $J$. Thus the triplet states lie lower than the corresponding singlet states and this is an example of a general rule, known as *Hund's rule*, that *states of the highest spin normally lie lowest*.

Putting all this together, we are now in a position to exhibit the spectrum of helium in its main features. This is done, rather schematically, in Figure 2. In the figure, the unperturbed configuration is given, as is the common spectroscopic designation of the states $^{2S+1}L_J$ and of the electronic configuration.

The difference in the sign of the contribution of the exchange energy $K_{nl}$ in singlet and triplet states can be expressed in the following interesting and suggestive way. Recall that the operator $\sigma_1 \cdot \sigma_2 = -3$ when acting upon singlet spin states and that $\sigma_1 \cdot \sigma_2 = 1$ when acting upon triplet spin states. The first-order energy can thus be *symbolically* written

$$E_{21}^+ = \mathcal{E}_{21} + J_{21} + K_{21}$$

$1s2p$ ——————— $^1P_1$
$1s2s$ ——————— $^1S_0$

$$E_{20}^+ = \mathcal{E}_{20} + J_{20} + K_{20}$$

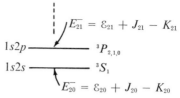

$$E_{21}^- = \mathcal{E}_{21} + J_{21} - K_{21}$$

$1s2p$ ——————— $^3P_{2,1,0}$
$1s2s$ ——————— $^3S_1$

$$E_{20}^- = \mathcal{E}_{20} + J_{20} - K_{20}$$

Triplet states (ortho-helium)

$(1s)^2$ ——————— $^1S_0$

$E_0$

Singlet states (para-helium)

FIGURE 2.   Spectrum of helium.

in the form

$$E_{nl} = \mathcal{E}_{nl} + J_{nl} - \left(\frac{1 + \sigma_1 \cdot \sigma_2}{2}\right) K_{nl},$$

because the factor in parentheses is $+1$ for triplet spin states, $-1$ for singlet. In this special sense, the energy depends explicitly upon the relative spins of the electrons, even though the Hamiltonian does not contain the spin. Because the electrons carry magnetic moments aligned with their spin angular momentum, we here discern the basis of ferromagnetism, including the *purely electrostatic* nature of the forces responsible for it.

Thus far we have neglected all spin-dependent interactions, and particularly the spin-orbit interaction, which is the dominant one of these. When such a term is included in the Hamiltonian, the following modifications result:

(1)   $L^2, L_z, S^2, S_z$ are no longer exact constants of the motion, although $J^2$ and $J_z$ are, of course. Thus the states are no longer strictly singlet or triplet spin states, and weak transitions between the (approximately) "singlet" and "triplet" series occur. These are called *intercombination lines*.

(2)   "Triplet" states are split into three components corresponding, for given $l$, to $j = l + 1, l, l - 1$. This classification of the states is only approximate, but the spin-orbit interaction is weak, the splitting is small and the approximation is good.

We conclude this section with some brief remarks about the periodic table.[1] We consider a highly approximate *independent particle* descrip-

[1] For a detailed treatment, see J. C. Slater, *Quantum Theory of Atomic Structure,* Vols. 1 and 2, McGraw-Hill (1960). See also, for a more elementary discussion, G. P. Harnwell and W. E. Stephens, *Atomic Physics,* McGraw-Hill (1955).

tion of an atom in which each electron is assumed to move in the Coulomb potential of the nucleus and in the *average* electrostatic field of all the other electrons. Further, this average field is assumed to be spherically symmetric. Neglecting spin-orbit forces, each electron can be assigned a principal quantum number $n$ and angular momentum quantum numbers $l$ and $m$, in analogy to hydrogenic states. The energy does not depend on $m$, of course, and hence each level of given $l$ is $(2l + 1)$-fold degenerate. Also taking into account the two possible orientations of electron spin, the total degeneracy is $2(2l + 1)$. For hydrogenic states, the energy depends only on $n$ and not on $l$, as a consequence of the special properties of the Coulomb potential. However, the average electrostatic interaction between the electrons alters the radial dependence of the potential energy in which each electron moves, and hence introduces a dependence upon $l$. Specifically, *states of smallest $l$ for a given $n$ have the lowest energy.* This follows because the smaller the angular momentum, the less effective is the centrifugal barrier and the more likely is an electron to be found close to the nucleus, where the attractive nuclear Coulomb potential is strong. The energy of states of given $n$ and $l$ depend upon the nuclear charge $Ze$, of course, and gradually shift downward as $Z$ increases from one element to the next.

The ground state of a given element is that in which its $Z$ electrons occupy the lowest possible set of independent particle states consistent with the Pauli exclusion principle. Generally speaking, the states are occupied in order of increasing principle quantum number $n$ and of increasing $l$ for each $n$. A state of given $n$ and $l$ can contain $2(2l + 1)$ electrons, each of which has the same energy. Such states are called *shells* and electrons in the same shell are called *equivalent electrons*. In large part the ground state configurations of atoms can be inferred from a knowledge of the ordering of these shells with energy. Using the standard spectroscopic notation in which the principle quantum number is given its proper numerical value and the $l$ value is denoted by a letter, the empirically observed order of the states is

$$1s, 2s, 2p, 3s, 3p, [4s, 3d], 4p, [5s, 4d],$$

$$5p, [6s, 4f, 5d], 6p, [7s, 5f, 6d].$$

As we have said, these states generally appear in order of increasing $n$ and of increasing $l$. Note, however, that for larger $l$ values the increase in energy with $l$ more than compensates the increase with $n$ and the states appear in reverse order. Thus, for example, the $5s(n = 5, l = 0)$ and $5p(n = 5, l = 1)$ shells have lower energies than the $4f(n = 4, l = 3)$. Similarly the $6s$ and $6p$ shells have lower energies than the $5f$. The brackets enclose shells where this compensation is almost exact, so that two or more shells have very nearly the same energy. The filling of such states is rather complicated, since it involves a detailed competition between the shells in question. The filling of the $3d$ shell is responsible for the first transition elements, the iron group, while the filling of the $4d$ shell produces the palladium group. The filling of the fourteen $4f$ states produces the rare earths and of the $5f$ states the actinide group.

The ground state electronic configuration for a given atom is specified by the number of electrons in each shell. This number is conventionally attached as a superscript to the shell designation. Thus, using the shell ordering scheme above, we can write the following examples:

$Z = 1$,　　H $: 1s$

$Z = 2$,　　He $: 1s^2$

$Z = 3$,　　Li $: 1s^2 2s$

$Z = 4$,　　Be $: 1s^2 2s^2$

$Z = 5$,　　B $: 1s^2 2s^2 2p$

$\vdots$

$Z = 11$,　　Na $: 1s^2 2s^2 2p^6 3s$

$\vdots$

$Z = 36$,　　Kr $: 1s^2 2s^2 2p^6 3s^2 3p^6 4s^2 3d^{10} 4p^6$.

To make the meaning clear, Kr has two electrons in its $1s$ shell, two in its $2s$ shell, 6 in its $2p$ shell, and so on up to 6 in its $4p$ shell. This notation is clearly redundant, since the filled shells need not be explicitly written out; only the last partially filled shell is required to uniquely specify the configuration. Thus one uses the abbreviation $3s$ for Na, $4p^6$ for Kr, $5d^{10}$ for Hg $(Z = 80)$, $5s4d^8$ for Rh $(Z = 45)$ or $5s^0 4d^{10}$ for Pd $(Z = 46)$. This last pair illustrates the complicated nature of the competition between the $5s$ and $4d$ states in the Pd group.

The chemical properties of atoms are determined primarily by the most loosely bound, or valence, electrons. The dominant factors are the number of such valence electrons and the energy interval to the next higher unfilled shell. The recurrence of similar sequences of shells produces a similar recurrence in chemical properties and hence is responsible for the periodic character of the table of elements.

## 2. THEORY OF SCATTERING

In our discussion of the stationary states characterizing the motion of a pair of particles interacting through a potential $V(\mathbf{r})$ we have thus far emphasized only discrete, bound states. We now consider states in the continuum. We shall assume that $V(\mathbf{r})$ vanishes at infinity sufficiently rapidly that

$$\lim_{r \to \infty} r V(\mathbf{r}) = 0, \tag{14}$$

so that continuum states exist for all positive energies, $E \geqslant 0$. Note that condition (14) rules out the Coulomb potential. We shall return to that special case later.

One immediate consequence of equation (14) is that for sufficiently large values of $r$, $V(\mathbf{r})$ becomes negligible and Schrödinger's equation reduces to that for a free particle. The state function at large $r$ is thus composed of familiar free particle states. In Chapter IX we obtained two alternative descriptions of such states, one in terms of eigenfunctions of the linear momentum, the other in terms of eigenfunctions of the orbital angular momentum. *Both* are involved in the description of the states we now seek if these states are to be physically significant.

In order to understand the question of physical significance, we turn for a moment to a time dependent description. Imagine a wave packet in which the particles are far apart initially, but are approaching each other with mean relative momentum $\mathbf{p}$. As time goes on the particles come close together, eventually interact and then finally separate with altered direction but with the magnitude of their mean relative momentum unchanged from its original value, as is required if energy is to be conserved. We have thus given a description of a collision or scattering process. Now imagine the initial wave packet to become broader and broader. As it does so its width in momentum space, and therefore also its energy spread, becomes narrower and narrower. Indeed, if the packet is made broad enough, the energy spread can be made infinitesimal. Further, the time required for the incident wave packet to complete its interaction with the potential becomes longer and longer. Hence, in the limit of an infinitely broad wave packet, the momentum (and energy) become precisely defined and the incident and scattered wave packets all coexist in time. In this limit, which corresponds to a stationary state description, the incident wave $\psi_{\text{inc}}$ becomes an eigenstate of the linear momentum,

$$\psi_{\text{inc}} = e^{i \mathbf{p} \cdot \mathbf{r}/\hbar} = e^{i \mathbf{k} \cdot \mathbf{r}}, \tag{15}$$

where we have introduced the wave vector $\mathbf{k} = \mathbf{p}/\hbar$.

On the other hand, the scattered wave $\psi_{\text{sc}}$, which emanates from the neighborhood of the origin, since that is where the interaction which produces it occurs, must have the form of an outward traveling generalized spherical wave. That is, at large distances, as $\mathbf{r}$ approaches infinity along some given direction $\hat{\mathbf{n}}$, we must have

$$\mathbf{r} \to \infty, \qquad \psi_{sc}\,(\mathbf{\hat{n}}r) \simeq f(\mathbf{\hat{n}})\,\frac{e^{ipr/\hbar}}{r} = f(\mathbf{\hat{n}})\,\frac{e^{ikr}}{r}. \qquad (16)$$

Note that $\mathbf{\hat{n}}$ is simply a unit vector along the direction of the radius vector $\mathbf{r}$. Putting equations (15) and (16) together, the complete field for large r has the form

$$\psi(\mathbf{\hat{n}}, r) \simeq e^{i\mathbf{k}\cdot\mathbf{\hat{n}}r} + f(\mathbf{\hat{n}})\,\frac{e^{ikr}}{r}. \qquad (17)$$

The coordinates used in this expression are shown in Figure 3, which also depicts the incident plane wave and the scattered spherical wave.

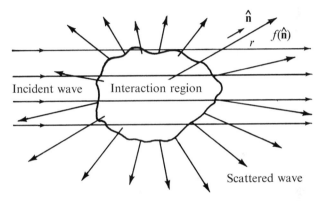

FIGURE 3.   The incident and scattered waves of equation (17).

The language of our subsequent analysis is considerably simplified if we temporarily imagine one of the particles to be infinitely massive and at rest. We shall call this particle the target particle. The states then involve only the motion of a single particle which is incident upon and scattered by the target particle. For now, we shall assume this to be the case.

The quantity $f(\mathbf{\hat{n}})$ is called the *scattering amplitude*. It is the probability amplitude that the incident particle will emerge along the direction $\mathbf{\hat{n}}$ as a result of the collision. To put this idea into more quantitative terms, note that the probability flux in the incident wave, normalized to unit amplitude, is

$$\mathbf{j}_{inc} = \frac{\mathbf{p}}{m}|\psi_{inc}|^2 = \frac{\mathbf{p}}{m}.$$

With the same normalization the probability flux in the outgoing spherical wave is

$$\mathbf{j}_{sc} = \frac{\mathbf{\hat{n}}p}{m}|\psi_{sc}|^2 = \frac{\mathbf{\hat{n}}p}{m}|f(\mathbf{\hat{n}})|^2/r^2.$$

Now the probability per second that the particle will emerge after the collision through some surface element $\mathbf{dS}$ is $\mathbf{j}_{sc} \cdot \mathbf{dS}$. If $\mathbf{dS}$ lies at a distance $r$ from the origin, then

$$\mathbf{j}_{sc} \cdot \mathbf{dS} = \frac{p}{m} |f(\hat{\mathbf{n}})|^2 \frac{\mathbf{n} \cdot \mathbf{dS}}{r^2} = \frac{p}{m} |f(\hat{\mathbf{n}})|^2 \, d\Omega,$$

since $\hat{\mathbf{n}} \cdot \mathbf{dS} = r^2 d\Omega$, where $d\Omega$ is the solid angle subtended at the origin by $\mathbf{dS}$. Finally, the relative probability $d\sigma$ that the particle will emerge into the solid angle $d\Omega$ is given by

$$d\sigma = \frac{\mathbf{j}_{sc} \cdot \mathbf{dS}}{|\mathbf{j}_{inc}|} = |f(\mathbf{n})|^2 \, d\Omega. \tag{18}$$

The quantity $d\sigma$ has the dimensions of an area, and it is called the differential cross section for scattering into the element of solid angle $d\Omega$ at $\hat{\mathbf{n}}$. The quantity $|f(\hat{\mathbf{n}})|^2$, which may be symbolically written as $d\sigma/d\Omega$, is usually simply called the *differential cross section*.[2]

The reason for calling these relative probabilities "cross sections" is the following. Think of the uniform flux of probability $\mathbf{j}_{inc}$ in the incident wave. Then $d\sigma$ is the effective cross-sectional area of the interaction region for intercepting probability flux from the incident wave and transferring it into the solid angle $d\Omega$. This will perhaps be clearer if we think not of a single incident particle and a single target particle but rather of a beam of particles incident upon an ensemble of target particles. Then, if the flux of incident particles is $J$ per cm$^2$ per second, and if the number of target particles irradiated by the beam is $N$, the number of particles $dN$ emerging per second into $d\Omega$ is

$$dN = JN d\sigma = JN |f(\hat{\mathbf{n}})|^2 \, d\Omega, \tag{19}$$

and each target particle has an effective area $d\sigma$ for intercepting an incident particle and scattering it into $d\Omega$. The quantity $dN$ is an experimentally observable quantity, and indeed, the observation of cross sections and of their dependence on energy and direction is the main source of information about the elementary particles and their inter-

---

[2] It is instructive to compare this quantum mechanical description, involving incident and scattered probability waves, with the classical description in terms of trajectories. Recall that the classical cross section is defined in terms of the number of particles deflected into $d\Omega$ per unit time and per unit incident flux, in complete analogy to the quantum definition. Recall also that for a given spherically symmetrical potential, the angle through which a classical particle of given energy is deflected depends only upon its *impact parameter* and hence only upon its angular momentum. The classical analysis thus conveniently proceeds through the isolation of states of definite angular momentum. As we shall see in a moment, the quantum analysis for such potentials is also conveniently treated in the same way, although the simple direct and precise connection between angular momentum and deflection angles is lost. A detailed presentation of classical scattering theory is given in Reference [15]; the subject is also treated, but more briefly, in Reference [14].

actions. The information is somewhat tenuous, of course, since one must essentially work backwards in order to draw inferences about the nature of the interactions from the observed cross sections.

For most of the discussion above we have treated the target particle as if it were infinitely massive. However, no serious complications occur when the mass of the target particle is taken into account. In fact all of our results apply as written, provided only that the center-of-mass and not laboratory coordinates are understood in all of the equations.

We must still supply a method for calculating scattering amplitudes and cross sections. We restrict our attention to spherically symmetric potentials, considering states of definite angular momentum $l$. The radial part of the wave function $R_{El}$ then satisfies equation (IX–49), which we rewrite here for reference,

$$-\frac{\hbar^2}{2m}\frac{1}{r^2}\frac{d}{dr}\left(r^2\frac{dR_{El}}{dr}\right)+\left[V(r)+\frac{l(l+1)\hbar^2}{2mr^2}\right]R_{El}=ER_{El}. \quad (20)$$

Now if $V(r)$ were zero, these radial functions would be the free particle functions $j_l(kr)$ which, according to equation (IX–63), take the form, for large $r$,

$$r\to\infty,\qquad j_l(kr)\simeq\frac{\sin\,(kr-l\pi/2)}{kr}.$$

Since $V(r)$ becomes negligible at large $r$, the presence of the potential cannot change this functional form but can, at most, alter the *phase* of the sinusoidal function. We thus write,

$$r\to\infty,\qquad R_{El}\simeq\frac{\sin\,(kr-l\pi/2+\delta_l)}{kr}. \quad (21)$$

The quantity $\delta_l=\delta_l(E)$ is called the *phase shift* of the $l$th *partial wave.* It can be calculated by solving the radial equation for $R_{El}$ and examining the asymptotic form of the solution for large $r$.

Let us suppose that this has been done and that the $\delta_l$ have been determined. Our final task is then to express the scattering amplitude and cross section in terms of the phase shifts. This is achieved by forming a general superposition of partial waves,

$$\psi=\sum_{l,m}C_{lm}R_{El}(r)Y_l^m(\theta,\phi),$$

and then choosing the expansion coefficients $C_{lm}$ in such a way that $\psi$ has precisely the form of equation (17) as $r$ becomes very large. Choosing the z-axis to coincide with the direction of incidence for simplicity, the incident plane wave term in equation (17) can be expressed in terms of spherical waves using equation (IX–68),

$$e^{ikz} = e^{ikr\cos\theta} = \sum_{l=0}^{\infty} i^l\,(2l+1)\,j_l(kr)\,P_l(\cos\theta)\,,$$

whence it is not too hard to show that (see, for example, Reference [24])

$$C_{lm} = 0 \qquad\qquad , \qquad m \neq 0$$

$$C_{lm} = \sqrt{4\pi\,(2l+1)}\ e^{i\delta_l}, \qquad m = 0.$$

With this result it then follows that

$$r \to \infty\,, \qquad \psi \sim e^{ikz} + f(\theta)\,\frac{e^{ikr}}{r}\,,$$

where

$$
\begin{aligned}
f(\theta) &= \frac{1}{k}\sum_{l=0}^{\infty}\sqrt{4\pi\,(2l+1)}\ e^{i\delta_l}\,\sin\delta_l\,Y_l^0(\theta)\\[2mm]
&= \frac{1}{k}\sum_{l=0}^{\infty}(2l+1)\,e^{i\delta_l}\,\sin\delta_l P_l(\theta)\,.
\end{aligned}
\tag{22}
$$

For given energy $E$, the scattering amplitude thus depends only upon the angle $\theta$ between the direction of incidence and the direction of scattering, and it is completely determined once the $\delta_l$ are known. The magnitudes of the $\delta_l$ are related to the strength of the interaction potential in a rather complicated way. We note, however, that if there is no interaction, so that the particles move freely, then by their definition the phase shifts are all zero and the scattering amplitude and cross section are seen to vanish, as they must.

The differential cross section $d\sigma$ is a measure of the probability for scattering into some particular infinitesimal element of solid angle $d\Omega$. It is also of interest to consider the *total cross section* $\sigma$, which is a measure of the probability for scattering into any and all elements of solid angle. From its definition, we have

$$\sigma = \int \frac{d\sigma}{d\Omega}\,d\Omega = \int |f(\theta)|^2\,d\Omega = \int |f(\theta)|^2\,\sin\theta\,d\theta\,d\phi \tag{23}$$

whence, using the first form of equation (22) and noting the orthonormality of the $Y_l(\theta)$, we obtain

$$\sigma = \frac{4\pi}{k^2}\sum_{l=0}^{\infty}(2l+1)\,\sin^2\delta_l. \tag{24}$$

The fact that there is a definite probability, proportional to the total cross section $\sigma$, for an incident particle to be scattered in some direction or other means that there must be a compensating decrease in the probability flux along the direction of incidence, if probability is to be conserved. This compensation is accomplished through interference between the incident state function and the scattering amplitude in the forward direction. This argument implies the existence of a general relationship between the total cross section and the forward scattering amplitude. We now establish this relationship. To do so, we examine the forward scattering amplitude $f(\theta = 0)$. Recalling that for all $l$

$$P_l(\theta = 0) = 1,$$

we have, from equation (22),

$$f(\theta = 0) = \frac{1}{k} \sum_{l=0}^{\infty} (2l + 1) \, e^{i\delta_l} \sin \delta_l$$

$$= \frac{1}{k} \sum_{l=0}^{\infty} (2l + 1) \cos \delta_l \sin \delta_l + \frac{i}{k} \sum_{l=0}^{\infty} (2l + 1) \sin^2 \delta_l.$$

The second term is seen to be porportional to the total cross section, and hence the relation we seek may be expressed in the form

$$\sigma = \frac{4\pi}{k} \, \mathrm{Im} \, f(\theta = 0). \tag{25}$$

In words, *the total cross section $\sigma$ is $4\pi$ divided by the reduced wave number $k$ times the imaginary part of the scattering amplitude in the forward direction.* This important result is known as the *optical theorem* or as the *cross section theorem.* For emphasis, we repeat that it expresses the *probability conservation* requirement that the amplitude of the incident wave must ultimately be reduced in proportion to the total probability that the particle is scattered in any way. We also emphasize that it is a far more general result than our method of derivation would seem to indicate; it holds for *arbitrary* potentials, spherical or not, and for *arbitrary* scattering processes.

The expressions we have obtained for the scattering amplitude and cross section involve summations over the partial wave phase shifts. Questions of the convergence of these infinite series expressions deserve some attention at this point. The role of the centrifugal barrier in keeping the particles apart plays the dominant role in these considerations. For sufficiently large $l$ values the particles do not approach each other closely enough to interact and the phase shifts corresponding to such $l$ values must vanish. To become a bit more quantitative, the distance of closest approach $r_0$ of two particles with relative energy $E$ and relative angular momentum $l$ is roughly that point at which the centrifugal barrier height is equal to the total energy $E$,

$$\frac{l(l + 1)\hbar^2}{2mr_0^2} \simeq E.$$

If we neglect unity in comparison with $l$ we obtain the somewhat simpler relation

$$l \simeq kr_0, \qquad k = \sqrt{2mE}/\hbar.$$

Now if the effective range of the interaction is $R$, we thus expect that $\delta_l$ will become very small when $l$ appreciably exceeds $l_{\max} \equiv kR$, because then the particles simply do not come close enough to each other to interact. Hence, the number of contributing terms in the summations over the partial wave phase shifts is of the order of $kR$. This means that at high energies, where $kR \gg 1$, the cross section involves many phase shifts and becomes a rapidly varying function of angle. On the other hand, at low enough energies that $kR \ll 1$, only the $l = 0$ or $S$-wave phase shift differs significantly from zero and the scattering is isotropic.

We conclude this section with some remarks on the special case of the Coulomb potential, which violates equation (14). Because of this, the continuum states do not reduce to equation (17) for large $r$, but are somewhat more complicated. Nonetheless, the analysis can be carried through and the scattering amplitude and differential cross section found. Indeed, using parabolic coordinates, a *complete* and *exact* solution can be obtained and the scattering amplitude can be written in closed form, the only known case for which this can be done. Specifically, it is found that if two particles of reduced mass $m$, carrying charges $Z_1 e$ and $Z_2 e$ respectively, are incident upon each other with relative momentum $\hbar k$, the amplitude for Coulomb scattering through the angle $\theta$ with respect to the direction of incidence is

$$f_C(\theta) = -\frac{\eta}{2k \sin^2 \theta/2} \exp\left[-i\eta \, ln(\sin^2 \theta/2) + 2i\beta\right], \qquad (26)$$

where the *Coulomb parameter* $\eta$ is given by

$$\eta = Z_1 Z_2 e^2 m/\hbar^2 k$$

and where the phase factor $\beta$ is such that

$$e^{2i\beta} = \Gamma(1 + i\eta)/\Gamma(1 - i\eta).$$

The function $\Gamma$ is called the *gamma function*[3] and is a generalized factorial function, defined by

$$\Gamma(\nu + 1) \equiv \int_0^\infty e^{-x} x^\nu \, dx.$$

When $\nu$ is an integer it is readily verified that $\Gamma(\nu + 1) = \nu!$.

The Coulomb parameter $\eta$ is frequently expressed in terms of the relative velocity $v$ of the particles,

---

[3] See, for example, Reference [8].

$$\eta = Z_1 Z_2 e^2 / \hbar v = \alpha \frac{Z_1 Z_2}{v/c},$$

where the dimensionless *fine structure constant* $\alpha$ is given by

$$\alpha = \frac{e^2}{\hbar c} \simeq 1/137.$$

It should be remarked that the expression we have given for the Coulomb scattering amplitude is that corresponding to the repulsive interaction of two like charges. For two unlike charges $\eta$ changes sign, which simply means that $f_C$ is replaced by the negative of its complex conjugate.

The differential cross section for Coulomb scattering is obtained in the usual way from the scattering amplitude and is thus given by

$$\frac{d\sigma}{d\Omega} = |f_C|^2 = \frac{\eta^2}{4k^2 \sin^4 \theta/2} = \frac{(Z_1 Z_2 e^2 / 2mv^2)^2}{\sin^4 \theta/2}.$$

This expression is in *exact* agreement with the *classical* result for the scattering of point charges, first obtained by Rutherford in 1911, and is commonly called the *Rutherford cross section*. No explanation can be offered for this unique and strange correspondence between the classical and quantum mechanical cross sections. It is important to observe, however, that this exact agreement occurs because the *magnitude* of the scattering amplitude happens not to contain $\hbar$ and hence neither does $d\sigma/d\Omega$, which must necessarily assume its classical value as a consequence. On the other hand, the phase of the scattering amplitude is an entirely quantum mechanical quantity so that any processes which depend upon that phase will exhibit quantum effects. An important and interesting example in which this is so is the scattering of identical particles, as we now show.

If two particles interact and if one of them emerges at angle $\theta$ in the center-of-mass system of coordinates, the other must emerge at angle $\pi - \theta$. On the other hand, if the first emerges at $\pi - \theta$ the second must emerge at $\theta$. If the particles are identical these two possibilities cannot be distinguished from each other in any way whatsoever. Classically, therefore, the cross sections for both processes must be added so that the classical result for identical particles is

$$\left( \frac{d\sigma}{d\Omega} \right)_{\text{classical}} = \frac{d\sigma(\theta)}{d\Omega} + \frac{d\sigma(\pi - \theta)}{d\Omega}.$$

Quantum mechanically, on the other hand, the scattering *amplitudes*, and not the cross sections, must be combined, with relative sign determined by the symmetry of the state function. For spinless particles

($\alpha$-particles, for example) the total wave function must be symmetric and the scattering amplitudes add with the same sign,

$$\frac{d\sigma}{d\Omega} = |f(\theta) + f(\pi - \theta)|^2,$$

For spin one-half particles (electrons or protons, for example) the wave function must be antisymmetric in the space and spin coordinates together. The spatially symmetric state combines with a singlet spin state, the spatially antisymmetric state with a triplet spin state. The former has statistical weight one, the latter three. Hence for unpolarized spin one-half particles

$$\frac{d\sigma}{d\Omega} = \frac{1}{4} |f(\theta) + f(\pi - \theta)|^2 + \frac{3}{4} |f(\theta) - f(\pi - \theta)|^2.$$

In both cases, the results contain interference terms which depend on the relative phase of the scattering amplitude at $\theta$ and at $\pi - \theta$ and which are clearly quantum mechanical in structure.[4] Specifically, for Coulomb scattering of identical particles these expressions become, upon substitution for $f_C$ from equation (26):

*spin zero*

$$\frac{d\sigma}{d\Omega} = \left(\frac{Z_1 Z_2 e^2}{2mv^2}\right)^2 \left[ \frac{1}{\sin^4 \theta/2} + \frac{1}{\cos^4 \theta/2} + \frac{2\cos\left(\eta ln \tan^2 \theta/2\right)}{\sin^2 \theta/2 \cos^2 \theta/2} \right]$$

*spin one-half*

$$\frac{d\sigma}{d\Omega} = \left(\frac{Z_1 Z_2 e^2}{2mv^2}\right)^2 \left[ \frac{1}{\sin^4 \theta/2} + \frac{1}{\cos^4 \theta/2} - \frac{\cos(\eta ln \tan^2 \theta/2)}{\sin^2 \theta/2 \cos^2 \theta/2} \right].$$

The first two terms in each case give the classical result; the last is the quantum mechanical interference term. In the classical limit the interference term must disappear, of course, and it is interesting to see how this comes about. As $\hbar$ tends toward zero, $\eta$ increases without limit and the interference term oscillates more and more rapidly. An infinitesimal spread in energy, or in angle of observation, thus causes this term to average to zero and hence to become unobservable.

## 3. GREEN'S FUNCTIONS FOR SCATTERING; THE BORN APPROXIMATION

In the preceding section we have given a rather formal treatment of the theory of scattering. The scattering amplitude was introduced, and in

---

[4] Note that these effects are far from small. At 90°, for example, $d\sigma/d\Omega$ is twice the classical result for spinless particles and half the classical result for spin one-half particles.

terms of it was defined the experimentally measurable quantity, the cross section. For the special case of spherically symmetrical potentials a method was given for evaluating this quantity, the method of partial waves. The actual calculation requires the solution of the set of radial equations (20). These are so complicated, even for the simplest inter-action, that no attempt was made to even discuss their properties.[5] As a result we were unable to present, to any significant degree, a qualitative description of the general character of quantum mechanical scattering processes.

In the present section we correct this defect by presenting an approxi-mation method first introduced by Born. The *Born approximation* is nothing more than perturbation theory applied to continuum states, as we shall see, and it is therefore restricted to sufficiently weak inter-actions.[6] Even so, it is the physicist's absolutely indispensable quali-tative guide to scattering and it effectively serves as the norm to which all is referred.

We shall derive the Born approximation by two quite distinct methods. In the first, we obtain an exact integral equation for the state function through the intermediary of the *free particle Green's function*. In the second we regard the scattering process as a transition, induced by the interaction potential, between states of definite initial and final momenta, and we thus use the methods of time dependent perturbation theory.

(a)   *Green's Function Method.*[7]   We seek a solution of the time in-dependent Schrödinger equation, which we write in the form

$$(\nabla^2 + k^2)\psi(\mathbf{r}) = \frac{2m}{\hbar^2} V(r)\psi(\mathbf{r}).  \tag{27}$$

Here

$$k^2 = 2mE/\hbar^2,$$

so that $k$ is the free particle wave number for energy $E$. We assume $E > 0$ and that $rV(\mathbf{r})$ vanishes at infinity, in accord with equation (14). The solution we seek is subject to the asymptotic boundary conditions for large $r$, given in equation (17), which we repeat here for convenience:

$$\psi(\mathbf{r}) = e^{i\mathbf{k}\cdot\mathbf{r}} + f(\hat{\mathbf{n}}) \frac{e^{ikr}}{r},$$

---

[5] In practice, these equations are almost exclusively solved numerically with the aid of a high-speed digital computer. See Section 10, Chapter VI.

[6] It is not, however, restricted to spherically symmetrical interactions. This is in contrast to the method of partial waves, which becomes completely intractable for non-spherical interactions, except in rather special cases.

[7] For a general discussion of Green's functions, see References [6] through [13].

where $\mathbf{r} = \hat{\mathbf{n}} r$.

We now obtain an integral equation for $\psi(\mathbf{r})$, which incorporates these boundary conditions, by introducing the Green's function $G(\mathbf{r}, \mathbf{r}')$ $= G(\mathbf{r}', \mathbf{r})$, defined as the purely *outgoing* wave solution of the inhomogeneous equation

$$(\nabla^2 + k^2) \, G(\mathbf{r}, \mathbf{r}') = - \delta(\mathbf{r} - \mathbf{r}'). \tag{28}$$

Here $\delta(\mathbf{r} - \mathbf{r}')$ is the three-dimensional Dirac $\delta$-function, which may be thought of as the product of three one-dimensional $\delta$-functions,

$$\delta(\mathbf{r} - \mathbf{r}') \equiv \delta(x - x') \, \delta(y - y') \, \delta(z - z')$$

and which therefore is such that

$$\int \delta(\mathbf{r} - \mathbf{r}') \, d^3r' = 1.$$

The Green's function is thus seen to be the wave function at the point $\mathbf{r}$ generated by a point source at $\mathbf{r}'$.

Deferring the actual construction of $G$ for a moment, we observe that any function $\psi$ which satisfies the *integral equation*

$$\psi(\mathbf{r}) = e^{i\mathbf{k}\cdot\mathbf{r}} - \frac{2m}{\hbar^2} \int_{\text{all space}} G(\mathbf{r}, \mathbf{r}') \, V(\mathbf{r}') \, \psi(\mathbf{r}') \, d^3r' \tag{29}$$

is a solution of Schrödinger's equation (27) and that this solution automatically satisfies the boundary conditions of equation (17). That $\psi$ is a solution follows by operating on equation (29) with $(\nabla^2 + k^2)$. This operator annihilates the first term on the right and, when taken under the integral sign to operate on $G$, the only quantity dependent upon $\mathbf{r}$, it yields the Dirac $\delta$-function according to equation (28). Hence we obtain

$$(\nabla^2 + k^2)\psi = - \frac{2m}{\hbar^2} \int (\nabla^2 + k^2) \, G(\mathbf{r}, \mathbf{r}') \, V(\mathbf{r}') \, \psi(\mathbf{r}') \, d^3r',$$

$$= \frac{2m}{\hbar^2} \int \delta(\mathbf{r} - \mathbf{r}') \, V(\mathbf{r}') \, \psi(\mathbf{r}') \, d^3r',$$

$$= \frac{2m}{\hbar^2} \, V(\mathbf{r}) \, \psi(\mathbf{r}),$$

which is indeed equation (27). That equation (17) is satisfied follows because $G$ contains only outgoing waves at $\mathbf{r}$ generated by a point source at $\mathbf{r}'$, whence the integral on the right of equation (29) is simply a superposition of such outgoing waves, and therefore assumes the required form for $r$ sufficiently large. We shall shortly demonstrate this aspect quite explicitly.

The next step is the construction of $G$ itself. We start with the relation,

$$\nabla^2 \frac{1}{r} = -4\pi\delta(\mathbf{r}),\qquad(30)$$

which is the transcription into the language of differential equations of the Coulomb expression for the electrostatic potential of a point charge. Recall that the electrostatic potential $\phi$ generated by a charge density $\rho$ is given by Poisson's equation

$$\nabla^2\phi = -4\pi\rho.$$

The potential of a point charge of unit strength is $1/r$, and its charge density $\rho$ is $\delta(\mathbf{r})$, whence equation (30) follows.[8] Next, using the readily verifiable identity

$$\nabla^2 \frac{e^{ikr}}{r} = e^{ikr}\,\nabla^2 \frac{1}{r} - k^2 \frac{e^{ikr}}{r},$$

we obtain

$$(\nabla^2 + k^2)\,\frac{e^{ikr}}{r} = e^{ikr}\,\nabla^2 \frac{1}{r}$$

$$= -4\pi\delta(\mathbf{r})\,e^{ikr}$$

$$= -4\pi\delta(\mathbf{r}),$$

where, in going to the last line, we have used the fact that, because of the properties of the $\delta$-function, $g(\mathbf{r})\delta(\mathbf{r}) = g(0)\delta(\mathbf{r})$ for arbitrary $g(\mathbf{r})$. Finally, shifting the origin to the point $\mathbf{r}'$, we have

$$(\nabla^2 + k^2)\,\frac{e^{ik|\mathbf{r}-\mathbf{r}'|}}{4\pi|\mathbf{r}-\mathbf{r}'|} = -\delta(\mathbf{r}-\mathbf{r}'),$$

so that, upon comparison with equation (27), we obtain

$$G(\mathbf{r},\mathbf{r}') = \frac{e^{ik|\mathbf{r}-\mathbf{r}'|}}{4\pi|\mathbf{r}-\mathbf{r}'|}.\qquad(31)$$

Substitution of this explicit expression for the Green's function into equation (29) then gives the *exact* integral equation,

$$\psi(\mathbf{r}) = e^{i\mathbf{k}\cdot\mathbf{r}} - \frac{m}{2\pi\hbar^2}\int \frac{e^{ik|\mathbf{r}-\mathbf{r}'|}}{|\mathbf{r}-\mathbf{r}'|}\,V(\mathbf{r}')\psi(\mathbf{r}')\,d^3r'.\qquad(32)$$

We now find an expression for the scattering amplitude by letting $\mathbf{r} = \hat{\mathbf{n}}r$ approach infinity. Under these circumstances

---

[8] This appeal to electrostatics can be avoided by direct analysis of equation (30). It is readily verified by carrying out the differentiations that the Laplacian of $1/r$ vanishes for $r \neq 0$. Integration of equation (30) over an infinitesimal volume containing the origin then yields $-4\pi$ for the right side, and the same result is obtained for the left after application of Green's theorem (Gauss's law). See Reference [18], especially pp. 543,4.

$$\frac{e^{ik|r-r'|}}{|r-r'|} = \frac{e^{ik|\hat{n}r-r'|}}{|\hat{n}r-r'|} = \frac{e^{ikr}}{r} e^{-ik\hat{n}\cdot r'} + 0\left(\frac{1}{r^2}\right),$$

where we have expanded $|r-r'|$ in a Taylor series according to

$$h(r-r') = h(r) - r' \cdot \nabla h(r) + \cdots.$$

We thus find, upon substitution of this result into equation (32), that for large $r$,

$$\psi(r) = e^{ik\cdot \hat{n}r} - \frac{m}{2\pi\hbar^2} \int e^{-ik\hat{n}\cdot r'} V(r')\psi(r') \, d^3r' \cdot \frac{e^{ikr}}{r},$$

which is of the form demanded by equation (17). Further, upon comparison with that equation, we see that the scattering amplitude is identified as

$$f(\hat{n}) = -\frac{m}{2\pi\hbar^2} \int e^{-ik\hat{n}\cdot r} V(r)\psi(r) \, d^3r, \qquad (33)$$

where we have dropped the prime on the integration variables for simplicity. Again, this expression is *exact* for any potential $V(r)$.

The Born approximation follows quickly by observing that if $V(r)$ is sufficiently small,[9] then the second term of equation (32) is a small correction to the incident wave term $e^{ik\cdot r}$. We can thus consider solving equation (32) iteratively, just as we did in perturbation theory. The zeroth-order solution is the incident wave, the first-order solution is obtained by replacing $\psi$ in the integral by its zeroth-order expression, the second-order by replacing it by the first-order solution, and so on. The result is a series in increasing powers of the interaction potential which we shall *assume* to be convergent if $V$ is small enough. We shall not carry the process beyond the first step. Looking only at the resulting expression for the scattering amplitude, obtained from equation (33) by replacing $\psi(r)$ by the zeroth-order expression $e^{ik\cdot r}$, we find

$$f(\hat{n}) = -\frac{m}{2\pi\hbar^2} \int e^{ik(\hat{n}_0-\hat{n})\cdot r} V(r) \, d^3r, \qquad (34)$$

where we have introduced the direction of incidence $\hat{n}_0$ by writing $k = \hat{n}_0 k$. The quantity $k(n_0 - n) \equiv \Delta k$ is just the change in the wave number vector between the direction of incidence and the direction of scattering, and we see that, to this approximation, the scattering amplitude is proportional to the *Fourier transform of the potential* with $\Delta k$ as transform variable. The intrinsically quantum mechanical nature of this result is readily apparent; contributing to the scattering through a given angle is the potential function at *every* point in space, not just the poten-

[9] We discuss the meaning of this condition later.

tial along a localized trajectory, as would be the case classically.

It is instructive to rewrite equation (34) in terms of momenta rather than wave number. We have, rearranging the terms slightly,

$$f(\hat{\mathbf{n}}) = -\frac{m}{2\pi\hbar^2} \int e^{-i\mathbf{p}_f \cdot \mathbf{r}/\hbar} V(r) \, e^{i\mathbf{p}_i \cdot \mathbf{r}/\hbar} \, d^3 r$$

$$= -\frac{m}{2\pi\hbar^2} < e^{i\mathbf{p}_f \cdot \mathbf{r}/\hbar} \, |V(r)| \, e^{i\mathbf{p}_i \cdot \mathbf{r}/\hbar} > ,$$

(35)

where we have introduced the initial and final momenta $\mathbf{p}_i$ and $\mathbf{p}_f$ by writing

$$\mathbf{p}_i = \hbar k \hat{\mathbf{n}}_0 = p\hat{\mathbf{n}}_0$$

$$\mathbf{p}_f = \hbar k \hat{\mathbf{n}} = p\hat{\mathbf{n}}$$

and where

$$p = \sqrt{2mE}.$$

We thus observe that $f(\hat{\mathbf{n}})$ is proportional to the matrix element of $V$ between unperturbed (free particle) final and initial momentum states. The differential cross section, given by

$$\frac{d\sigma}{d\Omega} = |f(\hat{\mathbf{n}})|^2 = \frac{m^2}{4\pi^2\hbar^4} \, | \, \langle \, e^{i\mathbf{p}_f \cdot \mathbf{r}} \, |V| \, e^{i\mathbf{p}_i \cdot \mathbf{r}} \, \rangle \, |^2,$$

(36)

is then proportional to the square of this matrix element and hence has the form of a transition probability. We next show how this result can be obtained directly, using the conventional methods of time dependent perturbation theory.

(b) *Perturbation Theory Method.* Consider now the interaction potential as a perturbation which induces transitions between a state of definite initial momentum $\mathbf{p}_i$ and a dense set of energy conserving final states, namely those with momentum $\mathbf{p}_f = p\hat{\mathbf{n}}$. According to the golden rule of time dependent perturbation theory, equation (VII-37), the transition rate is given by

$$W = \frac{2\pi}{\hbar} \, \rho(E) |V_{fi}|^2,$$

where $\rho(E)$ is the density of final states and $V_{fi}$ is the matrix element connecting the initial state and a typical final state.[10] We are here dealing with idealized, nonphysical continuum states of definite momentum, and we therefore adopt the conventional artifice of periodic boundary

[10] The precise meaning of typical with reference to the final states will be made clear in a moment.

conditions in a cube of side $L$ to permit us to properly normalize these states and to calculate their density. With this convention, the momentum states are written as

$$\psi_p = \frac{1}{L^{3/2}} \, e^{i\mathbf{p}\cdot\mathbf{r}/\hbar} \tag{37}$$

and are evidently correctly normalized over the periodicity volume $L^3$. We thus have at once

$$V_{fi} = \frac{1}{L^3} \langle e^{i\mathbf{p}_f\cdot\mathbf{r}} \,|V|\, e^{i\mathbf{p}_i\cdot\mathbf{r}} \rangle . \tag{38}$$

Observe now that $V_{fi}$ is not necessarily slowly varying, as assumed in the derivation of the golden rule, if $\mathbf{p}_f$ is permitted to range freely over *all* final states which conserve energy. Otherwise stated, the requirement that $V_{fi}$ represent the matrix element to a *typical* final state has no meaning if $\mathbf{p}_f$ is permitted to vary over all directions. Accordingly, $\mathbf{p}_f$ must be restricted to an infinitesimal range of orientations about some given direction $\hat{\mathbf{n}}$, which is to say that it must lie in an infinitesimal solid angle $d\Omega$ about $\hat{\mathbf{n}}$. This means that $\rho(E)$ is to be interpreted as the density of that particular fraction of all the states with energy between $E$ and $E + dE$ which have momenta in $d\Omega$. To symbolize this fact, we denote the relative density by $d\rho(E)$ and the corresponding transition rate by $dW$, whence, using equation (38),

$$dW = \frac{2\pi}{\hbar L^6} |\langle e^{i\mathbf{p}_f\cdot\mathbf{r}}|V| \, e^{i\mathbf{p}_i\cdot\mathbf{r}} \rangle|^2 \, d\rho(E) . \tag{39}$$

Because the distribution of momentum states for a free particle is isotropic in momentum space, $d\rho$ is easily expressed in terms of the density $\rho$ of all states; it is just $\rho$ itself multiplied by the ratio of $d\Omega$ to $4\pi$, the total solid angle encompassing all directions,

$$d\rho = \rho \frac{d\Omega}{4\pi} .$$

Now, according to equation (IX-18),

$$\rho(E) = \frac{(2m)^{3/2}}{4\pi^2\hbar^3} L^3 \sqrt{E} = \frac{mp}{2\pi^2\hbar^3} L^3 ,$$

whence

$$d\rho = \frac{mp}{8\pi^3\hbar^3} \, L^3 \, d\Omega ,$$

and equation (39) yields

$$dW = \frac{mp}{4\pi^2\hbar^4 L^3} \, |\langle e^{i\mathbf{p}_f \cdot \mathbf{r}} \, |V| \, e^{i\mathbf{p}_i \cdot \mathbf{r}}\rangle|^2 \, d\Omega.$$

This, then, is the probability per unit time of a transition from the initial state to any final state with momentum oriented in the solid angle $d\Omega$. Otherwise stated, $dW$ is the probability per unit time that scattering into $d\Omega$ has occurred. By the definition of the differential cross section $d\sigma$, equation (18), this is nothing more than $d\sigma |j_{\text{inc}}|$, where $|j_{\text{inc}}|$ is the magnitude of the incident probability flux,

$$|j_{\text{inc}}| = \frac{p}{m} \, |\psi_{\text{inc}}|^2 = \frac{p}{mL^3},$$

in virtue of the normalization of equation (36). Hence, finally,

$$d\sigma = \frac{mL^3}{p} \, dW = \frac{m^2}{4\pi^2\hbar} \, |\langle e^{i\mathbf{p}_f \cdot \mathbf{r}} \, |V| \, e^{i\mathbf{p}_i \cdot \mathbf{r}}\rangle|^2 \, d\Omega,$$

in agreement with equation (37).

The utility and simplicity of the Born approximation is best illustrated by considering some examples. As the first case, which we shall work through in some detail because of its importance, we examine scattering by the so-called *Yukawa pontential*,[11]

$$V(r) = V_0 R \, \frac{e^{-r/R}}{r}. \tag{40}$$

The constant $V_0$ is usually called the *strength* of the Yukawa potential, and $R$ is called the *range*. Substitution into equation (34) gives

$$f(\hat{n}) = -\frac{m V_0 R}{2\pi\hbar^2} \int e^{ik(\hat{n}_0 - \hat{n})\cdot \mathbf{r}} \, \frac{e^{-r/R}}{r} \, d^3r. \tag{41}$$

The integration is easily performed by choosing the polar axis to lie along $(\hat{n}_0 - \hat{n})$, as illustrated in Figure 4. Denoting the angle between $(\hat{n}_0 - \hat{n})$ and $\mathbf{r}$ by $\omega$, as indicated, we thus have

$$f(\hat{n}) = -\frac{m V_0 R}{2\pi\hbar^2} \int_0^\infty r^2 \, dr \int_0^\pi \sin\omega \, d\omega \int_0^{2\pi} d\phi \, e^{ik|\hat{n}_0 - \hat{n}|r \cos\omega} \, \frac{e^{-r/R}}{r}.$$

The integrand is independent of azimuthal angle so that integration over azimuth simply gives a factor of $2\pi$. The integration over $\omega$ is also readily carried out to give

$$f(\hat{n}) = -\frac{m V_0 R}{i\hbar^2 k|\hat{n}_0 - \hat{n}|} \int_0^\infty dr \, e^{-r/R} \left( e^{ik|\hat{n}_0 - \hat{n}|r} - e^{-ik|\hat{n}_0 - \hat{n}|r}\right)$$

$$= -\frac{2m V_0 R^3}{\hbar^2} \left( \frac{1}{1 + k^2 R^2 (\hat{n}_0 - \hat{n})^2}\right).$$

[11] The Yukawa potential is an essential ingredient in nucleon-nucleon interactions. It also is frequently used to describe a *screened Coulomb potential* (see Problem 8).

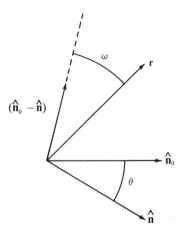

FIGURE 4.    Coordinate system for the integration of equation (40).

This result is readily expressed in terms of the scattering angle $\theta$, which is the angle between $\hat{\mathbf{n}}_0$ and $\hat{\mathbf{n}}$, as shown in Figure 4. We have at once

$$(\hat{\mathbf{n}}_0 - \hat{\mathbf{n}})^2 = 4 \sin^2 \theta/2,$$

whence, writing $f(\hat{\mathbf{n}})$ as $f(\theta)$,

$$f(\theta) = -\frac{2mV_0R^3}{\hbar^2} (1 + 4k^2R^2 \sin^2 \theta/2)^{-1}, \tag{42}$$

and the differential cross section is

$$\frac{d\sigma}{d\Omega} = |f(\theta)|^2 = \left(\frac{2mV_0R^3}{\hbar^2}\right)^2 (1 + 4k^2R^2 \sin^2 \theta/2)^{-2}. \tag{43}$$

Finally, the total cross section is obtained by integrating the differential cross section over all solid angles, as in equation (23). Upon performing the integration,[12] we find

$$\sigma = \left(\frac{2mV_0R^3}{\hbar^2}\right)^2 \left(\frac{4\pi}{1 + 4k^2R^2}\right). \tag{44}$$

Observe first that at energies sufficiently low that $4k^2R^2 \ll 1$, equation (43) shows the scattering to be isotropic, agreeing with our earlier prediction in terms of the properties of the phase shifts $\delta_l$. As $kR$ increases, the cross section becomes more and more peaked in the forward direction and, for $4k^2R^2 \gg 1$, the scattering is largely confined to the small angular domain $\theta \lesssim 1/kR$, which is called the main diffraction peak.

[12] The integral is readily evaluated by introducing the new variable of integration $z = (1 - \cos \theta) = 2 \sin^2 \theta/2$.

This behavior is illustrated in Figure 5. As shown in the figure, the forward scattering turns out to be independent of energy in Born approximation, whereas the total cross section decreases with increasing energy, ultimately inversely as the energy according to equation (44).

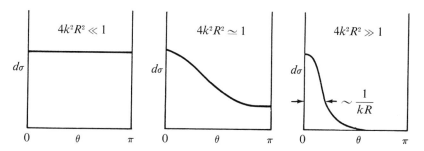

FIGURE 5.    Differential scattering for the Yukawa potential at different values of $kR$, according to the Born approximation, equation (43).

All of these features are quite general ones, with only the details differing from one interaction potential to the next. To make this explicit, let us consider as a second example a *Gaussian potential*,

$$V(r) = V_0 \, e^{-(r/R)^2}.\tag{45}$$

Omitting details, we obtain

$$f(\theta) = -\left(\frac{mV_0R^3}{2\hbar^2}\right) \sqrt{\pi} \; e^{-k^2R^2 \sin^2 \theta/2}\tag{46}$$

$$d\sigma/d\Omega = \left(\frac{mV_0R^3}{2\hbar^2}\right)^2 \pi \; e^{-2k^2R^2 \sin^2 \theta/2}\tag{47}$$

and

$$\sigma = \left(\frac{mV_0R^3}{2\hbar^2}\right)^2 \frac{2\pi^2}{k^2R^2} \left(1 - e^{-2k^2R^2}\right).\tag{48}$$

---

**Exercise 1.**    Derive equations (46), (47), and (48).

---

Again we find that the scattering in the forward direction, $\theta = 0$, is independent of energy, that the cross section is independent of angle at low energies but as the energy increases a diffraction peak develops which is of angular width $\theta \simeq 1/kR$ and, finally, that the total cross section decreases inversely as the energy at high energies. The explanation of these common general features follows:

*Forward Scattering.* According to equation (34), the scattering amplitude in the forward direction, $\hat{n} = \hat{n}_0$, is simply proportional to the volume integral of the potential. Specifically, denoting the forward scattering amplitude by $f(0)$, we have

$$f(0) = -\frac{m}{2\pi\hbar^2} \int V(r)\, d^3r. \tag{49}$$

Up to numerical factors of order unity, this can be expressed, in agreement with equations (42) and (46), as

$$f(0) \simeq -\frac{mV_0R^3}{\hbar^2}, \tag{50}$$

where $V_0$ is the strength of the potential and $R$ is its range.

*Angular Dependence.* At energies low enough that $kR \ll 1$, the exponential factor in the integrand of equation (34) never deviates significantly from unity over the effective domain of integration, which is simply a sphere of radius approximately $R$. Consequently, for all angles,

$$f(\theta) \simeq f(0),$$

whence the cross section is isotropic at low energies and the total cross section is

$$\sigma \simeq 4\pi|f(0)|^2 \simeq 4\pi \left(\frac{mV_0R^3}{\hbar^2}\right)^2. \tag{51}$$

Comparison with equations (44) and (48) in the low energy limit shows that we have again obtained the correct result, up to numerical factors of order unity.

We next consider the opposite limit in which the energy is high enough that $kR \gg 1$. Here the oscillations of the exponential factor in the integrand of equation (34) must be taken into account. Evidently this factor oscillates more and more rapidly as the scattering angle increases, thus causing the scattering amplitude to diminish sharply as one moves away from the forward direction. This diminution sets in at scattering angles such that

$$kR\,|\hat{n}_0 - \hat{n}| = 2kR \sin \theta/2 \simeq 1,$$

because at smaller angles than this the exponential deviates little from unity. Since $kR \gg 1$, this relation can be written in the simpler form

$$\theta \simeq 1/kR,$$

and we have thus demonstrated that the width of the high energy diffraction peak is of the order $1/kR$, as claimed.

*Total Cross Section at High Energies.* This last result permits us to quickly estimate the high energy total cross section. The essential feature is that only the main diffraction peak contributes significantly to the scattering, whence we have

$$\sigma = \int |f(\theta)|^2 \sin\theta \, d\theta \, d\phi \simeq 2\pi \int_0^{1/kR} |f(\theta)|^2 \sin\theta \, d\theta$$

$$\simeq 2\pi \int_0^{1/kR} |f(\theta)|^2 \, \theta \, d\theta \simeq \pi |f(0)|^2 / k^2 R^2$$

and, finally, using equation (48),

$$\sigma \simeq \left(\frac{mV_0R^3}{\hbar^2}\right)^2 \frac{\pi}{k^2R^2}, \tag{52}$$

in agreement with the high energy limit of equations (44) and (48), up to the usual numerical factors.

We conclude our discussion of the Born approximation with some remarks about its domain of validity. Most treatments of this subject are based on a consideration, which turns out to be rather complicated, of the magnitude of the first-order correction to the zero-order plane wave state function.[13] However, we give here a much simpler analysis in which the total cross section $\sigma$ plays the key role. The essence of the argument is that, in the spirit of time dependent perturbation theory, the cross section directly measures the effective rate for transitions out of the initial state. Now this rate must be small if the perturbation treatment is to be valid. To put this on a quantitative basis, we shall take the effective rate to be the ratio of the scattered probability flux to the maximum possible scattered flux, namely that incident on the *geometrical* cross section of the scatterer,[14] $\pi R^2$. The ratio in question is simply $\sigma/\pi R^2$ and hence our condition for the validity of the Born approximation is that this ratio be small,

$$\sigma/\pi R^2 \ll 1. \tag{53}$$

At low energies, using equation (51), the validity condition is thus

$$\left(\frac{2mV_0R^2}{\hbar^2}\right)^2 \ll 1 \tag{54}$$

and at high energies, using equation (52), it is

---

[13] A particularly complete treatment is presented in Reference [18].

[14] Strictly speaking, this is the maximum only at high energies. At low energies the maximum is considerably larger because of diffraction. As a result, our estimate is on the conservative side.

$$\frac{1}{k^2R^2}\left(\frac{mV_0R^2}{\hbar^2}\right)^2 \ll 1. \tag{55}$$

Observe that the low energy condition is much more stringent than the high energy one; whenever equation (54) is satisfied, equation (55) is automatically, and much more strongly, satisfied, because it contains the extra factor $(kR)^{-2}$, which is small at high energies. Moreover, at sufficiently high energies, equation (55) can always be satisfied, even if equation (54) is not. For this reason, the Born approximation is primarily regarded as a high energy approximation and it thus nicely supplements the phase shift method which, it is recalled, becomes quite unwieldy in the high energy domain.

Equations (54) and (55) cannot be thought of as particularly precise or rigorous conditions, as is evident from our mode of derivation. The following argument, however, may perhaps serve to increase our confidence in their reliability. Recall that the exact solution to any scattering problem satisfies the optical theorem of equation (25),

$$\sigma = \frac{4\pi}{k}\, \mathrm{Im}\, f(0).$$

Now the Born approximation can *never* satisfy this theorem because, as equation (49) makes clear, the forward scattering amplitude in Born approximation is *always* real. This implies that the Born approximation is not reliable unless the imaginary part of the true forward scattering amplitude is negligibly small compared to the entire forward scattering amplitude. In brief, we must have as a kind of self-consistency condition

$$\mathrm{Im}\, f(0) \ll |f(0)|$$

or, multiplying through by $4\pi/k$ and using the optical theorem,

$$\sigma \ll \frac{4\pi}{k}\,|f(0)|.$$

At low energies this gives, using the estimates of equations (50) and (51),

$$kR\left(\frac{mV_0R^2}{\hbar^2}\right) \ll 1. \tag{56}$$

This is seen to be a generally weaker condition than equation (54) because it contains the factor $kR$, which becomes vanishingly small in the low energy limit. On the other hand, at high energies we obtain

$$\frac{1}{4kR}\left(\frac{mV_0R^2}{\hbar^2}\right) \ll 1, \tag{57}$$

which is the square root of equation (55) up to numerical factors, and hence is a generally somewhat stronger condition.

In summary, then, the self-consistency requirement and the requirement that the total transition rate be small lead roughly, but not precisely, to equivalent conditions for the validity of the Born approximation. This may be taken as rather strong evidence that we have indeed identified the relevant features. As far as the application of these conditions in practice is concerned, the safest guide is to use whichever of the two sets of conditions is more restrictive in the energy domain under consideration.

## 4. MOTION IN AN ELECTROMAGNETIC FIELD

Consider a particle of positive charge $e$ moving in a given externally imposed electromagnetic field described by the vector potential $\mathbf{A}(\mathbf{r}, t)$ and scalar potential $\phi(\mathbf{r}, t)$. It is recalled that in Gaussian units the electromagnetic field strengths $\mathscr{E}$ and $\mathscr{H}$ are given in terms of these potentials by

$$\mathscr{H} = \nabla \times \mathbf{A}$$
$$\mathscr{E} = -\nabla\phi - \frac{1}{c}\frac{\partial \mathbf{A}}{\partial t}, \tag{58}$$

and the classical Hamiltonian is

$$H = \frac{1}{2m}(\mathbf{p} - e\mathbf{A}/c)^2 + e\phi + V, \tag{59a}$$

where $V$ is any additional potential which may be present.[15] It is not difficult to verify that Hamilton's equations

$$\frac{dx_i}{dt} = \frac{\partial H}{\partial p_i}$$
$$\frac{dp_i}{dt} = -\frac{\partial H}{\partial x_i} \tag{59b}$$

then yield the correct equations of motion

$$m\frac{d^2r}{dt^2} = e\mathscr{E} + \frac{e}{c}(\mathbf{v} \times \mathscr{H}) - \nabla V. \tag{60}$$

Note however, that from the first of equations (59b)

$$m\frac{dx_i}{dt} = p_i - \frac{eA_i}{c}, \tag{61}$$

[15] See, for example, Reference [14], especially Chapter I, Section 5, and Chapter VII, Section 3. See also Problem 11.

so that **p** is *not* the kinetic momentum. It is nonetheless the *canonical momentum*, that is, the formal momentum variable in the sense of Hamilton's equations.

Note also that the electromagnetic potentials have no direct physical significance; only the field strengths do. Indeed **A** and $\phi$ are not uniquely defined by equation (58). The class of *transformations* which leave $\mathcal{E}$ and $\mathcal{H}$ unchanged are called *gauge transformations* and are generated by arbitrary scalar functions $\chi$ according to

$$\mathbf{A}' = \mathbf{A} - \nabla\chi, \qquad \phi' = \phi + \frac{1}{c}\frac{\partial\chi}{\partial t}. \tag{62}$$

The choice of $\chi$ determines the gauge, but no physical result depends upon this choice. Otherwise stated, the potentials $\mathbf{A}'$, $\phi'$ yield the same field strengths, equations of motion, and so forth, as do $\mathbf{A}$, $\phi$. Note that while the canonical momentum *does* depend upon the gauge, the particle velocity does not, as may be seen from equation (60), which makes no reference to gauge dependent quantities.

As usual, the Hamiltonian operator is obtained by replacing the classical dynamical variables in the classical Hamiltonian by the quantum mechanical operators which represent them. Thus Schrödinger's equation becomes

$$\frac{1}{2m}\left(\mathbf{p} - \frac{e\mathbf{A}}{c}\right)^2\psi + (e\phi + V)\psi = -\frac{\hbar}{i}\frac{\partial\psi}{\partial t}. \tag{63}$$

The commutation rules between **p** and **r** are unaltered by the presence of the field and thus assume their usual form

$$(p_i, x_j) = \frac{\hbar}{i}\delta_{ij}.$$

In configuration space we still have, therefore,

$$\mathbf{p} = \frac{\hbar}{i}\nabla.$$

Note that **p** and $\mathbf{A}(\mathbf{r}, t)$ do not commute in general and hence that the meaning of the first term of equation (63) must be spelled out. Specifically, this term is to be understood as having the symmetrized form,

$$\left(\mathbf{p} - \frac{e\mathbf{A}}{c}\right)^2 = p^2 - \frac{e}{c}(\mathbf{p}\cdot\mathbf{A} + \mathbf{A}\cdot\mathbf{p}) + \frac{e^2}{c^2}A^2, \tag{64}$$

in which case it is not hard to see that it is properly Hermitian.

We have argued that the vector and scalar potentials are not uniquely defined but may be altered at will by a gauge transformation. Classically, no physical results are affected by such a transformation, and the same conclusion must hold in the quantum mechanical case as well. To see how this comes about, consider the gauge transformation of equation (62). Let $\psi'(x, t)$ denote the transformed state function, which is to say, let it be the solution to Schrödinger's equation in the new gauge,

$$\frac{1}{2m} \left( \mathbf{p} - \frac{e\mathbf{A}'}{c} \right)^2 \psi' + (e\phi' + V)\psi' = -\frac{\hbar}{i} \frac{\partial \psi'}{\partial t}.$$

Substitution of equation (62) then yields

$$\frac{1}{2m} \left( \mathbf{p} + \frac{e}{c} \nabla\chi - \frac{e}{c}\mathbf{A} \right)^2 \psi' + (e\phi + V)\psi' = -\frac{\hbar}{i} \frac{\partial \psi'}{\partial t} - \frac{e}{c} \frac{\partial \chi}{\partial t} \psi'. \quad (65)$$

Comparison of equations (63) and (65) now shows that $\psi$ and $\psi'$ are related by the expression

$$\psi'(x, t) = \psi(x, t)\, e^{-iex/\hbar c}, \quad (66)$$

as is readily verified by direct substitution. Otherwise stated, the gauge transformation of equation (62) is generated by replacing $\psi$ by $\psi\, e^{iex/\hbar c}$ in equation (63). We thus conclude that the arbitrariness in the definition of the electromagnetic potentials is reflected in a corresponding arbitrariness in the *phase* of the state function. The phase is actually determined only within an *undetectable* scalar function $ex/\hbar c$, and this is the meaning of gauge invariance for Schrödinger's equation.[16]

As an illustrative example, consider a charged particle in a uniform magnetic field $\mathcal{H}$. For this case the vector potential can be written in the form

$$\mathbf{A} = \frac{1}{2} (\mathcal{H} \times \mathbf{r}).$$

It may readily be verified that equation (58) is satisfied with this choice. Since $\mathbf{p}$ commutes with this particular vector potential, we can write $\mathbf{A} \cdot \mathbf{p} + \mathbf{p} \cdot \mathbf{A} = 2\mathbf{A} \cdot \mathbf{p}$ and hence

$$(\mathbf{p} - e\mathbf{A}/c)^2 = p^2 - \frac{e}{c} (\mathcal{H} \times \mathbf{r}) \cdot \mathbf{p} + \frac{e^2}{4c^2} (\mathcal{H} \times \mathbf{r})^2.$$

Now

$$(\mathcal{H} \times \mathbf{r}) \cdot \mathbf{p} = \mathcal{H} \cdot (\mathbf{r} \times \mathbf{p}) = \mathcal{H} \cdot \mathbf{L},$$

where $\mathbf{L}$ is the orbital angular momentum operator, and hence Schrödinger's equation becomes

$$\frac{p^2}{2m} \psi + \left[ V - \frac{e}{2mc} \mathcal{H} \cdot \mathbf{L} + \frac{e^2}{8mc^2} (\mathcal{H} \times \mathbf{r})^2 \right] \psi = -\frac{\hbar}{i} \frac{\partial \psi}{\partial t}. \quad (67)$$

---

[16] See Problem 10, Chapter V, for a discussion of the undetectability of a phase factor such as that referred to above.

The term quadratic in $\mathcal{H}$ is most easily understood by considering the motion of a *free* particle in a magnetic field. Classically, this motion proceeds along a helical orbit and is bounded in the plane perpendicular to $\mathcal{H}$. The quadratic term in $\mathcal{H}$ in Schrödinger's equation is responsible for a similar confinement of the state function with respect to motion in the transverse plane. Indeed, it is not hard to show, although we shall not do so, that the motion in this plane is equivalent to that of a two-dimensional harmonic oscillator.[17] The quadratic term is thus clearly essential for the description of such states. On the other hand, in discussing the *bound* states of a particle in some confining potential $V$, the contribution of the quadratic term is generally quite small and it can often be neglected entirely.

The term linear in $\mathcal{H}$ in Schrödinger's equation is at once seen to describe the interaction of a magnetic dipole of magnetic moment

$$\mu = \frac{e}{2mc}\,\mathbf{L}$$

with the magnetic field. If the quadratic term is neglected, if $V(\mathbf{r})$ is spherically symmetrical and if the $z$-axis is chosen to coincide with $\mathcal{H}$, then the stationary states can be classified as simultaneous eigenfunctions of $L^2$ and of $L_z$, as usual. However, the energy now depends upon the eigenvalues $m_l\hbar$ of $L_z$. Specifically, if we have a state $\psi_{nlm_l}$ with energy $E_{nl}$ in the absence of a magnetic field, then in the presence of the field we have

$$E_{nlm_l} = E_{nl} - \frac{e\hbar}{2mc}\,\mathcal{H}\cdot m_l,$$

and the intrinsic $(2l+1)$-fold degenerate state is split into $2l+1$ states with equal energy separations

$$\Delta E = \frac{e\hbar}{2mc}\,\mathcal{H}.$$

The quantity $e\hbar/2mc$ is called a Bohr magneton; it is the magnetic moment of a particle with unit orbital angular momentum. Because of the role of the quantum number $m_l$ in the above, $m_l$ is frequently called the *magnetic quantum number*. It is well to remark at this point that all of our equations have been written for a particle of positive charge $e$. For an electron, $e$ must be replaced by $-e$ everywhere.

We now briefly mention the effect of spin. Associated with the spin is a magnetic moment, which we write in the form

$$\mu_{\text{spin}} = g\,\frac{e\hbar}{2mc}\,\mathbf{S}/\hbar. \tag{68}$$

[17] See References [19] or [24].

The dimensionless quantity $g$ measures the ratio of the magnetic moment, in units $e\hbar/2mc$, to the angular momentum in units of $\hbar$. For orbital motion $g$ was seen to be unity. *For the intrinsic magnetic moment of the electron $g$ turns out to have the value two.* In a sense, the spin angular momentum is thus twice as effective as the orbital angular momentum in generating a magnetic moment.[18]

The existence of the spin magnetic moment means that the Hamiltonian must be supplemented by an additional magnetic energy term $\mu_{\text{spin}} \cdot \mathcal{H}$. Thus, for example, for an electron in a uniform magnetic field the term linear in $\mathcal{H}$ in the Hamiltonian has the form

$$(\mu + \mu_{\text{spin}}) \cdot \mathcal{H} = \frac{e}{2mc} (\mathbf{L} + 2\mathbf{S}) \cdot \mathcal{H} = \frac{e}{2mc} (\mathbf{J} + \mathbf{S}) \cdot \mathcal{H} ,$$

where $\mathbf{J}$ is the total angular momentum. The analysis of this term, leading to the so-called anomalous Zeeman effect, is considerably more complicated than for the case of a particle without spin, and we shall not carry it out.[19]

## 5. DIRAC THEORY OF THE ELECTRON

We now want to develop a relativistic version of Schrödinger's equation for the motion of an electron. We shall consider primarily the case in which no external forces act so that the electron can be treated as free. In that case, the classical relativistic Hamiltonian is

$$H = \sqrt{(pc)^2 + (mc^2)^2} ,$$

where $m$ is the rest mass of the electron and $\mathbf{p}$ is its momentum. We now see at once that if $\mathbf{p}$ is the usual quantum mechanical operator, $H$ is not well-defined because of the square root sign.[20] One way out of this difficulty was suggested by Klein and Gordon, who considered

---

[18] It is of interest to remark on the $g$-values for other particles, measuring magnetic moments always in the natural units $e\hbar/2mc$, where $m$ is the mass of the particle. For the $\mu$-meson, $g$ is again two, as for the electron. Both cases are in agreement with the predictions of the Dirac theory. On the other hand, the intrinsic magnetic moment of the proton is not one nuclear magneton, as it is called, $(g = 2)$, but 2.79 nuclear magnetons $(g = 5.59)$, while the neutron's magnetic moment is not zero (as for the neutrino) but is $-1.91$ nuclear magneton, the minus sign meaning that it is directed opposite to the spin. These anomalous magnetic moments clearly indicate the existence of some kind of charge structure for the proton and neutron. These are subjects of great current interest in the physics of elementary particles.

[19] See, for example, Reference [22].

[20] In configuration space the square root must be expanded in a power series, whence $H$ is equivalent to a differential operator of infinite order. This can be circumvented by working in momentum space, but the resulting equations are completely intractable except for the special case of a free particle.

$$H^2\psi = \left( -\frac{\hbar}{i}\frac{\partial}{\partial t} \right)^2 \psi$$

rather than

$$H\psi = -\frac{\hbar}{i}\frac{\partial\psi}{\partial t}.$$

This is a perfectly good relativistic wave equation but, because of the second time derivative, probability is not conserved if $\psi$ is interpreted as a probability amplitude. It turns out that this equation, the Klein–Gordon equation, can be interpreted in the quantum theory of fields, but it cannot be applied to the motion of a single particle.

The dilemma was resolved by Dirac using the following argument. If $\psi$ is to be a probability amplitude, then only a first-order time derivative can appear in Schrödinger's equation. Since time and space coordinates enter on an essentially equal footing relativistically, the space derivatives in Schrödinger's equation must also appear only linearly. Thus he wrote

$$H\psi = -\frac{\hbar}{i}\frac{\partial\psi}{\partial t},$$

where $H$ is now constrained to be linear in the momentum,

$$H = [\alpha \cdot \mathbf{p}c + \beta mc^2], \tag{69}$$

and where $\alpha$ and $\beta$ are to be determined by the requirement that

$$H^2 = (pc)^2 + (mc^2)^2. \tag{70}$$

Thus Dirac forced the issue by imposing the condition that

$$H = \sqrt{(pc)^2 + (mc^2)^2}$$

be a well-defined linear function of $\mathbf{p}$. It is clear, of course, that $\alpha$ and $\beta$ cannot be ordinary numbers if equations (69) and (70) are to be satisfied but must be (space and time independent) operators.

Keeping in mind the operator nature of $\alpha$ and $\beta$ we thus require

$$[\alpha \cdot \mathbf{p}c + \beta mc^2]^2 = (pc)^2 + (mc^2)^2$$

or, preserving the order of all relevant factors,

$$\sum_i \alpha_i^2 p_i^2 c^2 + \frac{1}{2}\sum_{i \neq j}(\alpha_i\alpha_j + \alpha_j\alpha_i)\,p_i p_j c^2 + \sum_i (\alpha_i\beta + \beta\alpha_i)mc^2 p_i c$$
$$+ \beta^2(mc^2)^2 = p^2 c^2 + (mc^2)^2.$$

Comparing terms we thus see that $\alpha_i^2 = \beta^2 = 1$ or

$$\alpha_x^2 = \alpha_y^2 = \alpha_z^2 = \beta^2 = 1 \tag{71}$$

and

$$(\alpha_i, \alpha_j)_+ = 0, \qquad i \neq j$$

$$(\alpha_i, \beta)_+ = 0. \tag{72}$$

In words, *the four matrices $\alpha_x$, $\alpha_y$, $\alpha_z$ and $\beta$ mutually anti-commute and the square of each is unity.*

The algebra of these Dirac operators is seen to be identical to that of the Pauli spin operators, except that there are four of them. Since the Pauli operators (along with the unit matrix) exhaust the independent two-by-two matrices, the four Dirac operators cannot be represented by two-by-two matrices. It turns out that three-by-three matrices do not suffice, either, and the smallest matrices that do are four-by-four. These are not uniquely defined by the commutation relations, but the conventional choice is

$$\beta = \begin{pmatrix} 1 & 0 & 0 & 0 \\ 0 & 1 & 0 & 0 \\ 0 & 0 & -1 & 0 \\ 0 & 0 & 0 & -1 \end{pmatrix} \qquad \alpha_x = \begin{pmatrix} 0 & 0 & 0 & 1 \\ 0 & 0 & 1 & 0 \\ 0 & 1 & 0 & 0 \\ 1 & 0 & 0 & 0 \end{pmatrix}$$

$$\alpha_y = \begin{pmatrix} 0 & 0 & 0 & -i \\ 0 & 0 & i & 0 \\ 0 & -i & 0 & 0 \\ i & 0 & 0 & 0 \end{pmatrix} \qquad \alpha_z = \begin{pmatrix} 0 & 0 & 1 & 0 \\ 0 & 0 & 0 & -1 \\ 1 & 0 & 0 & 0 \\ 0 & -1 & 0 & 0 \end{pmatrix} \tag{73}$$

or, in compressed notation,

$$\beta = \begin{pmatrix} 1 & 0 \\ 0 & -1 \end{pmatrix} \qquad\qquad \alpha = \begin{pmatrix} 0 & \sigma \\ \sigma & 0 \end{pmatrix}, \tag{74}$$

where each element now stands for a two-by-two matrix and $\sigma$ denotes the usual Pauli spin operator.

The Dirac operators commute with all external variables such as **p** and **r** and hence must act on some kind of internal degrees of freedom. Evidently $\psi$ must now be regarded as a four-component wave function, called a *spinor*, in view of the dimensionality of the Dirac matrices. We shall write such a four-component function as a column matrix, that is, as

$$\psi = \begin{pmatrix} \psi_1 \\ \psi_2 \\ \psi_3 \\ \psi_4 \end{pmatrix}, \tag{75}$$

with the understanding that when an arbitrary four-by-four matrix $A$, with elements $A_{ij}$, acts on $\psi$ it does so according to the rules of matrix multiplication to produce an altered column matrix. Explicitly written out, this means that

$$A\psi = \begin{pmatrix} A_{11} & A_{12} & \cdot & A_{14} \\ A_{21} & & & \cdot \\ \cdot & & & \\ A_{41} & \cdot & \cdot & A_{44} \end{pmatrix} \begin{pmatrix} \psi_1 \\ \psi_2 \\ \psi_3 \\ \psi_4 \end{pmatrix} = \begin{pmatrix} \Sigma A_{1i}\psi_i \\ \Sigma A_{2i}\psi_i \\ \Sigma A_{3i}\psi_i \\ \Sigma A_{4i}\psi_i \end{pmatrix}.$$

Thus, specifically,

$$\beta\psi = \begin{pmatrix} \psi_1 \\ \psi_2 \\ -\psi_3 \\ -\psi_4 \end{pmatrix} \qquad \alpha_x\psi = \begin{pmatrix} \psi_4 \\ \psi_3 \\ \psi_2 \\ \psi_1 \end{pmatrix}$$

$$\alpha_y\psi = \begin{pmatrix} -i\psi_4 \\ i\psi_3 \\ -i\psi_2 \\ i\psi_1 \end{pmatrix} \qquad \alpha_z\psi = \begin{pmatrix} \psi_3 \\ -\psi_4 \\ \psi_1 \\ -\psi_2 \end{pmatrix}. \tag{76}$$

The Dirac equation

$$c\left[\alpha \cdot \mathbf{p} + \beta mc\right]\psi = -\frac{\hbar}{i}\frac{\partial\psi}{\partial t} = E\psi \tag{77}$$

is now to be understood as an equation in these column matrices. Note that $E$ is the operator for the *total* energy, including the *rest energy* of the particle.

Recall now that if two matrices are equal, each element of the first must equal the corresponding element of the second. We thus see that the Dirac equation is a compact way to write a set of four coupled linear differential equations in the four components of $\psi$. Writing these out explicitly, using equation (76), we have

$$c(p_x - ip_y)\psi_4 + cp_z\psi_3 + mc^2\psi_1 = -\frac{\hbar}{i}\frac{\partial\psi_1}{\partial t}$$

$$c(p_x + ip_y)\psi_3 - cp_z\psi_4 + mc^2\psi_2 = -\frac{\hbar}{i}\frac{\partial\psi_2}{\partial t}$$

$$c(p_x - ip_y)\psi_2 + cp_z\psi_1 - mc^2\psi_3 = -\frac{\hbar}{i}\frac{\partial\psi_3}{\partial t} \tag{78}$$

$$c(p_x + ip_y)\psi_1 - cp_z\psi_2 - mc^2\psi_4 = -\frac{\hbar}{i}\frac{\partial\psi_4}{\partial t}.$$

We now consider a stationary state of definite linear momentum **p** and energy $E$. For simplicity, we shall take **p** to be directed along the $z$-axis. Writing

$$\psi = Ue^{ipz/\hbar} \tag{79}$$

where $U$ is a spinor with constant components,

$$U = \begin{pmatrix} U_1 \\ U_2 \\ U_3 \\ U_4 \end{pmatrix},$$

equation (78) then reduces to the *algebraic* equations

$$cpU_3 - (E - mc^2)U_1 = 0$$
$$-cpU_4 - (E - mc^2)U_2 = 0$$
$$cpU_1 - (E + mc^2)U_3 = 0 \tag{80}$$
$$-cpU_2 - (E + mc^2)U_4 = 0.$$

It is then easily verified that we obtain a solution, provided that

$$E = \pm\sqrt{(pc)^2 + (mc^2)^2}. \tag{81}$$

With either choice of sign, we then have

$$\frac{U_1}{U_3} = \frac{E + mc^2}{pc} = \frac{pc}{E - mc^2}$$

$$\frac{U_2}{U_4} = -\frac{E + mc^2}{pc} = -\frac{pc}{E - mc^2}. \tag{82}$$

Note first that equation (81) gives the correct relativistic relation between energy and momentum, but that states of both positive and negative energy are possible. For now we restrict our attention to positive energy states. We shall return later to the question of negative energies.

Next note that $U_1$ and $U_3$ are entirely independent of $U_2$ and $U_4$. This means that our solution can be expressed as an arbitrary linear combination of two independent states. Let us exhibit these states by writing

$$U = a_+\chi_+ + a_-\chi_-, \tag{83}$$

where

$$\chi_+ = \begin{pmatrix} U_1 \\ 0 \\ U_3 \\ 0 \end{pmatrix}, \qquad \chi_- = \begin{pmatrix} 0 \\ U_2 \\ 0 \\ U_4 \end{pmatrix}. \tag{84}$$

The coefficients $a_+$ and $a_-$ are arbitrary and, of course, $U_1$ and $U_3$ are related to each other by equation (82), as are $U_2$ and $U_4$.

We thus see that states of given energy $E$ and given linear momentum vector (which is directed along the $z$-axis in the case under consideration) are not unique but are doubly degenerate. This means $H$ and $\mathbf{p}$ do *not* form a complete set of commuting operators but must be supplemented by an additional operator which can be associated only with some kind of internal coordinate. This internal coordinate turns out to be the spin and the missing operator to be the component of the spin along the direction of $\mathbf{p}$, as we now demonstrate.

First define the four-component analog of the Pauli spin operator by writing

$$\hat{\sigma} = \begin{pmatrix} \sigma & 0 \\ 0 & \sigma \end{pmatrix}. \tag{85}$$

The notation means, for example, that

$$\hat{\sigma}_z = \begin{pmatrix} \sigma_z & 0 \\ 0 & \sigma_z \end{pmatrix} = \begin{pmatrix} 1 & 0 & 0 & 0 \\ 0 & -1 & 0 & 0 \\ 0 & 0 & 1 & 0 \\ 0 & 0 & 0 & -1 \end{pmatrix}, \tag{86}$$

and similarly for the remaining components. It is now seen at once that

$$\hat{\sigma}_z \chi_\pm = \pm \chi_\pm$$

where $\chi_\pm$ are given by equation (84). Since the Dirac operator $\hat{\sigma}$ has exactly the same algebraic properties as the Pauli operator $\sigma$, we thus tentatively conclude that $\chi_+$ describes a state in which the spin is oriented along the positive $z$-axis and $\chi_-$ a state in which it is oriented along the negative $z$-axis. As our notation indicates, these states are the spinor analogs of the two-component nonrelativistic states discussed in Chapter X. This relationship can be made clearer by passing to the nonrelativistic limit, which can be defined by

$$E - mc^2 \ll mc^2.$$

In that limit, it follows from equation (82) that $U_3$ and $U_4$ become negligible compared to $U_1$ and $U_2$. If these small components are neglected, the Dirac spinors $\chi_\pm$ can be collapsed to two-component states which are precisely the usual representations of nonrelativistic spin states $\chi_+$ and $\chi_-$, and the Dirac operator $\hat{\sigma}$ collapses to the Pauli $\sigma$.

The argument above makes it plausible that $\hat{\sigma}$ is related to the internal angular momentum of the electron, but it is by no means a proof. Indeed, it should be observed that when relativistic effects are important, a state function which is an eigenstate of $H$ and $\mathbf{p}$ can be simultaneously made an eigenfunction only of that component of $\hat{\sigma}$ which is directed along $\mathbf{p}$, and not of either perpendicular component. Otherwise stated, only the parallel component of $\hat{\sigma}$ commutes with $H$. Thus, relativistically, there is a coupling between the internal and external degrees of freedom and the separation into "internal" and "external" is no longer quite so sharp.

A genuine proof, instead of a plausibility argument, can be constructed in the following way. We first ask whether orbital angular momentum is a constant of the motion. To answer this, let us examine the commutation relations between $\mathbf{L}$ and $H$. After some algebra,[21] we find

$$(\mathbf{L}, H) = i\hbar c (\alpha \times \mathbf{p}),$$

so that orbital angular momentum is not conserved. This is no surprise; a glance at the Dirac Hamiltonian shows that it is *not* invariant under space rotations alone. Next examine the commutator of $\hat{\sigma}$ and $H$. After some more algebra we obtain

$$(\hat{\sigma}, H) = -2ic\ (\alpha \times \mathbf{p}),$$

whence it follows that $\mathbf{J} = \mathbf{L} + (\hbar/2)\hat{\sigma}$ *is* a constant of the motion,

$$\left[ \left( \mathbf{L} + \frac{\hbar}{2}\hat{\sigma} \right), H \right] = (\mathbf{J}, H) = 0.$$

Hence the quantity $\mathbf{J}$, which satisfies the requisite commutation relations for angular momentum, must be interpreted as the total angular momentum and $(\hbar/2)\hat{\sigma}$ as the operator representing intrinsic angular momentum. The Dirac equation thus has the automatic consequence that the electron, or any other particle it describes, has spin angular momentum one-half.

Since spinor wave functions have four components, we must extend our definition of expectation values and matrix elements. The meaning of an expression like $\langle \psi | \phi \rangle$ is

$$\langle \psi | \phi \rangle = \sum_{i=1}^{4} \langle \psi_i | \phi_i \rangle,$$

where $\psi_i$ and $\phi_i$ are the $i$th components of the spinor states $\psi$ and $\phi$. The quantities $\langle \psi_i | \phi_i \rangle$ have their normal meaning. Thus, for example, in configuration space

---

[21] The details are left to the problems.

$$\langle \psi_i | \phi_i \rangle = \int \psi_i^*(\mathbf{r}, t)\phi_i(\mathbf{r}, t)\ d^3r.$$

Note that this rule is all that is required for the evaluation of the matrix elements of an arbitrary operator $A$, arbitrary in both spinor and ordinary space, since we can always write $\langle \psi|A|\phi \rangle = \langle \psi|\phi' \rangle$, where $\phi'$ is simply the new spinor obtained by operating upon $\phi$ with $A$.

We turn now to the question of the negative energy states. According to equation (81), the free particle energy spectrum is a continuum ranging from plus infinity down to the rest energy $mc^2$. Below this energy there is a gap containing no states, and this gap extends to the negative energy $-mc^2$. At more negative energies still the spectrum again becomes continuous and extends to minus infinity. This spectrum is shown in Figure 6. In a certain sense it is no different than the classical spectrum, since the same negative energy solutions are *formally* possible in the classical case as well. Classically, however, no communication whatsoever is possible between the positive and negative energy portions of the spectrum. A particle cannot change from a state of motion with positive energy to one with negative energy, for to do so it would have to pass through an energy interval in which motion is impossible. The existence of such states thus causes no difficulty on the classical level; such states are simply inaccessible and therefore unobservable.

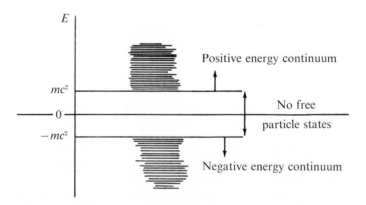

FIGURE 6.   Spectrum of allowed energies for a free Dirac particle.

On the quantum level, all is changed; discontinuous changes are the rule. Hence, transitions from positive to negative energy states would certainly be expected to occur, and with the release of an enormous amount of energy—in excess of $2mc^2$. Indeed, since the negative energy spectrum extends to indefinitely large values it would appear that no Dirac particle could ever be observed in a positive energy state. In other words, all of the electrons in the universe would quickly fall into the negative energy "sea," a sea more than deep enough to accept them.

It might appear that one way out of this difficulty would be to rule out negative energy states as physically inadmissible. It turns out, however, that if this is done the Dirac states do not form a complete set.[22] Dirac gave a brilliant solution to this problem by postulating that in the normal state of the universe all of these negative energy states are *completely* occupied. Under these conditions the exclusion principle then forbids any transitions of an electron from a state of positive to a state of negative energy. Furthermore, this vast, infinitely dense ensemble of negative energy state electrons would be electrically and mechanically inert, except for transitions a particle in the filled negative sea might make to the relatively empty states of positive energy. Such a transition would require an energy in excess of $2mc^2$. This huge energy is not available under normal conditions and the negative energy particles are not observable, in agreement with experience.

Suppose, however, that enough energy were supplied to produce a transition from the negative sea to a positive energy state. We would then observe the sudden appearance of an ordinary electron, originally unobservable as an occupant of a negative energy state. We would also observe a correspondingly sudden change in the properties of the negative energy sea. This sea would no longer be electrically and mechanically inert, and therefore unobservable, because it would no longer be completely filled; one state would be unoccupied. This hole state, as it is called, is associated with the removal of a negative charge of negative energy from the inert normal state of the universe, which we might call the vacuum. The *removal* of a negative charge of negative energy from the vacuum can be interpreted as the appearance of a positive charge of positive energy, called the positron, completely equivalent to an electron except for its opposite charge.

That this interpretation is reasonable can be seen by considering the effects of an applied electromagnetic force on the negative energy sea when the sea contains a hole state. The electrons would tend to move in the direction of the applied force, and transitions to the hole state, the *only* available state within the sea, would be favored if they supported this tendency. The hole would then tend to move oppositely to the electrons, and indeed to do so as freely as if it were an isolated object.

---

[22] This is one example, and a very important one, in which the mathematics initially forced the physical picture, rather than the other way around. We have given a physical interpretation of completeness by arguing, correctly, that the totality of physically admissible states must be complete. This leaves the question of how we are to identify all of the physically admissible states. In the past we have more or less taken this identification for granted and have then asserted completeness as a necessary consequence. In the present instance, even if our intuition about physical admissibility is inadequate, the completeness requirement forces us to admit negative energy states.

It is now seen that the transition of an electron from a negative to a positive energy state corresponds to the sudden appearance of an electron and positron at the same point in space and time and requires a minimum energy expenditure of $2mc^2$. This is the famous phenomenon of *pair production*. The reverse of this process also occurs, of course, when an electron happens to fall from a positive energy state into a hole state. This is the phenomenon of *electron–positron annihilation,* and the energy appears as electromagnetic energy, normally as a pair of photons.

So far we have considered only the free particle Dirac Hamiltonian. Suppose now that the electron moves in an external electromagnetic field or other potential. The Dirac Hamiltonian is then modified in the canonical way by writing

$$H = c[\alpha \cdot (\mathbf{p} - e\mathbf{A}/c)] + \beta mc^2 + e\phi + V,$$

where $\mathbf{A}$ and $\phi$ are the electromagnetic vector and scalar potentials and where $V$ is any other independent interaction which may exist. If one solves this equation for an electron in a uniform magnetic field, for example, the magnetic interaction energy turns out to be precisely that of a spin one-half particle carrying a magnetic moment of one Bohr magneton, just as required by observation. To mention a second example, the solution of the Dirac equation for the states of the hydrogen atom gives almost perfect agreement with experiment, including all fine structure corrections.[23]

Starting with only its mass and charge, the Dirac theory of the electron is thus seen to account for all of its intrinsic properties as well as for the existence of the positron and the phenomena of pair production and annihilation.

We have emphasized the electron in our discussion. However, every spin one-half particle is described by the Dirac equation. Muons and neutrinos, for example, are described by precisely the same equation, with appropriate rest mass, as that for electrons. On the other hand, protons and neutrons, as a consequence of their anomalous magnetic moments, are described by a Dirac Hamiltonian supplemented by an extra term, empirically adjusted to give the observed magnetic moment. Pair production and annihilation involving all of these particles and their anti-particles, as they are called, have been observed, in full agreement with the predictions of the theory.

---

[23] However, the Dirac equation as we have considered it does not give the so-called Lamb shift. This very small shift is generated by interactions with the fluctuating vacuum electromagnetic fields. Such field fluctuations, like the zero point motion of the harmonic oscillator, are entirely quantum mechanical in nature and do not appear in any treatment, such as ours, in which the electromagnetic field is taken to be classical.

## 6. MIXED STATES AND THE DENSITY MATRIX

Every *uniquely* defined state of a quantum mechanical system must be a simultaneous eigenfunction of some complete set of operators. Expressed in terms of observations, this means that some complete set of noninterfering measurements must be performed if a state is to be uniquely specified. Such a uniquely specified state is called a *pure state*. The evolution of a pure state with time is governed by Schrödinger's equation, and this evolution proceeds in a perfectly definite predictable way. The intrinsic statistical nature of quantum mechanics manifests itself in the distributions of observed values which are obtained when measurements are performed upon an ensemble of identically prepared systems, each in the same pure state. Only pure states have concerned us to the present. More often than not, however, our knowledge of complicated systems is not complete and a description in terms of pure states is impossible. This lack of knowledge introduces a second statistical element, which is not quantum mechanical in origin and which is familiar from classical statistical mechanics. We now briefly show how incompletely defined or *mixed states* can be treated quantum mechanically.

The description of a mixed state requires a statistical mixture of all those pure states which happen to be consistent with our incomplete knowledge, whatever it may be. Alternatively, we can think of an ensemble of systems, each of which is in one or another of such pure states. Consider now one such system of the ensemble described by the pure state $\psi$, and suppose $\psi$ to be expressed in terms of some complete orthonormal set,

$$\psi = \sum_n C_n \phi_n; \qquad |C_n| \leqslant 1. \tag{87}$$

Let $A$ be an operator representing some arbitrary observable. Then

$$\langle A \rangle = \sum_{n,m} C_m C_n{}^* A_{nm}, \tag{88}$$

where the $A_{nm}$ are the matrix elements of $A$,

$$A_{nm} = \langle \phi_n | A | \phi_m \rangle.$$

Averaging over the ensemble, in the usual sense of statistical mechanics, we then obtain

$$\overline{\langle A \rangle} = \sum_{n,m} \overline{C_m C_n{}^*} A_{nm}, \tag{89}$$

where a bar over a quantity denotes that an ensemble average has been taken. Note that the $A_{nm}$ are fixed known numbers so that the ensemble average on the right involves only the $C_n$, which play the role of statistical variables. Since $A$ is arbitrary, we thus see that the ensemble average for *any* physical operator whatsoever is uniquely specified by a knowledge of the quantity

$$\rho_{mn} \equiv \overline{C_m C_n^*}; \qquad |\rho_{mn}| \leq 1. \tag{90}$$

The operator $\rho$, with matrix elements $\rho_{mn}$, is called the *density matrix*. It contains *all* of the available information about the system, since if $\rho$ is known, so is the distribution of measured values for every measurement which might conceivably be performed upon the ensemble.

Expressing equation (89) in terms of $\rho$, we now obtain

$$\overline{\langle A \rangle} = \sum_{n,m} \rho_{mn} A_{nm}.$$

Since the right-hand side can be interpreted as a matrix product, this reduces to

$$\overline{\langle A \rangle} = \sum_m (\rho A)_{mm}. \tag{91}$$

The sum of the diagonal elements of a matrix is called the *trace* of the matrix and is written as

$$\mathrm{Tr}\, B \equiv \sum_m B_{mm},$$

whence equation (91) can be expressed in the representation independent form

$$\overline{\langle A \rangle} = \mathrm{Tr}\,(\rho A). \tag{92}$$

We have now achieved our main objective, namely, to develop a suitable characterization of mixed states. The density matrix for such states is seen to play the same role as does the state function for pure states. In particular, all of the information available is contained in each, and each defines the state it characterizes.

We next develop some of the properties of the density matrix. Note first, from equation (90), that $\rho$ is Hermitian,

$$\rho_{mn} = \rho_{nm}^*. \tag{93}$$

This follows because of the independence of ensemble averaging and complex conjugation. Next observe, setting $A$ equal to unity in equation (92), that

$$\mathrm{Tr}\, \rho = 1. \tag{94}$$

In words, *the diagonal elements of the density matrix sum to unity*. This result can be verified from equation (90), since we have

$$\mathrm{Tr}\, \rho = \sum_m \overline{C_m C_m^*} = \overline{\sum_m C_m C_m^*} = 1.$$

Observe also that the diagonal element $\rho_{mm}$ directly gives the probability that the system is in the state $\phi_m$. Hence $\rho_{mm}$ can never be negative, but must lie between zero and unity. Considering a representation in which $\rho$ is diagonal, the diagonal elements are then the eigenvalues of $\rho$, and we see that the eigenvalues of the density matrix all lie between zero and unity, so that $\rho$ is a positive definite, bounded operator.

Suppose now that $\rho$ is prescribed initially. What can we say about its development in time? Let $H$ denote the Hamiltonian of the system and consider the $C_n$ in equation (87) to be functions of time,

$$\psi(t) = \Sigma\, C_n(t)\phi_n.$$

Since $\psi$ satisfies

$$H\psi = -\frac{\hbar}{i}\frac{\partial\psi}{\partial t},$$

we find in the usual way that the $C_m$ satisfy the set of equations

$$-\frac{\hbar}{i}\frac{dC_m}{dt} = \sum_i H_{mi}C_i, \tag{95}$$

whence also,

$$\frac{\hbar}{i}\frac{dC_n{}^*}{dt} = \sum_i H_{ni}{}^*C_i{}^* = \sum_i C_i{}^*H_{in}, \tag{96}$$

where we have used the fact that $H$ is Hermitian in taking the last step. Multiplying equation (95) by $C_n{}^*$ and equation (96) by $C_m$ and subtracting, this gives

$$-\frac{\hbar}{i}\frac{d}{dt}(C_mC_n{}^*) = \sum_i (H_{mi}C_iC_n{}^* - C_mC_i{}^*H_{in}).$$

The operations of time differentiation and ensemble averaging are independent, and hence, upon averaging over the ensemble, we obtain

$$-\frac{\hbar}{i}\frac{d\rho_{mn}}{dt} = \sum_i (H_{mi}\rho_{in} - \rho_{mi}H_{in})$$

$$= (H, \rho)_{mn}.$$

Thus finally, the time development of $\rho$ is given by

$$-\frac{\hbar}{i}\frac{d\rho}{dt} = (H, \rho), \tag{97}$$

whence we observe that $\rho$ is stationary if it commutes with the Hamiltonian.

To illustrate some of these properties of $\rho$ we now give some simple examples. Consider a system in a state of definite energy $E$ and angular momentum $l$, but suppose its angular momentum to be oriented at random. Then each of its $2l + 1$ degenerate states of the given $E$ and $l$ are equally probable. These $2l + 1$ states are complete for the system, and in a representation using these states $\rho$ is simply the unit matrix divided by $(2l + 1)$. A similar example is that of an unpolarized particle of spin one-half. Then, in the usual representation in which $\sigma_z$ is diagonal,

$$\rho = \begin{pmatrix} 1/2 & 0 \\ 0 & 1/2 \end{pmatrix},$$

which describes a mixed state in which the spin is equally likely to be up or down with respect to any axis. Note that in both cases the trace of $\rho$ is properly unity and that $\rho$ is stationary.

A final and extremely important example is that of a system in thermal equilibrium at some temperature $T$. The probability that the system is in a given state is determined by the Boltzmann factor and hence

$$\rho = \frac{1}{Z} e^{-H/kT},$$

where $Z$, the partition function, is given by

$$Z = \mathrm{Tr}\ e^{-H/kT}$$

so that the trace of $\rho$ is unity. Note that since $\rho$ is a function only of $H$ it commutes with $H$ and hence is stationary, as it must be for equilibrium. All thermodynamic properties of the system are determined in the usual way in terms of the partition function.

We conclude by remarking that the density matrix formalism is applicable to the description of pure states as well as to mixed. If a state is pure, all systems of the ensemble are in the same state and the ensemble average has no effect. Thus $\rho_{mn}$ has the *product form*

$$\rho_{mn} = C_m C_n^*. \tag{98}$$

It is precisely this product form that identifies $\rho$ as describing a pure state. To obtain a representation-independent characterization, observe that

$$(\rho^2)_{mn} = \sum_i \rho_{mi}\rho_{in}$$

$$= \sum_i C_m C_i^*\ C_i C_n^*,$$

whence, since the $|C_i|^2$ sum to unity,

$$(\rho^2)_{mn} = \rho_{mn},$$

that is to say, $\rho^2 = \rho$ for a pure state. As a consequence

$$\mathrm{Tr}\ \rho^2 = 1, \tag{99}$$

which can be shown to be a necessary and sufficient condition that the state is pure.[24]

---

**Problem 1.** Evaluate the integral of equation (5) to find the first-order perturbation theory correction to the energy of helium-like atoms.

**Problem 2.** Starting with the trial function of equation (8) for the helium-like atom, derive equation (10).

**Problem 3.** Consider the helium-like atom with $Z = 1$, the *negative hydrogen ion*. Show that according to the variational estimate of equation (10) the system is unstable against dissociation by a little less than one eV. [The $H^-$ system is actually *stable*, but it requires a considerably more refined calculation to show this than that leading to equation (10).]

**Problem 4.**
(a) Treating the protons as infinitely massive, and neglecting spin-dependent terms, show that the Hamiltonian for the *hydrogen molecule* is

$$H = \frac{p_1^2}{2m} + \frac{p_2^2}{2m} - \frac{e^2}{r_1} - \frac{e^2}{r_2} + \frac{e^2}{R} + \frac{e^2}{|\mathbf{r}_1 - \mathbf{r}_2 - \mathbf{R}|} - \frac{e^2}{|\mathbf{r}_1 - \mathbf{R}|} - \frac{e^2}{|\mathbf{r}_2 + \mathbf{R}|},$$

using the coordinate system shown in Figure 7.

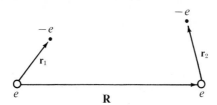

FIGURE 7.   Coordinate system for the hydrogen molecule.

(b) A reasonable trial function for a Rayleigh–Ritz calculation is one in which each electron is assumed to be in a hydrogenic ground state $\phi_0$ with respect to its "own" nucleus. Show that a *properly symmetrized* trial function based on this idea is

$$\psi(\mathbf{r}_1, \mathbf{r}_2) = \phi_0(r_1)\phi_0(r_2) \pm \phi_0(\mathbf{r}_2 + \mathbf{R})\phi_0(\mathbf{r}_1 - \mathbf{R}).$$

(c) Using this trial function show that

$$E_\pm(R) = \frac{J \pm K}{1 \pm \alpha},$$

where $J$ and $K$ are direct and exchange integrals of the total Hamiltonian $H$, and where

[24] See Reference [22].

$$\alpha = \int \phi_0(r_1)\phi_0(r_2)\phi_0(\mathbf{r}_2 + \mathbf{R})\phi_0(\mathbf{r}_1 - \mathbf{R}) \ d^3r_1 \, d^3r_2.$$

This result, when the integrals are evaluated, shows that only the symmetrical case leads to a bound state, $E_+(R)$ having a minimum of about the right value at about the right value of $R$. See reference [20].

**Problem 5.** Plot the ratio of quantum mechanical to classical differential cross sections for Coulomb scattering of electrons by electrons at a center-of-mass energy of 100 electron volts; for alpha particles on alpha particles at 5 MeV.

**Problem 6.** Use Born approximation to find the scattering amplitude plus the differential and total cross sections for the exponential potential

$$V = V_0 \, e^{-r/R}.$$

Compare the low and high energy behavior with that predicted in the text.

**Problem 7.** The same as problem 6 for a square well potential,

$$V = \begin{cases} -V_0, & r < R \\ 0, & r \geq R. \end{cases}$$

**Problem 8.** The screening effect of the electrons in a neutral atom modifies the electrostatic interaction between an atom and an incident charged particle. A reasonable approximation to this *screened Coulomb potential*, as it is called, is

$$V(r) = \frac{Zze^2}{r} e^{-r/R},$$

which is seen to be the same as the Yukawa potential in form. Here $Z$ is the atomic number of the atom, $ze$ is the charge of the incident particle, and $R$ is the effective atomic radius.

(a)  Show that the Born approximation gives the *exact* result for the Rutherford cross section in the limit $R \to \infty$.

(b)  Use Born approximation to estimate the angular domain over which the differential scattering cross section from a screened Coulomb potential differs from Rutherford scattering in the following cases:

(i)  a 5 MeV alpha particle scattered by gold;

(ii)  a 1 MeV proton scattered by carbon;

(iii)  a 100 eV electron scattered by carbon.

For simplicity take $R = 1\text{Å}$ in all cases and assume all energies to be in the center-of-mass system.

**Problem 9.**

(a)  At a center-of-mass energy of 5 MeV, the phase shifts describing the elastic scattering of a neutron by a certain nucleus have the following values:

$$\delta_0 = 32.5°, \qquad \delta_1 = 8.6°, \qquad \delta_2 = 0.4°.$$

Assuming all other phase shifts to be negligible, plot $d\sigma/d\Omega$ as a function of scattering angle. What is the total cross section $\sigma$? For simplicity take the reduced mass of the system to be that of the neutron.

(b) The same if the algebraic sign of all three phase shifts is reversed.

(c) The same if the sign of only $\delta_0$ is reversed.

(d) Using the results of part (a), calculate the *total* number of neutrons scattered per second out of a beam of $10^{10}$ neutrons per cm² per sec, of cross-sectional area 2 cm², incident upon a foil containing $10^{21}$ nuclei per cm². How many neutrons per second would be scattered into a counter at 90° to the incident beam and subtending a solid angle of $2 \times 10^{-5}$ ster-radians?

**Problem 10.** Find the states of the hydrogen atom in a magnetic field of strength $10^4$ gauss. For simplicity, take the electron to be spinless and neglect the quadratic terms in the magnetic field. Is the latter justified?

**Problem 11.** Show that the classical Hamiltonian, equation (59a), yields the correct classical equations of motion, equation (60). Note that

$$\frac{d\mathbf{A}(\mathbf{r}, t)}{dt} = \frac{\partial \mathbf{A}}{\partial t} + (\mathbf{v} \cdot \nabla)\mathbf{A}$$

$$= \frac{\partial \mathbf{A}}{\partial t} + \nabla(\mathbf{v} \cdot \mathbf{A}) - \mathbf{v} \times (\nabla \times \mathbf{A})$$

and that, for any vector $\mathbf{B}$,

$$\nabla(B^2) = 2(\mathbf{B} \cdot \nabla)\mathbf{B} + 2\mathbf{B} \times (\nabla \times \mathbf{B}).$$

**Problem 12.** Show that, for the Dirac equation just as for Schrödinger's equation,

$$\frac{d}{dt}\langle \psi|A|\psi \rangle = \langle \psi|\frac{\partial A}{\partial t}|\psi \rangle + \frac{i}{\hbar}\langle \psi|H, A|\psi \rangle,$$

where $A$ is an arbitrary (spinor) operator.

**Problem 13.** A particle satisfying the free particle Dirac equation is in an arbitrary state. Find explicit expressions for

(a) $\langle x \rangle$, $\langle \mathbf{p} \rangle$

(b) $\dfrac{d\langle x \rangle}{dt}$, $\dfrac{d\langle \mathbf{p} \rangle}{dt}$.

**Problem 14.** Show by direct calculation that $\mathbf{J} = \mathbf{L} + \hbar\hat{\sigma}/2$ commutes with the Dirac Hamiltonian.

**Problem 15.**

(a)   Starting with equation (80), derive equation (83).

(b)   Find the free particle Dirac state function for a particle of rest mass $m$ moving along the $x$-axis with momentum $p$.

**Problem 16.**

(a)   A system is known to be in one of three states. Its probability of being in the first is 1/2 and of being in the second is 1/3. What is the density matrix for the system?

(b)   Which, if any, of the following density matrices describe pure states?

$$\frac{1}{4}\begin{pmatrix} 2 & i \\ -i & 2 \end{pmatrix} \qquad \frac{1}{5}\begin{pmatrix} 1 & 2e^{i\phi} \\ 2e^{-i\phi} & 4 \end{pmatrix}$$

**Problem 17.**   The state of a certain system is described by the density matrix

$$\rho = \begin{pmatrix} 0.4 & 0.2\,e^{i\pi/4} \\ x & y \end{pmatrix}.$$

(a)   $x = ?,\ y = ?$

(b)   Verify whether or not the system is in a pure state.

**Problem 18.**

(a)   A system is known to be in one of two states. Show that the most general possible density matrix has the form

$$\rho = \begin{pmatrix} \alpha & \beta e^{i\phi} \\ \beta e^{-i\phi} & 1-\alpha \end{pmatrix}; \qquad \begin{cases} ? < \alpha < ? \\ |\beta| < ? \end{cases}$$

(b)   Show that $\beta = \pm\sqrt{\alpha(1-\alpha)}$ if the system is in a pure state.

(c)   Show that $\alpha = \frac{1}{2}$, $\beta = 0$ for an unpolarized beam of particles of spin one-half.

(d)   Find $\alpha$ and $\beta$ (in a representation in which $\sigma_z$ is diagonal) for a partially polarized beam of particles such that spin parallel to the $z$-axis occurs twice as often as spin anti-parallel to the $z$-axis.

**Problem 19.**   The state of polarization of a beam of spin one-half particles is described by the polarization vector $\mathbf{P}$ defined in terms of the density matrix $\rho$ as follows:

$$\mathbf{P} = \langle \sigma \rangle = \mathrm{Tr}\,(\rho\sigma).$$

Supposing $\mathbf{P}$ to be determined experimentally, show that

$$\rho = \tfrac{1}{2}\,(1 + \mathbf{P}\cdot\sigma).$$

**Problem 20.**   Recalling that the spin-dependent part of the Hamiltonian for an electron in a magnetic field $\mathscr{H}$ is

$$H_{\text{spin}} = -\frac{e\hbar}{2mc}\, \sigma \cdot \mathcal{H},$$

(a)   show that the polarization vector **P** (see Problem 19) of a beam of electrons in such a field satisfies the equation of motion

$$\frac{d\mathbf{P}}{dt} = -\frac{e}{mc}\, \mathcal{H} \times \mathbf{P}.$$

(b)   Assuming the magnetic field to be uniform, solve this equation of motion and discuss your results.

# APPENDIX I

# *Evaluation of integrals containing Gaussian functions*

## 1. THE BASIC INTEGRAL

We seek to evaluate the integral

$$I_0 \equiv \int_{-\infty}^{\infty} e^{-(a+ib)x^2} \, dx.$$

To do so we consider the square of this expression, which we write in the product form

$$I_0^2 = \int_{-\infty}^{\infty} e^{-(a+ib)x^2} \, dx \int_{-\infty}^{\infty} e^{-(a+ib)y^2} \, dy$$

and which is then equivalent to the double integral

$$I_0^2 = \int\!\!\!\int_{-\infty}^{\infty} dx \, dy \, e^{-(a+ib)(x^2+y^2)}.$$

If $x$ and $y$ are regarded as rectangular coordinates in a plane, this double integral is readily evaluated by transforming to polar coordinates $r$, $\phi$. We have at once

$$I_0^2 = \int_0^{2\pi} d\phi \int_0^{\infty} r \, dr \, e^{-(a+ib)r^2} = \frac{\pi}{a+ib}$$

and, therefore,

$$I_0 = \sqrt{\frac{\pi}{a+ib}}.$$

Note that, in the limit $a \to 0$,

$$I_0 = \sqrt{\pi/ib} = \sqrt{\pi/b} \; e^{-i\pi/4}.$$

For $a = 0$, the Gaussian integral can be given a meaning in the sense of this limit.

## 2. GAUSSIAN MOMENT INTEGRALS

We next generalize our result by considering the integral, for integer $n$,

$$I_n \equiv \int_{-\infty}^{\infty} x^n \, e^{-(a+ib)x^2} \, dx.$$

Observe first that

$$I_{2m+1} = 0,$$

because the integrand is odd. Next observe that

$$I_{2m} = (-1)^m \left(\frac{d}{da}\right)^m \int_{-\infty}^{\infty} e^{-(a+ib)x^2} \, dx$$

$$= (-1)^m \frac{d^m I_0}{da^m},$$

whence

$$I_{2m} = (-1)^m \left(\frac{d}{da}\right)^m \sqrt{\frac{\pi}{a+ib}}.$$

As an example,

$$I_2 = \frac{1}{2} \sqrt{\pi} \, (a+ib)^{-3/2} = \frac{1}{2(a+ib)} \, I_0.$$

## 3. GENERALIZED GAUSSIAN INTEGRALS

Finally we consider an integral of the form

$$J_0 = \int_{-\infty}^{\infty} e^{-\alpha x^2 - \beta x} \, dx,$$

where both $\alpha$ and $\beta$ may be complex but where the real part of $\alpha$ must be greater than zero. To evaluate the integral we complete the square in the exponent, writing

$$\alpha x^2 + \beta x = \alpha \left(x + \frac{\beta}{2\alpha}\right)^2 - \beta^2/4\alpha.$$

Replacing $x + \beta/2\alpha$ by a new integration variable, say $y$, the integral reduces to the standard form considered in Section 1, and we obtain at once

$$J_0 = \sqrt{\frac{\pi}{\alpha}} \; e^{\beta^2/4\alpha}.$$

Again, this result holds for $\alpha$ pure imaginary, if it is regarded as a limit. Specifically, if $\alpha = a + ib$, $a > 0$, it is the limit attained as $a \to 0$.

Moment integrals are again readily evaluated by the method of Section 2. If we define

$$J_n \equiv \int_{-\infty}^{\infty} x^n \; e^{-\alpha x^2 - \beta x} \; dx,$$

then we obtain at once, as perhaps the simplest of many possible expressions,

$$J_{2m+1} = (-1)^{2m+1} \frac{\partial^{2m+1} J_0}{\partial \beta^{2m+1}}$$

$$= (-1)^{2m+1} \left( \frac{\partial}{\partial \beta} \right)^{2m+1} (\sqrt{\pi/\alpha} \; e^{\beta^2/4\alpha})$$

and

$$J_{2m} = (-1)^m \frac{\partial^m J_0}{\partial \alpha^m}$$

$$= (-1)^m \left( \frac{\partial}{\partial \alpha} \right)^m (\sqrt{\pi/\alpha} \; e^{\beta^2/4\alpha}).$$

As examples,

$$J_1 = -\frac{\beta}{2\alpha} J_0 \quad \text{and} \quad J_2 = \left[ \frac{1}{2\alpha} + \left( \frac{\beta}{2\alpha} \right)^2 \right] J_0.$$

---

**Exercise 1.** $J_3$ can be evaluated by the prescription given above, but it can also be evaluated from the relation

$$J_3 = -\frac{\partial J_2}{\partial \beta}.$$

Verify that both methods give the same result. Show that, in general,

$$J_{n+1} = -\frac{\partial J_n}{\partial \beta}.$$

---

# APPENDIX II

# *Selected references*

## A. HISTORICAL AND EXPERIMENTAL BACKGROUND

1. G. Holton and D. H. D. Roller, *Foundations of Modern Physical Science,* Addison–Wesley (1958). An interesting but exclusively elementary and descriptive introduction.

2. F. K. Richtmyer, E. H. Kennard and T. Lauritsen, *Introduction to Modern Physics,* fifth edition, McGraw–Hill (1955).

3. R. M. Eisberg, *Fundamentals of Modern Physics,* Wiley (1962).

4. M. Born, *Atomic Physics,* seventh edition, Blackie, Glasgow (1962). A classic book on the background and origins of quantum mechanics.

5. R. B. Leighton, *Principles of Modern Physics,* McGraw–Hill (1959).

## B. MATHEMATICAL BACKGROUND

6. E. A. Kraut, *Fundamentals of Mathematical Physics,* Mc-Graw–Hill (1967).

7. H. Margenau and G. M. Murphy, *The Mathematics of Physics and Chemistry,* second edition, Van Nostrand (1956).

8. P. Dennery and A. Krzywicki, *Mathematics for Physicists,* Harper and Row (1967).

9. J. S. Sokolnikoff and R. M. Redheffer, *Mathematics of Physics and Modern Engineering,* McGraw–Hill (1966).

10. J. D. Jackson, *Mathematics for Quantum Mechanics,* Benjamin (1962).

11. G. Arfken, *Mathematical Methods for Physicists,* Academic Press (1966).

12. A. Sommerfeld, *Partial Differential Equations in Physics* (translated by E. G. Straus), Academic Press (1949). A classic book for physicists on differential equations and related topics.

13. P. W. Berg and J. L. McGregor, *Elementary Partial Differential Equations,* Holden–Day (1966).


## C. CLASSICAL MECHANICS

14. H. Goldstein, *Classical Mechanics,* Addison–Wesley (1950). Emphasizes those aspects of classical mechanics of particular importance for quantum mechanics. A standard reference, if somewhat advanced for undergraduates.
15. J. B. Marion, *Classical Dynamics of Particles and Systems,* Academic Press (1965).
16. R. A. Becker, *Introduction to Theoretical Mechanics,* Mc-Graw–Hill (1954).
17. J. C. Slater and N. H. Frank, *Mechanics,* McGraw–Hill (1947).


## D. QUANTUM MECHANICS (UNDERGRADUATE LEVEL)

18. D. Bohm, *Quantum Theory,* Prentice–Hall (1951). A physically motivated, discursive treatment, with emphasis on the relations between classical and quantum concepts.
19. R. H. Dicke and J. P. Wittke, *Introduction to Quantum Mechanics,* Addison–Wesley (1960).
20. L. Pauling and E. B. Wilson, *Introduction to Quantum Mechanics,* McGraw–Hill (1935).
21. D. Park, *Introduction to the Quantum Theory,* McGraw–Hill (1964).


## E. QUANTUM MECHANICS (GRADUATE LEVEL)

22. A. Messiah, *Quantum Mechanics* (in two volumes), North–Holland, Amsterdam (1961). A remarkably careful and readable text.
23. L. I. Schiff, *Quantum Mechanics,* McGraw–Hill (1955).
24. L. D. Landau and E. M. Lifshitz, *Quantum Mechanics, Non-relativistic Theory* (translated by J. B. Sykes and J. S. Bell), Addison–Wesley (1958).
25. A. S. Davydov, *Quantum Mechanics* (translated by D. ter Haar), Addison–Wesley (1965).
26. G. L. Trigg, *Quantum Mechanics,* Van Nostrand (1964).
27. P. A. M. Dirac, *Quantum Mechanics,* fourth edition, Oxford (1958). An absolutely classic book. No student should be satisfied until he has read and understood it.
28. P. Stehle, *Quantum Mechanics,* Holden–Day (1966).

APPENDICES

**29.** D. ter Haar, editor, *Selected Problems in Quantum Mechanics,* Academic Press (1964). A collection of excellent problems. The solutions to most are fully worked out.

**30.** J. D. Bjorken and S. D. Drell, *Relativistic Quantum Mechanics,* McGraw–Hill (1964).

# APPENDIX III

# *Answers and solutions to selected problems*

---

## CHAPTER I

**Problem 2.** Perhaps the simplest procedure is to write, after solving for $p$ in terms of $(T/mc^2)$,

$$\lambda = \lambda_c[(T/mc^2)^2 + 2T/mc^2]^{-1/2},$$

where $\lambda_c = \hbar/mc$ is the reduced Compton wavelength of the particle. We find in this way,

| $T$ | $\lambda$ (cm) | |
|---|---|---|
| | electron | proton |
| 30 eV | $4 \times 10^{-9}$ | $10^{-10}$ |
| 30 keV | $10^{-10}$ | $4 \times 10^{-12}$ |
| 30 MeV | $7 \times 10^{-13}$ | $10^{-13}$ |
| 30 GeV | $7 \times 10^{-16}$ | $7 \times 10^{-16}$ |

Note that at ultra-relativistic energies, $T \gg mc^2$, the de Broglie wavelength is approximately $\hbar c/T$ and is *independent* of the mass of the particle.

---

## CHAPTER II

**Problem 1.**    (a)   $|A| = \pi^{-1/4} a^{-1/2}$;    (b)   $\langle x \rangle = x_0$

**Problem 2.**    (b)   $|c_\pm| = [2(1 \pm e^{-1/4})]^{-1/2}$

---

## CHAPTER III

**Problem 2.**

(a) $\quad -\dfrac{2L}{\pi} \sum_{1}^{\infty} (-1)^n \dfrac{\sin (n\pi x/L)}{n}$

(d) $\quad \dfrac{\alpha}{L} \sum_{-\infty}^{\infty} \dfrac{1 - (-1)^n \, e^{-\alpha L}}{\alpha^2 + n^2 \pi^2/L^2} \, e^{in\pi x/L}$

**Problem 3.**

(c) $\quad g(k) = \sqrt{\dfrac{1}{2\pi}} \left( \dfrac{\sin k/2}{k/2} \right)^2$

**Problem 4.**

(a) $\quad A = L^{-1/2}$

(b) $\quad \phi(p) = (2 \, \hbar^3/\pi L^3)^{1/2} \, [(p - p_0)^2 + \hbar^2/L^2]^{-1}$

**Problem 7.** The ground state energies are roughly

(a) $\quad \hbar^2/mL^2$

(b) $\quad \hbar\omega$

(c) $\quad (\hbar^2 mg^2)^{1/3}$.

---

## CHAPTER IV

**Problem 1.**

(a) $\quad v_g = \dfrac{1}{2} \sqrt{\dfrac{g\lambda}{2\pi}} = v_p/2$

(b) $\quad v_g = c \sqrt{1 - (\omega_0/\omega)^2} \leqslant c$

**Problem 3.**

(a) $\quad \langle \psi_{E_n}|x|\psi_{E_n} \rangle = \dfrac{L}{2}$

$\langle \psi_{E_n}|x^2|\psi_{E_n} \rangle = \dfrac{L^2}{3} \left( 1 - \dfrac{3}{2n^2\pi^2} \right)$

$\langle \psi_{E_n}|p|\psi_{E_n} \rangle = 0$

$\langle \psi_{E_n}| \dfrac{p^2}{2m} |\psi_{E_n} \rangle = \dfrac{1}{2m} \left( \dfrac{n\pi\hbar}{L} \right)^2 = E_n$

(b) Consider a wave packet initially centered about $x = x_0$ with mean momentum $p_0$ and described by

$$\psi(x, t = 0) = f(x - x_0)\, e^{ip_0(x - x_0)/\hbar}. \tag{1}$$

According to equations (IV–43) and (IV–45), the development of such a state with time is given by

$$\psi(x, t) = \frac{2}{L} \int_0^L dx' \sum_{n=1}^{\infty} f(x' - x_0)\, e^{ip_0(x' - x_0)/\hbar} \sin \frac{n\pi x'}{L} \sin \frac{n\pi x}{L}$$

$$\times \exp\left[-in^2\pi^2\hbar t/2mL^2\right]. \tag{2}$$

Note that any such state is *exactly* periodic, with period

$$T = \frac{4mL^2}{\pi\hbar},$$

but that this period obviously has nothing to do with the classical period (why?). In fact, it is of the order of $10^{27}$ seconds for a macroscopic object in a macroscopic box!

We now consider the classical limit of equations (1) and (2). To settle orders of magnitude, think of a particle with mass a gram or so, in a box a few centimeters in length. Suppose its initial momentum to be one gm–cm/sec and its initial position to be defined within one micron ($10^{-4}$ cm). This last means that the amplitude function $f$ is sharply peaked about the point $x = x_0$ for which its argument is zero, with a width of about $10^{-4}$ cm. Because $p_0/\hbar$ is of the order of $10^{27}$, this in turn means that the exponential factor in equation (1) oscillates about $10^{23}$ times over the width of the wave packet. Accordingly, for the $n$th term in the summation of equation (2), the integral yields a negligible contribution unless $n$ is such that the oscillations of $\sin n\pi x'/L$ are almost *exactly* in phase with the oscillations of the exponential factor. To make this explicit, introduce a new summation index $s$ by writing

$$n = \frac{p_0 L}{\pi\hbar} + s \tag{3}$$

$$\simeq 10^{27} + s.$$

That the main contribution to the summation comes for values of $s$ which are relatively small is now readily established. The relevant factor in the $n$th term of equation (2) now becomes

$$f(x' - x_0)\, e^{ip_0(x' - x_0)/\hbar} \sin\left(\frac{p_0 x'}{\hbar} + \frac{\pi x'}{L}\, s\right)$$

$$= -\frac{1}{2i} f(x' - x_0)\, e^{-ip_0 x_0/\hbar}\left[e^{-i\pi x's/L} - \exp\left(2ip_0 x'/\hbar + i\pi x's/L\right)\right].$$

The second term in the bracket oscillates so rapidly over the wave packet that it yields a negligible contribution for all values of $s$. The first term oscillates approximately $s\Delta x\pi/L$ times over the one micron width, $\Delta x$, of the packet. Hence, the effective domain for the summation is

$$|s| \le \frac{L}{\pi \Delta x} \simeq 10^4. \tag{4}$$

To establish the classical periodicity of the motion, we now need examine the consequences of these results only for the *time dependent* exponential factor in each term of equation (2). Upon substitution of equation (3), this factor for the $n$th term becomes

$$\exp\left[-in^2\pi^2 ht/2mL^2\right] = \exp\left[-i\frac{p_0^2 t}{2m\hbar} - i\frac{s\pi p_0 t}{mL} - i\frac{s^2\pi^2\hbar t}{2mL^2}\right]. \tag{5}$$

Now according to equation (4), the term quadratic in $s$ does not exceed $\hbar t/m(\Delta x)^2 \simeq 10^{-19}\, t$ in magnitude, and hence it is negligible over time intervals comparable to the age of the universe. [Compare this analysis with that leading to equation (IV–9).] Omitting the quadratic term, therefore, we have

$$\exp\left[-in^2\pi^2\hbar t/2mL^2\right] \simeq \exp\left[-\frac{p_0^2 t}{2m\hbar} - i\frac{\pi p_0 t}{mL}s\right]. \tag{6}$$

Hence, except for the phase factor $\exp\left[-i\left(p_0^2 t/2m\hbar\right)\right]$, which multiplies the entire wave packet and thus plays no significant role (why?), we see that the solution is indeed periodic with just the classical period

$$T = \frac{2mL}{p_0} = \frac{2L}{v}.$$

We conclude by remarking without proof that, using equation (6), first the summation and then the integration in equation (2) can be evaluated in closed form without further approximation, the result being that the wave packet travels without distortion and alternately bounces off the walls at $x = 0$ and $x = L$. In short, the center of the undistorted wave packet displays a periodic saw-toothed form when plotted against $t$, in exact agreement with the classical solution. The details, which are not entirely trivial, are left as an exercise.

**Problem 5.** Suppose, for generality, the particle to be in the $n$th state of the box when the walls are dissolved. We thus have, as the initial state,

$$\psi(x, 0) = \begin{cases} \sqrt{2/L}\,\sin n\pi x/L, & 0 \le x \le L \\ 0 & , \quad \text{otherwise}. \end{cases}$$

After the walls are dissolved, we have, in momentum space,

$$\phi(p, t) = \phi(p)\, e^{-ip^2 t/m\hbar},$$

where

$$\phi(p) = \frac{1}{\sqrt{2\pi\hbar}} \int_{-\infty}^{\infty} \psi(x, 0) \; e^{-ipx/\hbar} \; dx.$$

The integral is elementary, and we obtain at once

$$|\phi|^2 = \frac{\hbar p_n^2}{\pi L} \frac{|1 - (-1)^n \; e^{-ipL/\hbar}|^2}{(p^2 - p_n^2)^2}$$

where

$$p_n = n\pi\hbar/L = \sqrt{2mE_n}.$$

For large $n$, the momentum distribution is readily seen to be sharply peaked about $p = \pm p_n$, in agreement with the classical result.

---

## CHAPTER V

**Problem 2.**

(a)   $A$ is Hermitian if and only if $n$ is even.

(b)   For $n = 3$, $\psi_1 = x^2 - L^2$ (unnormalized). For $n = 4$, the general eigenfunction is a cubic in $x$. However, the eigenfunction identified above for $n = 3$, which we have called $\psi_1$, is automatically an eigenfunction for $n = 4$. We thus have a pair of degenerate eigenfunctions. The cubic eigenfunction can readily be made orthogonal to $\psi_1$, thereby determining its coefficients, and we thus find the degenerate orthogonal pair (unnormalized)

$$\psi_1 = x^2 - L^2$$

$$\psi_2 = x(x^2 - L^2).$$

For $n = 5$, the general eigenfunction is a quartic in $x$, and $\psi_1$ and $\psi_2$ are also eigenfunctions. Choosing the coefficients of the quartic so that it is orthogonal to *both* $\psi_1$ and $\psi_2$, we thus identify a set of three-fold degenerate and orthogonal eigenfunctions which are

$$\psi_1 = x^2 - L^2$$

$$\psi_2 = x(x^2 - L^2)$$

$$\psi_3 = x^4 - \frac{8}{7} L^2 x^2 + \frac{1}{7} L^4 = (x^2 - L^2)\left(x^2 - \frac{L^2}{7}\right).$$

Note that, except for $n = 3$, these eigenfunctions are not unique. For example, for $n = 4$, one can choose any linear combination of $\psi_1$ and $\psi_2$ as a new eigenfunction, $\bar{\psi}_1$ say, and then use orthogonalization to find a second and linearly independent combination $\bar{\psi}_2$. Similarly for

$n = 5$. The particular orthogonal eigenfunctions given above are merely the easiest to construct.

**Problem 6.**

    (a)  ii, iv, v;    (b)  ii;    (c)  iii;    (d)  none

**Problem 10.**

    (a)  The new momentum eigenfunctions are defined by

$$\left[\frac{\hbar}{i}\frac{d}{dx} + f(x)\right]\psi_p = p\psi_p,$$

and hence are given by

$$\psi_p = \frac{1}{\sqrt{2\pi\hbar}}\, e^{i[px - g(x)]/\hbar},$$

where

$$g(x) = \int^x f(x)\,dx.$$

These states thus differ by a phase factor from the conventional momentum states, which we denote by $\bar{\psi}_p$ to make the distinction clear, and which are given by

$$\bar{\psi}_p = \frac{1}{\sqrt{2\pi\hbar}}\, e^{ipx/\hbar}.$$

    (b)  $\psi(x) = \int \phi(p)\psi_p(x)\,dp$

$$= e^{-ig(x)/\hbar}\int \phi(p)\bar{\psi}_p\,dp$$

$$= e^{-ig(x)/\hbar}\,\bar{\psi}(x),$$

where $\bar{\psi}(x)$ is the state function in the conventional representation. We emphasize that $\psi$ and $\bar{\psi}$ represent the same physical state, namely that corresponding to the particular momentum state function $\phi(p)$.

    (c)  Observe first that

$$\langle\psi|x|\psi\rangle = \langle\bar{\psi}|x|\bar{\psi}\rangle,$$

because the phase factor $e^{-ig(x)/\hbar}$ plays no role in the evaluation of the left side. Next, observe that

$$\langle\psi|\frac{\hbar}{i}\frac{\partial}{\partial x} + f(x)|\psi\rangle = \langle\bar{\psi}|\frac{\hbar}{i}\frac{\partial}{\partial x}|\bar{\psi}\rangle$$

because

$$\left[\frac{\hbar}{i}\frac{\partial}{\partial x} + f(x)\right]\psi = e^{-ig(x)/\hbar}\frac{\hbar}{i}\frac{\partial\bar{\psi}}{\partial x}$$

and the phase factors on the left then cancel. Finally, taking into account the fact that the commutation relation is unaltered, it follows that $\langle f(x, p) \rangle$ yields the same result in either representation. Hence the expectation values of any and all observables are independent of $f$, as was to be proved.

**Problem 12.**

$$\text{(a)} \quad \psi(x, 0) = \psi_1 \frac{e^{i\delta_1}}{\sqrt{2}} + \psi_2 \frac{e^{i\delta_2}}{2} + \psi_3 \frac{e^{i\delta_3}}{2},$$

where the $\delta_i$ are arbitrary constants.

$$\text{(b)} \quad \psi(x, t) = \psi_1 \frac{e^{i\delta_1 - iE_1 t/\hbar}}{\sqrt{2}} + \psi_2 \frac{e^{i\delta_2 - iE_2 t/\hbar}}{2} + \psi_3 \frac{e^{i\delta_3 - iE_3 t/\hbar}}{2}$$

(c)   iv

(d)   all but v

---

# CHAPTER VI

**Problem 2.**

(a)   $\delta_{0s}$

$$\text{(c)} \quad \text{zero;} \quad m\omega\hbar \left( n + \frac{1}{2} \right); \quad 3 \left( \frac{m\omega\hbar}{2} \right)^2 (2n^2 + 2n + 1)$$

$$\text{(d)} \quad \sqrt{\frac{(n + s)!}{n!}} \, \delta_{m, n+s}; \quad \sqrt{\frac{n!}{(n - s)!}} \, \delta_{m, n-s}$$

**Problem 3.**

$$\text{(a)} \quad \psi(x, 0) = \left( \frac{m\omega_0}{\hbar\pi} \right)^{1/4} e^{-m\omega_0 x^2 / 2\hbar}$$

(b)   Taking $\omega$ to be arbitrary, we treat both cases together. We then have

$$\psi(x, t) = \int \psi(x', 0) \, K(x', x; t) \, dx',$$

where the propagator $K$ is given by equation (68). The integral is of standard Gaussian form and, after some algebra, yields

$$\psi(x, t) = \frac{(m\omega_0/\hbar\pi)^{1/4}}{[\cos \omega t + (i\omega_0/\omega) \sin \omega t]^{1/2}}$$

$$\times \exp \left\{ -\frac{m\omega x^2}{2\hbar} \left[ \frac{\omega_0 \cos \omega t + i\omega \sin \omega t}{\omega \cos \omega t + i\omega_0 \sin \omega t} \right] \right\},$$

which is seen to reduce to the correct result when $\omega = \omega_0$. It also yields, for $|\psi|^2$, the expression to which equation (73) reduces for the particular Gaussian initial packet of part (a).

(c)   We have

$$\phi(p, t) = \frac{1}{\sqrt{2\pi\hbar}} \int \psi(x, t) \, e^{-ipx/\hbar} \, dx,$$

which is seen to be of standard Gaussian form. After evaluation of the integral one finds the momentum space probability density to be

$$|\phi(p, t)|^2 = \frac{(\pi m\omega_0\hbar)^{-1/2}}{[\cos^2 \omega t + (\omega^2/\omega_0^2) \sin^2 \omega t]^{1/2}}$$

$$\times \exp\left[-\frac{p^2/m\hbar\omega_0}{\cos^2 \omega t + (\omega^2/\omega_0^2) \sin^2 \omega t}\right].$$

(d)  $P_n \equiv |\langle\psi|\psi_n\rangle|^2,$

which yields, using the indicated representation for $\psi_n$,

$$P_{2S} = \frac{2\sqrt{\omega_0\omega}}{\omega_0 + \omega} \frac{(2S)!}{2^{2S}(S!)^2} \left(\frac{\omega_0 - \omega}{\omega_0 + \omega}\right)^{2S}$$

$$P_{2S+1} = 0.$$

Although it is not a trivial exercise, it can be verified that, as it must,

$$\sum_{S=0}^{\infty} P_{2S} = 1.$$

**Problem 4.**

(d)  Schrödinger's equation is

$$-\frac{\hbar^2}{2m} \frac{d^2\psi_E}{dx^2} - g\,\delta(x)\psi_E = E\psi_E.$$

Thinking of the δ-function as the limit of a sharply peaked function, it is not hard to see that $\psi_E$ is continuous but that $d\psi_E/dx$ changes dramatically over the width of the potential, by an amount proportional to the area under the potential. In the limit, $d\psi_E/dx$ actually becomes discontinuous. To display this discontinuity quantitatively, first integrate Schrödinger's equation from a point just to the left of the origin $(0_-)$ to a point just to the right $(0_+)$, and then consider the limit as each of these points approaches the origin (and thus each approaches the other from its respective domain). We find

$$-\frac{\hbar^2}{2m}\left\{\frac{d\psi_E}{dx}\bigg|_{0_+} - \frac{d\psi_E}{dx}\bigg|_{0_-}\right\} - g\psi_E(0) = 0,$$

where the right side vanishes, as indicated, because $\psi_E$ is continuous. The discontinuity in $d\psi_E/dx$ is thus

$$\frac{d\psi_E}{dx}\bigg|_{0_+} - \frac{d\psi_E}{dx}\bigg|_{0_-} = -\frac{2mg}{\hbar^2} \psi_E(0).$$

The prescription for solving Schrödinger's equation is now the following: Call the region to the left of the origin, $x < 0$, region I, to the right, $x > 0$, region II. Noting that the origin is excluded from each region, we see that in *both* regions Schrödinger's equation becomes that for a *free* particle. Hence, writing a general solution which satisfies the correct boundary condition at infinity on the right, say, the solution on the left is then fixed by requiring that $\psi_E$ be continuous and that $d\psi_E/dx$ have the correct discontinuity. We now construct solutions for bound and continuous states.

(i)  *Bound States, $E < 0$.*  Writing $E = -\epsilon$, we have in region II,

$$\psi_E^{II} = A \exp \left( - \sqrt{2m\epsilon/\hbar^2}\, x \right),$$

the positive exponential being inadmissible, of course. Similarly, in region I,

$$\psi_E^I = B \exp \left( \sqrt{2m\epsilon/\hbar^2}\, x \right)$$

the negative exponential now being inadmissible. The condition that $\psi_E$ be continuous at the origin then yields $A = B$. The discontinuity condition on $d\psi_E/dx$ can now only be satisfied if $\epsilon$ has the unique value,

$$\epsilon = mg^2/2\hbar^2,$$

which is the binding energy of the single bound state of a $\delta$-function potential, in agreement with the result quoted in part (a). The normalized bound state function is readily seen to be that given in part (b).

(ii)  *Continuum States, $E > 0$.*  To construct a solution corresponding to the conventional case of a wave incident *only* from the left with amplitude, say, $A$, we take the free particle solutions in each region to be

$$\psi_E^I = A\, e^{i\sqrt{2mE}\, x/\hbar} + B\, e^{-i\sqrt{2mE}\, x/\hbar}$$

$$\psi_E^{II} = C\, e^{i\sqrt{2mE}\, x/\hbar}.$$

Application of both boundary conditions at the origin then gives, almost at once,

$$\frac{C}{A} \equiv \tau = \frac{1}{1 - ig\sqrt{m/2E\hbar^2}}$$

$$\frac{B}{A} \equiv \rho = -\frac{ig\sqrt{m/2E\hbar^2}}{1 - ig\sqrt{m/2E\hbar^2}}.$$

Note that $|\rho|^2 + |\tau|^2 = 1$, as demanded by probability conservation.

**Problem 6.**

(a)  *Region I, $x < 0$.*  Letting $z = e^{x/2L}$ we obtain a form of Bessel's equation, and the general solution, for energy $E$, can be expressed in the form

$$\psi_E{}^{\mathrm{I}} = A J_\gamma(\alpha e^{x/2L}) + B J_{-\gamma}(\alpha e^{x/2L}),$$

where

$$\gamma = i\sqrt{2mE}\ (2L/\hbar)$$

$$\alpha = \sqrt{2mV_0}\ (2L/\hbar).$$

*Region* II, $x > 0$. Letting $z = e^{-x/2L}$, in the same way we find the general solution

$$\psi_E{}^{\mathrm{II}} = C J_\gamma(\alpha e^{-x/2L}) + D J_{-\gamma}(\alpha e^{-x/2L}).$$

*Boundary Conditions at* $x = 0$: Both $\psi_E$ and $d\psi_E/dx$ must be continuous.

(b)   *Bound States, $E < 0$.* Writing $E = -\epsilon$, we see that $\gamma$ is real,

$$\gamma = \sqrt{2m\epsilon}\ (2L/\hbar),$$

where we have arbitrarily chosen the sign of the square root to make $\gamma$ positive (why do we have the freedom to do this?). Observe now, looking at $\psi_E{}^{\mathrm{I}}$, that for $x \to \infty$, the argument of the Bessel functions approaches zero. Because

$$J_\nu(z) = \frac{(z/2)^\nu}{\Gamma(\nu + 1)} + \cdots$$

for $z \to 0$ (see references [6] – [13]), we see that only the $J_\gamma$ term is physically admissible. Similarly, in region II we see that again only the $J_\gamma$ term is physically admissible. Hence, we have

$$\psi_E{}^{\mathrm{I}} = A J_\gamma(\alpha e^{x/2L})$$

$$\psi_E{}^{\mathrm{II}} = B J_\gamma(\alpha e^{-x/2L}).$$

Both $\psi_E$ and $d\psi_E/dx$ must be continuous at the origin and hence we require

$$A J_\gamma(\alpha) = B J_\gamma(\alpha),$$

but also

$$A \frac{dJ_\gamma(\alpha)}{d\alpha} = - B \frac{dJ_\gamma(\alpha)}{d\alpha}.$$

The solutions are thus given by either $A = B$, $dJ_\gamma/d\alpha = 0$ or by $A = -B$, $J_\gamma(\alpha) = 0$. The former are seen to be even, the latter odd.

**Problem 8.**    (c)   $K(p', p; t) = e^{i(p'^3 - p^3)/6\hbar mF}\ \delta(p - p' - Ft)$

**Problem 18.**

(a)   Writing $a = b - \dfrac{\epsilon_2}{\epsilon_1}$, $a\dagger = b\dagger - \dfrac{\epsilon_2}{\epsilon_1}$, we find $H = \epsilon_1 b\dagger b - \epsilon_2{}^2/\epsilon_1$;

$(b\dagger, b) = 1$. Comparison with the harmonic oscillator Hamiltonian then shows $E_n = n\epsilon_1 - \epsilon_2^2/\epsilon_1$, $n = 0, 1, 2, \ldots$ .

## CHAPTER VII

**Problem 3.**
  (a) $E_0 = \hbar\omega/2$ (exact! why?)
  (b) $E_1 = \frac{33}{13}\hbar\omega \approx 1.6\,\hbar\omega$

**Problem 4.**

  (a) $E = 5\,\dfrac{\hbar^2}{mL^2}$, about 1% greater than the correct value $\dfrac{\pi^2}{2}\,\dfrac{\hbar^2}{mL^2}$

**Problem 7.**

  (c) $\qquad A^2 = \begin{pmatrix} 5 & 8 \\ 8 & 13 \end{pmatrix} \qquad B^2 = \begin{pmatrix} 5 & -3 \\ -3 & 2 \end{pmatrix}$

$A + B = \begin{pmatrix} 3 & 1 \\ 1 & 4 \end{pmatrix} \qquad AB = \begin{pmatrix} 0 & 1 \\ 1 & 1 \end{pmatrix} = BA$

**Problem 9.**
  (a) $b \gg 2m\hbar V_0/\,[2m(E - V_0)]^{3/2}$
  (b) $\tau = e^{i\delta(E)}, \qquad \delta(E) \approx -2V_0 bm/\hbar\sqrt{2mE}$

**Problem 15.**

  (a) $P(n) = \dfrac{2e^2 \mathscr{E}^2}{\hbar m\omega^3}\sin^2\dfrac{\omega\tau}{2}, \qquad n = 1$

$\qquad\qquad = 0, \qquad\qquad\qquad n \neq 1$

**Problem 16.**·

  (a) $\psi(t) = \left[\cos\left(\dfrac{\epsilon\tau}{\hbar}\right)\psi_1 - i\sin\left(\dfrac{\epsilon\tau}{\hbar}\right)\psi_2\right]e^{-iEt/\hbar}$

## CHAPTER VIII

**Problem 5.**

  (a) $\dfrac{1}{2}\hbar\omega - \dfrac{e^2\mathscr{E}^2\mu}{2m^2\omega^2}$, where $m$ is the mass of the charged constituent.
  (b) $\alpha = \mu e^2/m^2\omega^2$
  (c) uniform acceleration under the force $e\mathscr{E}$

**Problem 6.**
  (a) $\Delta E = \langle\phi_{10}(x_1)\phi_{20}(x_2)|H'|\phi_{10}(x_1)\phi_{20}(x_2)\rangle$, where $\phi_{10}$ and $\phi_{20}$

are the harmonic oscillator ground state eigenfunctions of particles 1 and 2, respectively. Thus

$$\Delta E = \sqrt{m_1 m_2} \, \frac{\omega V_0}{\hbar \pi} \int dx_1 \int dx_2$$

$$\exp \left( - \frac{\omega}{\hbar} \left( m_1 x_1{}^2 + m_2 x_2{}^2 \right) - (x_1 - x_2)^2 / a^2 \right)$$

$$= V_0 \left( \frac{\mu \omega a^2}{\mu \omega a^2 + \hbar} \right)^{1/2},$$

where $\mu$ is the reduced mass $m_1 m_2 / (m_1 + m_2)$.

(b)  The result is unchanged, but the analysis is much simpler. In center of mass coordinates, the Hamiltonian becomes

$$H = \frac{p^2}{2M} + \frac{1}{2} M\omega^2 X^2 + \frac{p^2}{2\mu} + \frac{1}{2} \mu \omega^2 x^2 + V_0 \, e^{-x^2/a^2},$$

and hence is separable. The problem thus reduces to an equivalent single-particle problem which, in fact, is just one of the examples discussed in detail in Section 3 of Chapter VII.

**Problem 12.**

(a)  $\psi_{pqrs} = \phi_p(x_1) \phi_q(x_2) \phi_r(x_3) \phi_s(x_4)$

$E_{pqrs} = (2 + p + q + r + s) \hbar \omega,$

where $p, q, r, s$ can take on all integral values starting with zero and where $\phi_n(x)$ is the $n$th harmonic oscillator eigenfunction.

(b)  Writing $p + q + r + s = N$, we see that

$$E_{pqrs} = (N + 2) \hbar \omega.$$

The degeneracy of the $N$th state is thus the number of ways four non-negative integers can be chosen so that their sum is $N$. It can be shown that this number is $(N + 1)(N + 2)(N + 3)/6$. The first four states then have degeneracies $1, 4, 10, 20$, respectively. Table I gives the quantum numbers of these degenerate states.

(c)  For spinless particles, only those states totally symmetric under exchange are physically realizable. The ground state is automatically symmetrical. The four states for $N = 1$ are an exchange degenerate set and a symmetric combination of them is the only realizable state. For $N = 2$, the states in the group (a) are an exchange degenerate set, as are separately those in group (b). Because these two groups of states are independent, there are two physically realizable states of $N = 2$, the symmetric combination of the states of type (a) and of the states of type (b). For $N = 3$, by the same argument, there are three physically realizable states, namely the symmetric combinations of the exchange degenerate states of types (a), (b) and (c), respectively.

| N | Degeneracy | | pqrs |
|---|---|---|---|
| 0 | 1 | | 0000 |
| 1 | 4 | | 1000, 0100, 0010, 0001 |
| 2 | 10 | (a) | 2000, 0200, 0020, 0002 |
| | | (b) | 1100, 1010, 1001, 0110 |
| | | | 0101, 0011 |
| 3 | 20 | (a) | 3000, 0300, 0030, 0003 |
| | | (b) | 1110, 1101, 1011, 0111 |
| | | (c) | 2100, 2010, 2001, 1200, |
| | | | 0210, 0201, 1020, 0120, |
| | | | 0021, 1002, 0102, 0012 |

TABLE I.  Quantum numbers and degeneracies of lowest four states

(d)  The ground state for spin 1/2 particles is easy to identify using the Pauli exclusion principle. It is the state $N = 2$ and is the totally anti-symmetric combination of single-particle states in which two particles with opposite spin are in the ground state and two with opposite spin are in the first excited state. It is not degenerate.

## CHAPTER IX

**Problem 3.**
  (a)  $\alpha = e^2/m\omega^2$
  (b)  same as with zero field

**Problem 4.**

  (a)  $V(r) = -\dfrac{Ze^2}{2R_0}[3 - (r/R_0)^2]$,    $r \leqslant R_0$

$$= -\frac{Ze^2}{r} \quad , \quad r \geqslant R_0$$

  (b)  $\Delta E \simeq \frac{2}{5} Z^4 e^2 R_0^2/a_0^3$

$$\simeq 7 \times 10^{-9} Z^{14/3} \text{ eV}$$

**Problem 6.**

  (a)  $\psi_m = \dfrac{1}{\sqrt{2\pi}} e^{im\phi}$,    $m = 0, \pm 1, \pm 2, \ldots$

$$E_m = \frac{\hbar^2 m^2}{2MR^2}$$

(b)  $\Delta E_m{}^{(1)} = 0$

$\Delta E_m{}^{(2)} = g^2 M^3 R^4 / (4m^2 - 1)\hbar^2$

(c)  WKB:  $E_0 \simeq \dfrac{\hbar}{2} \sqrt{g/R} - MgR$

**Problem 8.**

(a)  $\psi_{lm} = Y_l{}^m (\theta, \phi)$

$E_l = \hbar^2 l(l + 1)/2MR^2$

**Problem 11.**  Probability for 1s state $\simeq 0.70$, for 2s state $= 0.25$, for any state other than an s-state $= 0$. (See Reference [29], pp. 250-2.)

**Problem 17.**

(a)  $\phi(\mathbf{p}, t) = \dfrac{1}{(2\pi\hbar)^{3/2}} \displaystyle\int \dfrac{e^{-r/a_0}}{(\pi a_0{}^3)^{1/2}} \exp\left[-i\,\dfrac{\mathbf{p} \cdot \mathbf{r}}{\hbar} - i\,\dfrac{p^2 t}{2m\hbar}\right] d^3 r$

$= \dfrac{1}{\pi} \left(\dfrac{2a_0}{\hbar}\right)^{3/2} \dfrac{e^{-ip^2 t/2m\hbar}}{[1 + (pa_0/\hbar)^2]^2}$

$\rho(\mathbf{p}, t) = \phi^*\phi = \dfrac{8a_0{}^3}{\pi^2\hbar^3} [1 + (pa_0/\hbar)^2]^{-4} = \rho(p)$

(b)  $\rho_E(E) = 2\pi(2m)^{3/2} \sqrt{E}\, \rho(p = \sqrt{2mE}\,)$

**Problem 19.**

(a)  Yukawa:  $\epsilon = \dfrac{V_0}{8} - \dfrac{\hbar^2}{8\mu R^2}$

Exponential:  $\epsilon =$ same (coincidence!)

Square Well:  $\epsilon = \left(1 - \dfrac{5}{2} e^{-1}\right) V_0 - \dfrac{\hbar^2}{8\mu R^2}$

---

# CHAPTER X

**Problem 1.**

(a)  $E_0(l = \tfrac{1}{2}) = \tfrac{4}{9} E_0(l = 1)$

(b)  $\psi_{1, \frac{1}{2}, \pm\frac{1}{2}} = \sqrt{r}\, \sin\theta\, e^{\pm i\phi/2}\, e^{-2r/3a_0}$

**Problem 3.**  Some typical commutators are the following:

$(L_x, x) = 0; \qquad (L_x, y) = i\hbar z; \qquad (L_x, p_y) = i\hbar p_z$

**Problem 7.**  The pairs (i), (ii) and (v) commute; the other pairs do not. $T_a$ and $R_{\hat{n}}(\beta)$ commute if and only if $\hat{n}$ and $\mathbf{a}$ are co-linear.

**Problem 11.**

(a)  $J_z = (L_z + S_z)$

$L_z\psi = \hbar R(r) \sqrt{2/3} \, Y_1^{\,1} \, \chi_-$

$S_z\psi = \dfrac{\hbar}{2} R(r) \{ \sqrt{1/3} \, Y_1^{\,0} \chi_+ - \sqrt{2/3} \, Y_1^{\,1} \chi_- \}$

and hence

$$J_z\psi = \frac{\hbar}{2}\psi.$$

(b)  Probability density for spin-up:

$$\rho_+ = \tfrac{1}{3} \, |R(r)|^2 |Y_1^{\,0}|^2 .$$

Probability for spin down:

$$\rho_- = \tfrac{2}{3} \, |R(r)|^2 |Y_1^{\,1}|^2 .$$

(c)  $\rho_+ + \rho_- = \tfrac{1}{3} \, |R(r)|^2 \, (|Y_1^{\,0}|^2 + 2|Y_1^{\,1}|^2)$

$$= \frac{1}{3} \, |R(r)|^2 \left( \frac{3}{4\pi} \cos^2 \theta + \frac{3}{4\pi} \sin^2 \theta \right)$$

$$= \frac{1}{4\pi} \, |R(r)|^2$$

**Problem 13.**
  (a)  All are constants of the motion.
  (b)  All but the components of linear momentum are constants of the motion.
  (c)  the energy and the parity
  (d)  same as (b)
  (e)  $E$ and $L_z$
  (f)  only $E$
  (g)  $E, L_z, p_x, p_y$
  (h)  $L_z, p_x, p_y$.

**Problem 22.** The probability that the hydrogen atom will remain in its ground state is $[1 + (mva_0/2\hbar)^2]^{-4}$. (See Reference [29], pp. 310–14.)

---

## CHAPTER XI

**Problem 3.** According to the variational estimate of equation (10), the binding energy of $H^-$ is

$$E_0 = -\frac{121}{256} \frac{e^2}{a_0}.$$

This must be compared with the energy of the dissociated system, a neutral hydrogen atom plus a free electron at infinity. The ground state energy of the dissociated system is just the binding energy of the hydrogen atom, $-e^2/2a_0$. Accordingly, the $H^-$ system is unstable to dissociation in this approximation by $\frac{7}{256}\, e^2/a_0$, or by about 0.7 eV. Of course, the fact that the $H^-$ system is actually stable does not violate the minimum principle for ground state energies (why?).

**Problem 7.** After a straightforward evaluation of the integral, the Born approximation scattering amplitude for a square well can be expressed in the form

$$f(\theta) = \frac{2mV_0R^3}{\hbar^2} \frac{j_1\left(2kR\sin\dfrac{\theta}{2}\right)}{2kR\sin\dfrac{\theta}{2}},$$

where $j_1$ denotes the spherical Bessel function of order unity [see equation (IX–61)] and is thus given by

$$j_1(z) = \frac{\sin z}{z^2} - \frac{\cos z}{z}.$$

The differential cross section is then $d\sigma/d\Omega = |f(\theta)|^2$ and the total cross section is

$$\sigma = \left(\frac{2mV_0R^3}{\hbar^2}\right)^2 \frac{\pi}{2k^2R^2}\left(1 - \frac{1}{4k^2R^2} + \frac{\sin 4kR}{8k^3R^3} - \frac{\sin^2 2kR}{16k^4R^4}\right).$$

That the low and high energy behavior is in agreement with expectations can now be shown with relative ease. (See Reference [23], pp. 168–9, for a detailed discussion.)

**Problem 9.**
    (a)  $\sigma \approx 2 \times 10^{-25}$ cm$^2$
    (b)  Both $\sigma$ and $d\sigma/d\Omega$ are unaltered.
    (c)  $\sigma$ is unchanged but $d\sigma/d\Omega$ is altered.
    (d)  Total number scattered per second $\approx 4 \times 10^6$; the number scattered into the counter at 90° is about six per second.

**Problem 10.** Taking the $z$-axis along the magnetic field direction, the energy $E_{nlm_l}$ of a state of principal quantum number $n$, total angular momentum $l$ and $z$-component $m_l$ is

$$E_{nlm_l} = -\frac{e^2}{2n^2a_0} - \frac{e\hbar\mathcal{H}}{2mc}\, m_l$$

$$\approx -\left(\frac{13.5}{n^2} + 5 \times 10^{-5}\, m_l\right) \text{eV}.$$

Thus $s$-states are unaffected and states of $l \neq 0$ are split into $2l+1$ components of equal spacing, this spacing being about $5 \times 10^{-5}$ eV in magnitude.

We now consider the neglected (quadratic) term in $\mathcal{H}$, which we denote by $\Delta$. According to equation (67),

$$\Delta = \frac{e^2}{8mc^2} \, (\mathcal{H} \times \mathbf{r})^2.$$

Because $r$ is of the order of $n^2 a_0$, where $n$ is the principal quantum number, we thus have

$$\langle \Delta \rangle \simeq \frac{e^2 a_0{}^2 \mathcal{H}^2}{8mc^2} \, n^4.$$

Comparing this to the first-order splitting, we then have

$$\frac{\langle \Delta \rangle}{(e\hbar\mathcal{H}/2mc)} \simeq \frac{e a_0{}^2 \mathcal{H}}{4\hbar c} \, n^4 \simeq 10^{-6} n^4,$$

whence this term is seen to be negligible for almost all principal quantum numbers of interest.

**Problem 13.**

(a)  $\langle x \rangle = \langle \psi | x | \psi \rangle = \sum_{i=1}^{4} \langle \psi_i | x | \psi_i \rangle, \qquad \langle \mathbf{p} \rangle = \frac{\hbar}{i} \sum \langle \psi_i | \nabla \psi_i \rangle$

(b)  $\dfrac{d\langle x \rangle}{dt} = \dfrac{i}{\hbar} \, \langle \psi | H, x | \psi \rangle$

$\dfrac{d\langle \mathbf{p} \rangle}{dt} = \dfrac{i}{\hbar} \, \langle \psi | H, \mathbf{p} | \psi \rangle$

For a free particle, $(H, \mathbf{p}) = 0$ and $(H, x) = \dfrac{\hbar}{i} c\alpha_x$, whence, as expected,

$$\frac{d\langle \mathbf{p} \rangle}{dt} = 0,$$

but, surprisingly,

$$\frac{d\langle x \rangle}{dt} = c \, \langle \alpha_x \rangle.$$

Noting that this last result also holds when the electron is not free but moves in a static potential, we see that $c\alpha$ plays the role of a velocity operator in Dirac theory. Because the eigenvalues of each component of $\alpha$ are $\pm 1$, the eigenvalues of the velocity operator are $\pm c$, which is to say that a measurement of velocity can yield only the magnitude $c$. This seemingly paradoxical result can be understood intuitively in the following way. A measurement of instantaneous velocity implies an absolutely precise measurement of position at two infinitesimally separated time intervals. However, any precise determination of position makes the momentum completely uncertain, of course, so that the mean momentum between measurements becomes indefinitely large. Hence the velocity measurement must always yield the value $c$. For further discussion see Reference [30].

**Problem 18.**

(a)   $0 \leqslant \alpha \leqslant 1$,     $|\beta| \leqslant \sqrt{\alpha(1 - \alpha)}$

(d)   $\alpha = 2/3$   ,     $\beta = 0$

# *Index*